The Evolution of Human Sociability

Desires, Fears, Sex and Society

How do desires and fears motivate human sociability? What effect do these motivators have on reproductive, social and political behaviour? And, crucially, how might we understand them separate from preconceived notions of design, purpose or higher morality?

Taking these questions as a focus, this book examines human evolution with the emphasis on sexual selection and the evolution of a number of human psychological processes. Exploring evolutionary, sexual and maturational processes, along with primate, fossil and geological evidence, Professor Vannelli argues that human nature can be conceptualized as species-typical desires and fears, derived from sexual selection during human evolution, and that these can be treated as major motivators of human behaviour. Presenting additional evidence from the anthropology of band societies, along with material from group behaviour, Vannelli highlights the importance of pair-bonding, friendship, alliance behaviour, vengeance seeking and interpersonal and 'tribal' politics in human social behaviour, providing a unique and interdisciplinary framework for understanding human nature and the evolution of human sociability, including in its modern forms.

Ron Vannelli is Professor Emeritus at Birmingham City University, UK, where he taught epistemology, the psychology of personhood, social theory and political sociology for over 25 years. His PhD focused on accusation processes in human politics (interpersonal, sexual and political), and his research interests have spanned human evolution, the psychology of personhood, cultural anthropology and sociology. Developments in brain neuroscience, the study of emotions, sociobiology and evolutionary psychology fuelled his desire to explore the links between biology and human behaviour.

The Evolution of Human Sociability

Desires, Fears, Sex and Society

RON VANNELLI
Birmingham City University, UK

CAMBRIDGE
UNIVERSITY PRESS

University Printing House, Cambridge CB2 8BS, United Kingdom

Cambridge University Press is part of the University of Cambridge.

It furthers the University's mission by disseminating knowledge in the pursuit of education, learning and research at the highest international levels of excellence.

www.cambridge.org
Information on this title: www.cambridge.org/9781107114760

© R. Vannelli 2015

This publication is in copyright. Subject to statutory exception
and to the provisions of relevant collective licensing agreements,
no reproduction of any part may take place without the written
permission of Cambridge University Press.

First published 2015

A catalogue record for this publication is available from the British Library

Library of Congress Cataloguing in Publication data
Vannelli, Ron, 1938–
The evolution of human sociability : desires, fears, sex and society / Ron Vannelli.
  pages cm
Includes bibliographical references and index.
ISBN 978-1-107-11476-0
1. Human evolution. 2. Human behavior. 3. Behavior evolution. 4. Social evolution.
5. Cognition and culture. I. Title.
GN281.V348 2015
303.4–dc23
                                                                                    2015011609

ISBN 978-1-107-11476-0 Hardback

Cambridge University Press has no responsibility for the persistence or accuracy of
URLs for external or third-party internet websites referred to in this publication,
and does not guarantee that any content on such websites is, or will remain,
accurate or appropriate.

For
Liz, Alex and Mia

Contents

	Acknowledgements	*page* ix
1	**Introduction**	1
	In search of a science of humans	7
	Searching for desires and fears	21
2	**Human evolution: the background**	24
	Problems with traditional scenarios of human evolution	25
	Evolutionary processes	32
	The physical environment and human evolution	32
	Genetic and embryological processes	36
	Principles of sexual selection	42
	The phylogenetic background to hominin evolution	49
	Primates	49
	Chimpanzees	56
	Bonobos	60
3	**The evolution of human species-typical desires and fears**	64
	Human mating patterns	64
	Human sexuality	66
	Repressed sexuality	73
	Sex sublimated into human civilization	75
	Scenarios of human evolution	76
	The emergence of youth-apes	78
	The emergence of puberty and the taming of sexuality	83
	The evolution of human emotional-cognitive processes	92
	Offspring: evolutionary 'problematics'	92
	Females: evolutionary 'problematics'	100
	Males: evolutionary 'problematics'	103
	Evolution and human consciousness: some conclusions	107

4	**Bipedalism, brain growth, language and the development of human sociability**	111
	Bipedalism	111
	Brain growth	114
	Language	124
	Human sociability	140
	Human groups	141
	Hedonic relationships	155
	Creative subordination	158
	Status	159
5	**Desires, fears and the evolution of human politics**	163
	Male politics	163
	Alliances	165
	Alliance behaviour, politics and the rise of civilization	175
	Vengeance politics and the search for justice	179
	Patronage politics	183
	Modern politics	188
	Female politics	192
	Emotional power	193
	Female hedonic politics	199
	Human evolution and species-typical desires and fears	204
6	**Human fears**	206
	Primordial fears	206
	Freaks	226
	Witchcraft	230
	Struggling against evil	237
7	**A human science, justice and politics**	245
	Human species-typical desires and fears	246
	A human nature basis for a social (human) science	258
	In search of justice	259
	Human nature and social/political life	270
	References	273
	Index	304

Acknowledgements

This work has had a long gestation, with early versions being revised and built up along the way, during which numerous colleagues, students and friends made observations and comments which informed and otherwise influenced the final version. Vic Heatherington read a very early draft and made many useful observations concerning the problem of getting a balance between desires and fears in a view of humans. S. L. Washburn also commented on that draft, noting the problems of moving from evolutionary principles to human social behaviour without it being 'two books'. Harry Bauer read chapters of the current draft as they reached finality and made useful comments about its overall thrust and drew attention to a number of problematic issues. Mark Piney read Chapters 1, 2 and 3, drawing attention to a number of points which were not clear and making suggestions concerning academic points. My daughter Liz McKeown read the entire manuscript and with her lawyer mind let me know what made sense and what did not, as well as calling attention to specific issues that required a bit more thought. I thank all of them with heartfelt gratitude. In the final stages, Katrina Halliday and Megan Waddington and their colleagues at Cambridge University Press have been extremely friendly, helpful and efficient in steering this work to publication.

I think that it is important in this day and age of technology to acknowledge the vital role the internet plays in interdisciplinary scholarship of this kind. To put together this work a number of specialties had to be consulted, from the geology of Africa at the time of human differentiation to some of the legends and plays coming out of ancient Athens. Most academic journals are now available either free to anyone or via access through an institution of learning. It would take a great deal of time to find and travel to all the right libraries, wait for interlibrary loans and get all the notes required in relatively short sittings if this had to be done in pre-internet times. And information from a wide variety of sources, such as the Smithsonian Institute, World Health Organization, UNESCO or the CIA *World Factbook*, is instantly available, as well as from a number of online encyclopaedias, including such open and democratic sites as Wikipedia. The point is: 'times are changing,' and even we in academia are not immune; indeed we should embrace the opportunities provided by the new means of information storage and retrieval afforded to us to move beyond the confines of extremely narrow specialisms to more interdisciplinary approaches.

Most of all, I thank my wife, Susan Vannelli, who struggled with rewrites, always sharpening up the presentation, questioning the validity of many points, suggesting new and better ways of expressing certain things – postulating ideas all along the way. She knows the arguments as well as I do. This book would be dedicated to her if my other work on evolutionary theory and human nature had not already been (Vannelli, 2001); so for now I will just have to thank her for her work, patience and insights and dedicate the book to my offspring, Liz, Alex and Mia.

1 Introduction

As a species we have long wondered if there is a grand plan – or design – for our history, and for our individual part in it. We have sought answers in mythology, religion, philosophy and science, with a wide range of answers coming from these. Whatever the specifics, however, there has been a human-nature inspired tendency to come up with characterizations which suggest that there are elements of design or direction in our universe. Aspects of nature, spirits, ancestors, stars and gods, for example, have appeared to have a degree of control over events, a control that is not completely random in its manifestations or consequences concerning human fortunes. With the advent of human civilizations (city living, commerce, trade, world religions and relatively complex political administrations) ancestors, spirits and gods tended to become God, and this was a God of order and purpose, if not clear design. With the development of complicated theologies and philosophies, a grand design came to seem ever more likely, and it has been proposed that our purpose is to submit to it for the good of our souls, our good fortune, if not our fellow humans. With the philosophical notion of 'rationality' it has been argued that part of our *purpose* in being here is to discover the nature of the greater design. After all, design suggests a higher rationality and we are uniquely the 'rational' animal – *so we must have been put here to find that design and purpose.*

Yet, if we take the full implications of Darwin on board there is no grand design, purpose or higher universal moral imperative in human existence. We are animals that exist and reproduce and die, that is all, just like all plants and animals. We are nothing special. There is no evidence from nature that we are. Moreover, in Darwin's formulation *chance* is very important in the evolutionary process – as it also appears to be in human affairs. Or, if there is a grand plan, a master design, it is a very bad one indeed. We fight with each other, we kill each other, we torture each other, we ethnic cleanse, often in the name of a higher morality or rationality; we like food, drink and sex much more than reason (although we often do not admit it); we thrill at the conflicts, anxieties and double-dealing seen in mythological stories, soap operas and literature; we almost never read philosophies, and anyway they usually advocate just the opposite from each other, each claiming to be more rational and 'truth revealing' than all the others. We rarely end up in life as we – or our parents – originally planned. Our desires and fears take us ways we have not anticipated (for better and for worse). History is full of surprises, unintended consequences, good and bad luck. The enlightenment worship of reason, rationality and science has made little apparent difference, nor has the development of industrialization with its emphasis on mechanical explanations and material comforts.

Does this make an objective science of humans impossible or, indeed, a meaningless waste of time and energy? No. Human behaviour and human history are far from being completely random processes – we follow the principles of biology with its powerful urges and drives (as do all creatures), and we also follow the impulses contained in our *species-typical* human nature. Human nature is a complex mixture of unconscious and conscious drives, instincts, emotions, cognitions and learning abilities, derived largely from natural and sexual selection during human evolution; here I will argue that these mixtures can be expressed as 'species-typical desires and fears'. Desires and fears represent a surfacing of both the unconscious and conscious bioelectric activities (neural processes) during the gestation and growth of our brains. Just as brain bioelectrics can be seen to act as an interface between genetic processes and emotional-cognitive processes (including the formation of desires and fears), desires and fears act as an interface between emotional-cognitive processing and human social behaviour (the workings of these as mental processes and motivators are examined in some detail in my *Evolutionary Theory and Human Nature* [Kluwer 2001]).

Desires and fears thus act as the major motivators of human behaviour and, as a result, there is a noticeable degree of what might be called 'order-in-process' in our existence. And this (it will be argued) makes a non-teleological and/or non-ideological scientific approach to humans possible if we can discover desires and fears separate from any preconceived notions of direction, purpose, design or higher morality. Such an approach to humans, however, has not been especially popular, even among professional thinkers who believe in science and/or disclaim the existence of God. This may be because, as noted, we have a powerful *human nature* tendency to feel that there is meaning and purpose, if not direction, in our lives, and that progress is being made through the employment of 'rationality'. It may also be because non-teleological approaches not only greatly downplay the role of gods/God, but also the notion of a higher guiding morality.

It, therefore, raises the question: if we are largely emotional creatures, operating unconsciously a great deal of the time, and there is no predetermined direction or moral purpose in our existence, is it *worth* studying ourselves? Should we not let humans strive for moralistic perfection 'with hope' rather than impose a 'cold' scientific understanding which offers no specific utopia or human perfectibility? Unfortunately for such a view a great deal of human suffering has been caused by competing moralities designed to purify and perfect humans and human societies, each one just as ideological and dogmatic as the next. This has also been the case with 'pseudo-scientifically' discovered social utopias which, in fact, often have had even more efficient means of eliminating dissenters than moralistic endeavours have ever been able to employ.

So, in opposition to the above questions, it might be argued that: yes, we should study human nature *as it is* because, while anthropology and history tell us that a great deal of suffering has taken place, the argument here is: if we can more fully understand species-typical emotions, desires and fears – using our capacity for reason and science – we might be able to reduce (or at least evaluate) the fundamental causes of suffering and increase the number of emotionally rewarding times through reducing the circumstances in which fears are likely and by identifying and creating the conditions in which desires are most fully met. This approach would seek to be unhindered by the baggage of trying to spiritualize humanity and be free from the moralistic compulsion to perfect and purify humans (and to re-educate – re-programme – or eliminate the impure ones). The task would be to start from 'what humans are' and work to make social life as congenial to self and other humans as possible rather than to start from 'what humans should be'. It would mean that we make assessments of changes and ideas and possibilities in terms of their compatibility with species-typical desires and fears, and with the dangers these might present to humans physically, emotionally and politically rather than in terms of dogmatic moralistic or spiritual calculations. In this vein I will also suggest that a system of justice based on an innate sense of fairness can be greatly aided by a non-teleological, non-moralistic understanding of species-typical desires and fears.

However, emphasizing emotions, desires and fears leaves most of us with a sense that we are not in control of our lives, including of our long-term destinies. It also suggests that sex and emotionally based playing might be more natural and motivating to humans than thinking or reasoning. For some it implies that existence is simply 'survival of the fittest', 'dog eat dog', 'nature red in tooth and claw'. And it seems to go against a long tradition of thinkers who have painted a very different picture of the nature of the universe and of humankind's role in it as they searched for ways of perfecting both nature and human life. For example, thinkers such as: Plato, Aristotle, Augustine, Galileo, Newton, Hegel, Comte and Marx (together with classical economists, functionalist and social-constructionist social scientists) have set out to generate a picture of a designed, or at least designable, universe, with humans as basically reasoning, purposeful creatures who can bring instincts and emotions fully under control, if not completely eliminate them, and who can, through reason, seek perfection for individuals and society through discovery of the underlying rationality/purpose of the universe.

A flavour of some of these arguments, through time and across disciplines, can show us what a non-teleological science of humans is up against. For Aristotle, with his notion of a Final Cause of ultimate perfection, the path of history was, inevitably, guided by the *necessity* of making progress to it (Aristotle, 1885; see also, Ross, 1995). He argued that, 'Nature like a good householder throws away nothing of which anything useful can be made. Nature does nothing in vain, nothing superfluous ... Nature behaves as if it foresaw the future' (cited in Ross, p. 81). Galileo famously put forward the view that the Universe

Cannot be understood unless one first learns to comprehend the language and to understand the alphabet in which it is composed. It is written in the language of mathematics, and its characters are triangles, circles and other geometric figures, without which it is humanly impossible to understand a single word of it (cited in Gribbin, 2002, p. 95; see also, Drake, 1994).

This is a tradition of physics which held sway for some considerable time. Rene Descartes saw the universe as being mechanical in nature and set out to discover the laws of its working. He made a clear separation between body and mind (Morris, 1991). Isaac Newton carried on with this approach and Albert Einstein reacted negatively to the randomness and uncertainty built into quantum theory, saying (also famously) that: 'God does not play dice' (Hawking, 1995; see also, Gribbin and Gribbin, 1997; Gribbin, 2002).

Among theologians and philosophers there has been a tendency to see a great predetermining/designing power (God) as the driving force in nature. St Augustine, for example, saw evolution as moving from the 'City of Man' to the 'City of God' in which the communion of saints – the *predestined* elect – would be united in peace and enjoyment of God's complete goodness. He made it clear that there was a predetermined path for it all; it was God's Plan; no amount of 'free will' or merit could make one part of the elect unless one loved God at the expense of self and all earthly beings. An individual's capacity in this respect, however, was a predetermined gift of God and only afforded to a very small number of people (Augustine, 1950, 2012; see also, McClelland, 1996; Most, 2013).

By the time we get to Thomas Aquinas (1224–1274) the state, or at least political society, was 'the milieu for mental and moral development' (Aquinas, 1966; Black, 1984, p. 65). But nevertheless any policies developed were to be based on 'natural laws' provided by God: 'Thomas's God is a rational God. Like Aristotle's nature, God does nothing without a purpose' (McClelland, 1996, p. 112). All that was left was for humans (secular and theological) to find these laws and abide by them. In this tradition John Locke (1975 [1690], 1980 [1662]) firmly believed that God had provided the basis for a just commonwealth and it was up to humans to use their God-given reason to discover how it could be implemented and administered. For Locke: 'Like the rest of creation, God had set humankind a purpose, a telos, and it was not limited to this world' (Spellman, 1997, p. 120).

Early western anthropology set out to discover a 'natural order of existence', or at least the principles of such an existence, by observing a multitude of hunting and gathering societies. They came back from observing bands of conflicting, bickering individuals – usually with very little social stability, or even solidarity (Marwick, 1965; Harris, 1968; Douglas, 1970; Colson, 1974; Ambrose, 1975; Kuper, 1983; Jarvie, 1984; Woodburn, 1991) – and saw within them 'well-structured, synchronic social *systems*' full of rational, naturally cooperating actors' (Cf., Kroeber, 1917; Radcliffe-Brown, 1952; Gluckman. 1963; Boas, 1965 [1938]; Durkheim, 1965 [1912]; Levi-Strauss, 1966). Using 'science' ('social physics' – Saint-Simon), early sociologists had the very same aim with regard to discovering the laws of social evolution and of social *structure* to be able to establish principles for how societies should be organized to provide *social solidarity*, and thus social harmony and justice, and for guiding individual moral choices in daily life. There was an implication that societies evolved through distinct intellectual and 'structural' stages (Ward, 1897, 1906; Parsons, 1937, 1966; Spencer, 1967 [1874, 1896]; Marx and Engels, 2011 [1939]; see also, modernization theorists).

There are some very powerful thinkers here, and their intentions to improve to perfection humans and human social life can be hard to argue with. Yet their abiding faith in reason and teleology has had a long history of being questioned and even sometimes ridiculed. For example, the Sophists, the author of Ecclesiastes, Machiavelli, Hobbes, Voltaire, Hume, Swift, Darwin, Nietzsche and Freud considered humans somewhat less than rational creatures, often not in control of their own destinies and bounced about by Fate, Fortune, events, sex, emotions, desires and fears. This view of human nature is also frequently the stuff of stories and myths, ranging from the work of ancient Greek playwrights to numerous modern authors. 'Fortune lifts and Fortune fells the lucky and the unlucky every day. No prophet on earth can tell a man his fate' (*Antigone*, Sophocles – Fagles trans., 1984, p. 119). Jocasta rails to Oedipus against prophesy: 'What should a man fear? It's all chance, chance rules our lives. Not a man on earth can see a day ahead ... Better to live at random, best we can' (*Oedipus the King*, Sophocles – Fagles trans., 1984, p. 215).

Oedipus, of course, fought against this view in his desperate attempt to discover 'the truth' but his efforts only led to the mental demons of guilt and shame that accompany patricide and incest. In the process Sophocles raised the possibility that we might have *unconscious* desires to kill our fathers and sleep with our mothers. Jocasta tries to calm Oedipus well before he knows the truth: 'Many a man before you, in his dreams, has shared his mother's bed' (p. 215). The importance of both fantasy and delusion in human existence is noted in *Oedipus*: in the words of the chorus 'does there exist, is there a man on earth who seizes more joy than just a dream, a vision? And the vision no sooner dawns than dies blazing into oblivion' (Fagles, 1984, trans., p. 233).

Shakespeare (1977) was concerned with the arbitrary outcomes of much of human life. Lear bemoans: 'Let copulation thrive; for Gloucester's bastard son was kinder to his father than my daughters Got 'tween the lawful sheets.' (p. 908). And when his fool has been hanged he laments: 'No, no, no life! Why should a dog, a horse, a rat have life, And thou no breath at all?' (p. 915). And the doomed Cordelia – the non-schemer, the loyal daughter – acknowledges to her father, the king, as the end was near, 'We are not the first, Who, with best meaning, have incurr'd the worst' (p. 912). The novelist George Eliot put 'difficult choices before her characters, showing their human frailty and the sometimes disastrous consequences of dubious actions undertaken out of mixed and confused motives' (Ashton, 1996, p. 381). These elements were common in the novels of the great Russian writers: Turgenev, Tolstoy and Dostoevsky.

In the Bible book of *Ecclesiastes* (King James) the preacher asks: 'What profit hath man of all his labour wherein he taketh under the sun? *One* Generation passeth away and *another* generation cometh: and the earth abideth for ever' (chap. 1, verses 3–4). And as far as meaningful labour is concerned, what is the point? 'Then I looked on the works that my hands had wrought, and on the labour that I had laboured to do: and, behold, all was vanity and vexation of the spirit, and *there was* no profit under the sun' (chap. 2, verse 11). Moreover, it is not an especially good idea to seek wisdom, 'For in much wisdom *is* much grief: and he that increaseth knowledge increaseth sorrow' (chap. 1, verse 18). The Confucian master Xun Qing, argued in the third century BC that '"Fate" is not determined by Heaven but by chance' (*Treatise on Heaven*, in O'Grady, 2012, epigraph, chap. 15). A key doctrine of Buddhism emphasizes the relativity of human life. Impermanence is a major theme in Buddhism (and to a lesser extent Hinduism); 'all human striving is no more than a vanishing hand clutching at clouds' (Watts, 1962, p. 7). While the 'afterlife' might be different, the *Qur'an* points out: 'Know ye (all), that the life of this world is but play and amusement, pomp and mutual boasting and multiplying (in rivalry) among yourselves, riches and children' (chap. 57: sec 3:20).

Voltaire is renowned for (among other things) being scathing of the then current philosophical/rationalist notion that 'all was for the best in the best of possible worlds'

(See, *Candide* – 1759). His contemporary David Hume pointed out that, 'Necessity is regular and certain. Human conduct is irregular and uncertain. The one, therefore, proceeds not from the other' (Hume, 1958, p. 403). Hume's view of thoughts and experiences as they made up consciousness was that they were 'fluctuating, uncertain, fleeting ... in a perpetual flux ... succeed[ing] each other with an inconceivable rapidity' (Hume, 1958, p. 252; see also, Hume, 1963; Morris, 1991; Dollimore, 1998; Eagleman, 2011). Hume and the polemical writer Swift also noted some of the dangers and injustices that can be created in the seeking of rational answers to what are largely emotional and political issues. Nietzsche summed up the world: 'That which we now call the world is the result of a host of errors and fantasies which have gradually arisen in the course of the total evolution of organic nature' (Nietzsche, 1886: Section One, 16). In the whole of his work he (Nietzsche, 2010) generally held that reason and passion could not be separated; and it was not just the congenial passions that would necessarily predominate. For him the 'Will to Power' was the driving force of all human actions. The will to power was a combination of a multitude of instinctual forces that consciously and unconsciously, purposefully and subliminally, allowed humans to survive and proliferate. There was no grand plan behind it all.

And according to a number of philosophers with a political focus, the will to power clearly operated in human social life, and a 'natural order' was not assumed. Hobbes, for example, argued that the natural state of humans was a war of every man against every other. He considered that humans were naturally egotistical, competitive, aggressive and power-seeking (Hobbes, 1968; Morris, 1991). Before Hobbes, Machiavelli had warned us of the dangers of unconstrained humans when natural human emotions are given a free reign and means of providing order are absent. They both strongly suggested that issues of order were political rather than moral or ethical in nature. Machiavelli attributed about half of human affairs to 'Fortune', against which even the wisest and strongest (and most ruthless) of princes would have to work very hard to have any influence at all (Machiavelli, 1950). As 20th century British Prime Minister Harold Macmillan famously answered, when asked by a reporter what he thought would be his major problems over the coming period: 'events, dear boy, events'.

Even with a strong emphasis on reason and rationality within the evolving social sciences not all thinkers were convinced that human social life was amenable to following exact laws and structures. For the sociologist Max Weber, traditions, beliefs, emotions, charisma, status, economic positions and power politics, in a variety of contingent interrelationships, all played a part in explaining particular human social outcomes (Weber, 1978, 1981). And in sociology's sister discipline, economics, where rational, calculating man was the underpinning of the classical model, predictions of the business cycle were so often contradictory and proven wrong that the economist Ezra Solomon (1984) wryly observed 'the only function of economic forecasting is to make astrology look respectable' (see also, Galbraith, 1975). In social policy, Christopher Jencks (1975) concluded from a major statistical analysis of the effects of home and schooling on economic success in modern America (measured over two generations) that 'luck' accounted for about 50 per cent of predictability. By luck he did not mean a supernatural force. Rather, he argued that the variables accounting for success were so many, and so different for each individual, and *not following* any historical or clumping patterns, that particular mixes could not be easily identified, classified into useful categories or be seen to repeat themselves in a predictable manner.

Indeed, even at the heart of the scientific search for cast-iron laws and predictability in nature, uncertainty and a lack of complete predictability came to have an important role. In modern physics, for example, the development of quantum electrodynamics (QED) postulates that the position and direction of an electron is 'subject entirely to chance, selected at random from the options open to it' (Gribbin and Gribbin, 1997, p. 39). In Heisenberg's words, 'we *cannot* know, *as a matter of principle*, the present in all its details' (Hawking, 1995; Gribbin and Gribbin, 1997; Gribbin, 2002, p. 520). And, 'Heisenberg's uncertainty principle is a fundamental, inescapable property of the world' (Hawking, 1995, p. 61). 'Nothing is certain in the quantum world ... Quantum processes obey the same rules of chance as dice at a crap table in Las Vegas' (Gribbin and Gribbin, 1997, p. 39). Despite this QED is arguably 'the most successful scientific theory yet developed' (Gribbin, 2002, p. 526; see also, Hawking, 1995; Gribbin and Gribbin, 1997).

Nevertheless, the above scientists would not argue that the universe is a chaotic mess, unliveable in and not understandable by science. Just the opposite – they spend their lives searching for 'order in process' within it. While a number of the above thinkers, who were suspicious of rationality, design and purpose, suggested that human consciousness is often at war within itself (such 'wars' not being products of clear-cut factions struggling for supremacy, but more like bar-room brawls in which it is hard to tell what the issues are, who is on what side and what might be a satisfactory outcome), a disordered, conflict-prone picture of human nature and randomness in human life does not preclude the possibility of a science of human behaviour. Writers such as Hobbes, Machiavelli, Voltaire, Hume and Max Weber, for example, did not believe so. The task is to incorporate randomness, chance and conflict into the science. I suggest that we can undertake this through an understanding of species-typical emotions, desires and fears as being *evolutionarily derived* motivations of human behaviour (without *inventing* meaning and purpose for humans and then using these as primary *motivations* for human behaviour).

In search of a science of humans

A significant move in this direction was made when Charles Darwin provided a scientific paradigm free from a grand design, predictable direction and purpose in nature, including in human existence. Darwin's approach placed the causative processes – natural and sexual selection – prior to the outcome, with the outcome given no retrospective causal power whatsoever, and random chance being a significant factor. Publication of *The Origin of Species* in 1859, thus, presented a blow to any notion of a Grand Design or Final Cause in nature, reducing God's involvement – if not eliminating it altogether. Following the *Origin's* argument that organic life was largely a product of mindless forces unconnected to their outcome, he set in motion an approach to human consciousness (human nature) which did not presume the evolution of a unique higher rationality. In *The Descent of Man and Selection in Relation to Sex* (1874/1879) he emphasized the continuity of animal and human characteristics – including mental ones – and the significance of sexual selection in human evolution, implying that *sexual choice*, not man the big-brained/tool-using rationalist, might be the key to human evolution (see also, Ridley, 1993; Miller, 2000; Jones, 2002). In *The Expression of the Emotions in Man and Animals* (1872) Darwin considered emotional expressions in humans as representing modified inherited *animal* tendencies – and not always that much modified. He observed, for example, that, 'A young female chimpanzee, in a violent passion, presented a curious semblance to a child in the same state' (Darwin, 1872, pp. 139–140).

His argument was that physical capacities could take on new functions as evolution proceeded, and that this was very significant in the evolution of human emotions. For instance, faint smiles in humans are muscular facial reactions of recognition and non-threat, derived from the facial expressions of apes when signalling submission. And an automatic, or semi-automatic, smile remains in humans a means of saying 'hello, I will not harm you, I do not come to challenge you; we can have a peaceful chat'. In apes, 'laughter' is a common spontaneous result of social playing, including being tickled; tickling generates 'pleasure' at least as far back as dogs. In the case of laughter, a particular behaviour generates a physical reaction that provides pleasure and a sense of fun/happiness. And in apes and humans the pleasure is related to a social relationship – which can be seen from the fact that neither an ape nor a human can get pleasure or laughter from self-tickling however much they realize that tickling causes laughter and that laughter is pleasurable. In his search for a link between phylogenetically inherited physical tendencies and the development of human emotional responses, he was able to come to similar conclusions concerning the evolution of emotional expressions of such things as fear, terror, rage, grief, anxiety, joy and anger – these being set in motion by inherited animal reactions such as trembling, the grinding of teeth, changes in blood flow, perspiration and averting eyes. The significance is that it is the inherited physical reaction that generates the emotional feeling, rather than an abstract meaning given to a particular action or relationship generating the feeling.

Most of our emotions are so closely connected with their expressions, that they hardly exist if the body remains passive – the nature of the expression depending in chief part on the nature of the actions which have been habitually performed under this particular state of mind (Darwin, 1872, p. 234).

The link with 'habitual performance' and mind brings us to another level. Darwin speculated that actions of the body were linked to 'nerve force' – perhaps akin to the actions of neurotransmitters in today's physiology – which, at the same time, was linked to habitual behaviour – habit having a very similar feel to the current notion of the establishing of neural networks.

To establish the connection between behavioural expressions and the evolution of mind in humans he spent considerable time considering blushing, because: 'Blushing is the most peculiar and the most human of all expressions' (p. 310). Why? Because, although blushing is a result of very physical activity – the relaxation of muscular coats of small arteries by which capillaries in the face become filled with blood – it requires mind for its expression; 'this theory rests on mental attention having some power to influence the capillary circulation' (p. 336). Yet it is unconscious mind. We cannot cause blushing by any physical means (as tickling can cause a laugh), or by willing it, and we cannot suppress it by physical or mental means (attempting to only makes it worse) – but in certain mind states it is inevitable (and seemingly universal). What may have been a physical response to fear, anxiety and agitation came, in humans, to have personal and social meaning. We can see distress in others when they blush, and feel that we must be giving away our anxiety/distress/shame in the same way. So, when the human face became 'the chief seat of beauty and of ugliness, and throughout the world is the most ornamented' part of the human body (p. 356), humans became extremely conscious of the potential effects of the appearance of their faces. This awareness can cause involuntary blushing when attention seems extensively focused on self's face.

This we can assume was even more the case when human faces and facial expressions began to represent not only the sexual desirability of a person, but also their character, intentions, trustworthiness and value as a friend or long-term mate. Blushing, then, depends on 'a sensitive regard for the opinion, more practically for the depreciation of others, primarily in relation to our personal appearance, especially of our faces; and secondarily, through the force of association and habit, in relation to the opinion of others on our conduct' (p. 334). So when we feel under scrutiny we often blush and: 'We thus see that close attention [of our own minds] certainly affects various parts and organs which are not properly under the control of the will' (p. 340). It is not a big step from here for a human, especially when in the presence of others, to be able to imagine a 'moral' failing and blush because of the linkages between blushing and mind that have been established.

At this point consciousness, unconsciousness and concept formation are linked to a phylogenetic capacity for blood to flow to the face in certain mental states. And there is feedback with regard to these mind states.

> If, then, there exists, as cannot be doubted, an intimate sympathy between the capillary circulation in that part of the brain on which our mental processes depend, and in the skin of the face, it is not surprising that the moral causes which induce intense blushing should likewise induce, independently of their own disturbing influences, much confusion of the mind (pp. 323–324).

These 'confused' mind states come to be related to a powerful tendency to feel anxious from too much attention directed at self from others, especially teasing, questioning or deprecating attention, being part of a package of fear, embarrassment, shame and shyness which emerged during the evolution of the human mind. It is a mind derived from human phylogenetic history in which shyness and embarrassment evolved from what was already there. It is a mind evolving in a context of ever-changing – sometimes congenial, sometimes threatening – social worlds.

Darwin may have somewhat underestimated the possible reproductive advantages of shyness given the nature of human sexual evolution (see Chapter 3), but his comment about shyness and design is clear and to the point. 'Those who believe in design, will find it difficult to account for shyness being the most frequent and efficient of all the causes of blushing, as it makes the blusher to suffer and the beholder uncomfortable, without being of the least service to either of them' (p. 335).

Building upon this approach to emotions, William James – considered to be among the founders of modern psychology (1890) – put forward the hypothetical example of how we came to fear bears (1884). A fear of bears, according to James, does not cause us to run from bears; but rather a bio-electric tendency, phylogenetically inherited, to automatically run from big, dangerous creatures emotionally teaches us, through the physical changes generated from the running away from bears, to be afraid of bears specifically. Physical bodily action, thus, teaches us emotional feelings and understanding. James was criticized in that it was suggested his 'system' was too slow to explain the relatively rapid reactions of humans in the face of danger. However, this was based on the notion that James was arguing for visceral feedback as the 'teacher' of emotions when, in fact, he was also arguing for somatic feedback (as in the case of running from a bear), which is fast enough (LeDoux, 1998; see also, Damasiò, 2000).

Current evidence from neuroscience suggests that danger stimuli go to different sections of the brain via two or more pathways. Some stimuli – with minimal information – travel very fast to emotional and quick reaction centres, and others, with more information, to the cortex where slower, but more elaborate, analysis can take place (LeDoux, 1998; Brown, 2003; see also, Kahneman, 2011). The evolution of these mechanisms supports the approach taken by Darwin and James, although it suggests that, while running, other information is being stored to later be used in analysis. Whatever the details, it cannot be denied that this perspective does make considerable sense from an evolutionary point of view. It is based on the principle that 'The cost of treating a stick as a snake is less, in the long run, than the cost of treating a snake as a stick' (LeDoux, 1998, p. 165; see also, Hamilton, 2012), but that eventually it is useful to be able to know the difference, and to be able to assess the different dangers each might generate (in different circumstances).

Any evolving human that had to wait on the evolution of a relatively complex fear of bears (originally it would have been of something like lions, cheetahs, baboons or some big ape) would not have left very many copies of their genetic blueprint behind. But, if the act of running like hell with the image of a bear in one's head activated a 'charging-up' of an image of a bear as dangerous, with additional stimuli concerning bears and bear country being stored at the same time, a considered fear of bears would have become established. As humans became more self-aware, and capable of thinking and consciously remembering, a fear of bears might result in staying out of bear country – after all very few humans can outrun an angry bear. With increased emotional-cognitive capacities, however, a fear of bears could lead to using bear country when bears were away, or when it was a specific advantage, but being extremely vigilant about the presence of bears, and becoming very alert to signs of bears. (Who knows, the brave people who did this might then take the bear as a totem animal.)

In this approach, the interrelationship between body and mind is so intricate that it is not useful to consider them as separate entities. Reactions, emotion and reason evolved together, generating motivational packages (ones that are to a significant extent unconscious) that exist because they gave survival and reproductive success during the evolution of what eventually became the species *Homo sapiens*. We have an inherited innate propensity not only for actions but also for mood changes in response to certain somatic and visceral reactions, reactions that are themselves responding to specific external stimuli. James (and Darwin) were arguing that, 'we feel sorry because we cry, angry because we strike, afraid because we tremble' (James, 1892, pp. 375–376). Indeed, recent research (Meikle, 2003) that suggests that unborn babies can cry, smile and blink before they are born supports this possibility.

Darwin became somewhat Lamarckian in that he considered various linkages of body/mind-states/behaviour to have become biologically heritable 'habits'. These habits were originally derived from ancestral somatic/visceral reactions but then became socially learned habitual behaviour that became heritable as humans evolved. This is now considered to be wrong in that there is no evidence that learned behaviour is biologically inherited. Nevertheless, Darwin did have a strong sense of a biological base for emotions and the social behaviour they were related to. He clearly understood that many reactions, especially muscular ones, were related to what we now call emotions. Both Darwin and James challenged the over-reliance on rationality with which most of their contemporaries worked, as well as any notion that humans had some special quality that allowed a claim for them to be completely above biology. However, their hope of a non-teleological seamless transition from natural selection to human nature to social behaviour was not realized with the idea of inherited habits.

Nevertheless, there were some who saw promise in the work of Darwin and James for scientifically discovering biological mechanisms for the transition from biology to behaviour. Not least among them was one of the founders of modern psychology, Sigmund Freud (1974), a keen admirer of 'the great Darwin' (Freud, 1973, Lec. 4, pp. 104–105; Gay, 1989, p., 36; Ritvo, 1990; Badcock, 1994). Besides fully taking on board the principles of natural and sexual selection with the aim of developing an explanatory framework for the transition from biology to human behaviour, Freud anticipated some of the current thinking concerning the 'selfish gene' notion when he argued that:

the individual organism, which regards itself as the main thing and its sexuality as a means, like any other, for its own satisfaction, is from the point of view of biology only an episode in a succession of generations, a short-lived appendage to a germ-plasm endowed with virtual immortality – like the temporary holder of an entail which will outlive him (1973, p. 463).

With his strong interest in reason and science, Freud had an intellectual 'problem'. So much of human behaviour seemed not to be 'rational', indeed, not even intended. Certainly, humans often were hard pressed to explain why they felt or did certain things. The notion of unconscious motivations/behaviours made a lot of sense to him – as it had to Schopenhauer, Nietzsche and James – but if the human mind and behaviour were derived from animal minds and behaviour, acting *unconsciously*, how was modern human consciousness to be examined? And how could this be made into the basis of a science if a lot of key human motivations are unconscious? How are we to understand the unconscious if we cannot see, measure or ask people about it? Freud suggested that we look at the behaviour of children, neurotics and hysterics, and at dreams and slips of the tongue (parapraxes), and at jokes, myths, fantasies and, above all, at human sexual desires, for evidence of our basic, evolutionarily derived motivators (Freud, 1905, 1917, 1973, 1976).

The problem for many is that, to accept Freud's supposition, we have to give up any notion that evolution has completely eliminated instincts (or any sort of unconscious motivations) in favour of conscious reason and 'rationality'. From an evolutionary point of view Freud was on the right track (given that it is highly unlikely that drives and instincts would be completely eliminated and replaced by a new form of totally conscious motivation in the relatively short evolution of one particular primate species) and it is little wonder that Ernest Jones considered him to be the 'Darwin of the mind' (Morris, 1991). But this did not (and does not) generally make a lot of friends among religionists, humanists, rationalists or social constructionists.

The reaction to Freud was – and still remains for many, including scientists – not positive. Quite apart from a degree of ridicule of his suggested sources of evidence (such as children, neurotics, dreams and hysterics) his apparent attack on rationalism seemed out of place to enlightenment-inspired thinkers (Morris, 1991; Moore, 1994). Within the scientific community, behaviourists such as Watson, 1913 and Skinner, 1971, 1988, social constructionists and positivists were especially scathing (Morris, 1991). And since the 1960s cognitive scientists have generally tried to abolish not only instincts but also emotions altogether (LeDoux, 1998). In more recent times, evolutionary psychologists have continued this disregard (and a degree of disdain) for Freud and his 'unnecessary' messy middle between brain (as they see it as computer-like calculating processes) and behaviour (Cf., Cosmides, Tooby and Barkow, 1992; Pinker, 1998; Blackmore, 1999; Miller, 2000). Rather than being seen as a developer of Darwin's ideas, Freud was portrayed by most social scientists as a sort of 'sex-obsessed' preacher of irrationalism.

At the time, undaunted by his own critics – at least in terms of his work – Freud dug deeper into what his enemies considered to be his hole. If sexual selection was a major factor in human evolution, as Darwin had suggested, does this not mean that unconsciously held instincts and early learning potentials were all, at base, sexual? Well, yes, indeed. Freud, in fact, became quite insistent on this (Freud, 1962, 1971, 1973). His critics clamoured, 'look around, there is no evidence for such a claim. Humans pray, and show great modesty and love and care for children, and work hard at jobs, and build cathedrals. They do not run around all the time trying to have sex'. Freud's solution to this problem was, in a major sense, the heart of his complex formulation of the human mind/consciousness.

He recognized that if the behaviour of young children largely represents a number of the untamed instincts of human phylogenetic history, as he imagined it does, human society, as we know it, would be impossible. Famously, babies appeared to Freud to be 'polymorphously perverse' (Freud, 1973, Lec. 13, p. 246: 1962). That means they were a bundle of undifferentiated sexual impulses. Children do not mind which sex holds and cuddles them; they lovingly smile and coo for attention, stroking and kisses (from either sex); they love being tickled; they like sucking for pleasure (breasts, thumbs, dummies, lollypops – and one assumes a male penis or female vagina if allowed or encouraged); they are not ashamed to be naked; they do not mind their parents (of either sex) washing and powdering their sexual parts; they play with their own, and, if allowed, with their moms' and dads' sexual parts; brothers and sisters love taking baths together; they enjoy anal pleasure from defecation. Babies (and children in general), then, were seen as being wildly undiscriminating sexual creatures (1962, 1973, Lecs. 13–14, 20).

The biophysical mechanisms that underlie the pleasure of the above behaviours probably existed in the human ancestral line as they are often on public display as grooming, kissing, genital play, genital display, sexual coyness, pseudo-copulation, copulation, cuddling and infant care in primate species (see Chapter 2). If it is from this base that humans emerged, it is unlikely that when they did so all of the above would have completely disappeared. At the same time, it is equally unlikely that the modern human ideal of a totally non-sexual form of intense attraction to protecting 'innocent child-ness' would have appeared out of the blue. So, the question became for Freud: if in adults 'perverse sexuality is nothing else than a magnified infantile sexuality split up into its separate impulses' (Freud, 1973, Lec. 20, p. 353), how did this polymorphously perverse primate/child sexuality become adult human sexuality as we know it, and how did so much of childhood sexuality come to be considered perverse in adults?

This takes place, Freud argued (1962, 1971, 1973, Lec. 20), through a number of *developmental* psychosexual stages in which it can be said that the body and brain 'teach' the mind 'civilized living'. Primarily, during child development various pleasure-sensitive parts of the body become identifiable as specific erogenous zones (the mouth, breasts, anus and genitals) which we like to indulge with pleasure producing stimuli. During the oral stage, for example, 'sensual sucking' develops the already sensitive mouth into an erogenous zone which remains for life. So we suck breasts when very young but still lip-kiss as romantically and sexually excited adults and still suck breasts as adults; we enjoy giving/receiving oral sex; we paint our lips to make them look like the sexually receptive parts of a number of primates. And we smoke, suck lollypops, eat sweets, chew gum, lick our lips and suck our thumbs: all because they pleasure our mouths.

However, with increased self-awareness, contact with and exposure to specific erogenous zones can also communicate danger. So, we learn a degree of deferred gratification in that we do not always get our way; we are removed from the breast; we are weaned, sometimes forcibly. We learn that if we cry the breast might come back – or it might not. We learn to be strategic, too much demanding can drive away mother love (along with holding, cuddling), too little and we are hungry. So we realize that we are separate from the breast/mother. We learn that we have a degree of control over some events some of the time. This sense of individual agency carries on into later life. But by then we are extremely restrained and selective in our approach to oral gratification compared to what our primate ancestors got/get up to or how we were as babies. We decide when to paint our lips, who to kiss, who to allow to suck our breasts, who to provide oral sex for, and all this is done very sparingly. We also decide what to smoke, when to have a lollypop and when to suck our thumb. We become individuals with our own selective uniqueness in more than sexual matters.

Human sexual impulses have been made more *controllable* and more easily *repressed* in a self-aware creature in which controlled sexuality seems to be the norm or – as one might say today – 'to be what has provided reproductive success among evolving humans' (see Chapter 3). But a lot of this has been done unconsciously (manifest as shyness, embarrassment, shame) and the sexual impulses and the energy they generate do not completely go away. They become repressed, lurking in the unconscious, providing a potential for sublimation into a variety of more or less socially acceptable behaviours, such as defecation in private, modesty, cleanliness, secret sex, romantic love, elaborate nest building, hard work, religious devotion and political commitment.

Next is the anal stage. Defecation becomes increasingly positively pleasurable. We are proud of the results. We offer these as gifts to our loved ones. They progressively begin to react with horror. They clean us up thoroughly, over and over again, and they train us to do it in a special, hidden place. So we learn that cleanliness is next to godliness and tidiness is good and that we should be ashamed of our bodily expulsions, pleasurable though they at first seemed to be. But why has our gift been rejected so dramatically? Are we bad; are we rotten inside? Is this condition one we will have to monitor, control and hide all our lives? Is pleasure bad? We learn to save 'our unwanted gift' until we are in the secret place to dispose of it, and we clean ourselves up thoroughly, and we do not tell about the great pleasure (the bad pleasure) still derived from expulsion. But we feel a little guilty and we learn to organize ourselves and the things and those around us in a clean, protective manner. We may even learn to save our money for a rainy day. We are ready for taboos and laws.

At about four years of age we enter the phallic stage. Here the genitals become more than a means of expulsion, fun to explore and pleasurable to touch. They increasingly become a focus of attention, but also of restriction. We are told to not do that; leave them alone. Boys especially find this hard to do ('it being so easy to grab'), so they play with their genitals in secret. Punishment and negative attention generate shame, guilt and embarrassment concerning our genitals despite the pleasure of playing with them, preparing the basis for feelings of guilt when we secretly play with them. Boys and girls begin to notice the differences between themselves and are told to hide their sexually related differences from each other. Thus, reproductive characteristics become a sort of secret (if not shameful) thing, making it relatively easy for sexual intercourse, and even the actual birth of a child, to become highly restricted/regulated areas of human existence, often done out of sight, allowed only in the right circumstances, with the right person. In this whole process autoeroticism is overcome and transferred to external objects, with some of the sexual energy sublimated into rules of courtship behaviour, child rearing and civilized living in general.

Current brain research (Cf., Sternberg and Kaufman, 2002; Strauch, 2003) supports the idea of repression being a part of the way the brain develops. As a child grows there is a lot of 'pruning' of neurons and elimination of existing connections. 'Experience during this time may determine which neurons will live, their functionality, and their dedication to various brain operations' (Bjarklund and Kipp, 2002, p. 36). In Freud's formulation, after the phallic stage a period of latency occurs during which social restrictions (modesty), taboos (especially an incest taboo) and 'civilized behaviour' are more energetically taught. Brain studies suggest that during adolescence the process of 'pruning' continues. There appears to be a degree of refinement of the pathways and bridges so far developed in the brain. A certain number of excess interconnections, or partial connections, are eliminated. This is exactly at the time when parents, peers and social authorities might be looking to impose not just modesty or an incest taboo, but also some sense of 'responsibility' and guidance for a 'direction in life' onto emerging adults.

However, at adolescence (genital stage) new hormones (and most likely 'pre-knowledge' brain functions/pre-formed networks) kick in. At this stage a degree of exuberant emotionality and risk-taking among adolescents seems to be related to increased dopamine and a lack of melatonin (Strauch, 2003). Whatever the exact causes, a major conflict between parents and offspring usually emerges at this point. During this stage a person moves from the narcissistic orientations of autoeroticism (and perhaps the lust/love for the opposite sex parent) to the 'reality principle' represented by restricted and sub-limated sexuality, thus achieving an ego that will allow them to function in society (or not, possibly suffering neuroses and psychoses when it does not all work smoothly). But it is not easy. Mother love is hard to break away from having been established through sensual sucking, cuddling, eye-to-eye-gazing, close care and protection, with a good dose of separation anxiety to retard the process of breaking away. Most individuals do break away but guilt is often generated, keeping ambivalence alive when adolescents move towards separation (Freud, 1936, 1957; Laing, 1965; Carroll, 1985). So, from Freud's work we would not expect a finely balanced, harmonious self to emerge.

Neurosciences and elements of modern psychology have given a degree of support to Freud that he did not have in his own day. Yet, in the end, there was a tinge of teleology in his explanation of civilization. Repression existed because civilization *required* group living and group living *needed* repression in order to exist; civilization was a product of group living in which unfulfilled sexual and aggressive urges and parent/offspring conflicts were repressed and sublimated into such things as politics, art, architecture and democracy, among other things (Freud, 1921, 1961). Had he had some of the benefits of modern research into evolutionary processes (especially sexual selection) he may not have had to dip into teleology. The move clearly seen in human evolution from a degree of r selection (whereby males are as promiscuous as possible, and females nearly as much – at least when in oestrus – and females do all the child rearing) to k selection (with females becoming much more selective and less promiscuous, and males having to provide a degree of loyalty, protection and support in child rearing if they are to be selected by females) goes a long way to set the stage for the selection of repressed sexuality all round (see Chapters 2 and 3).

And the existence of very immature care-requiring offspring – competing for parental attention at the expense of parents paying attention to each other and to siblings – can go some way to explain parental/offspring conflict as well as sibling rivalry. In terms of the selfish gene argument, there is a natural conflict *among* offspring competing for parental attention, and *between* offspring and parents as the former break away to serve their own reproductive interests. There is plenty of scope here for the selection of the conflicting emotions, feelings of guilt and ambivalence that so interested Freud. And these emotionally charged ambivalences and potentials for guilt never leave us. So we live with the agonies of sex/love, the pressures of kinship, the fickleness of trust and loyalty, the potential treachery in friendships and in inter-group politics, but we call it 'being human' and very often we deem it 'civilization'.

So, the possibility of a non-teleological science of human motivation was inherent in the work of Freud, but for whatever reason he did not develop it. Perhaps he did not want to. Perhaps he did not have enough time. And no one else really took it up. This may have been because, however compatible with evolutionary theory Freud's work might have been, it did strongly question the notion of 'rational, calculating man'. 'Free will' seemed to disappear. And it suggested that human consciousness might not be a stable balanced whole with perfection just around the corner as soon as superstitions and irrationality had been banished. Statements like, 'it is simply the pleasure-principle which draws up the programme of life's purpose' (Freud, 1957, p. 27) did not draw the admiration of humanists and rationalists who wanted 'to progress' humans along to rationality and utopia.

Even in the study of biological evolution the teleology of 'adaptationism' began to creep back in and to dominate (Williams, 1966). Adaptation sounds rational; it sounds scientific; it almost sounds purposeful. In the social sciences the teleology of a postulated functional synchronicity of social institutions adapting into stable 'social structures' flourished, alongside cultural determinism and social constructionism as maintaining mechanisms. Psychology moved to behaviourism, then on to modern cognitive psychology, and then to evolutionary psychology, picturing human consciousness as rationally machine-like in its stimulus/response characteristics. For the behaviourist B. F. Skinner, for example, in terms of motivations of behaviour, 'feelings are incidental and irrelevant' and self is simply a 'mode of action ... [a] ... functionally unified system of responses' (Skinner, 1953, p., 285, cited in Morris, 1991, p., 120; Skinner, 1971, p., 544). Social psychology, desperate to escape from any form of 'biological determinism', looked to see how individuals were influenced, if not formed, by social processes rather than innate tendencies (Cf., Cooley, 1902; Mead, G., 1962; Goffman, 1971; for a discussion see Morris, 1991).

For cognitive psychologists humans are problem-solving creatures; they see problems, approach them with reason, set goals for their solution and come up with rational means to attain them (Forgas, 1981; Eysenck and Keane, 1990; Benjafield, 1992; LeDoux, 1998). Latching on to the so-called (or the self-labelled) 'cognitive revolution' of the 1950s and 1960s, (Tooby and Cosmides, 1992, 1995; Pinker, 1998, p. 23), certain evolutionary psychologists have increasingly avoided, indeed rejected, notions of emotionality in favour of the teleological notions of 'adaptationism' and 'mind-as-computer', 'which explains the mechanics of thought and emotion in terms of information and computation' (Pinker, 1998, p. 20). As Joseph LeDoux (1998) has critically pointed out, they generally like the 'idea of the human mind as a carefully engineered machine' (p. 39, see also, p. 280) in which emotions become (at best) aids to information processing (for additional critique see also, Vannelli 2001; Miller, 2000; Ruse, 2012). Furthermore, according to them the claim for adaptation is 'a claim about design' – design for very specific goals (Symons, 1992, p. 140), 'designed' by a 'target family of problems'; our 'psychological architecture can now be mapped... as a system of computational relationships' (Tooby and Cosmides, 1995, p. xv; see also, Dawkins, 1986).

In these approaches (as noted) instincts, drives, emotions, desires and fears became relegated to a significance well behind stimulus/response, stimulus/calculating/response, pure learning, rationality and assumptions of intellectual and social progress. Indeed, for many of the above (especially social constructionists), talking of instincts, emotions, desires and fears as important motivators of behaviours came to be considered 'essentialism' or 'reductionism', if not fascism and/or racism – and as such to be both academic and moral failings. So the human sciences banished potential conflicts, confusions, ambiguities and ambivalences (unconscious, pre-conscious and conscious) of the human mind – not to mention the oedipal conflicts, polymorphous perversions, neuroses and common-or-garden miseries which Freud (and before him a number of great storytellers and philosophers) struggled to make sense of.

And gone too was the role of chance in evolution as natural selection was seen to be a response to precise 'environmental problems'. Indeed, in recent times, *purpose* has been brought back into the evolutionary equation. For example, it has been argued that humans and animals have a purpose in 'achieving the greatest possible success of their genes ... our mission ... is to promote those genes that made us' (Barash, 2001, pp. 20, 23). He may mean that we have no supernatural purpose. He may mean that genes that promote their own survival better than others, and spread more rapidly in a population, predominate, and that is that. But it does not sound like this is a non-purposeful (or non-meaningful) process when the above quotes are read and words like 'our mission' appear. It almost sounds like there is a higher purpose in the universe that has given us genes that compel us to fulfil 'our mission'.

Richard Dawkins, of the pioneering work, *The Selfish Gene*, which brought back to scientific (and public) attention the possibility of considering humans in a non-teleological way, nevertheless concluded by claiming that humans can escape their genes, that we can escape from natural and sexual selection (Dawkins, 1976, 1989). In a 2004 television broadcast for Channel 5, 'The Big Questions' (January 7), he strongly implied that evolution had a driving force: a continuous move to greater and greater 'adaptation'. This is, of course, debatable. But what was more significant was the claim that humans had taken control of their evolution. Humans generate goals, science and a sense of purpose, and so have *introduced purpose* into biology. Humans have risen above the normal evolutionary process; they have circumvented natural selection and sexual selection; instead of struggling for survival humans create great works of art, architecture, science and civilization. He did not say, 'among other things' a *minority* of humans create the above (while most slave at extremely boring tasks, getting their kicks from drink, sport, sex and family). He said that is *what humans do*. There was a clear implication that he was saying 'that is what humans *are*'. More recently Dawkins has modified this stance somewhat, speaking of the evolution of 'neo-purpose'. 'In humans, the capacity to set up neo-purposes has evolved to such an extent that the original archi-purpose [the Darwinian selected functionality of a characteristic] can be eclipsed and even reversed. The subversion of [archi] purpose can be a curse, but there is some reason to hope that it might become a blessing' (Dawkins, 2009; see also, Elsberry, 2009; Moran, 2009).

If the human sciences have not escaped the teleology of postulating meaning, design and purpose as explanations of human behaviour, it is little wonder that idealist philosophers and policy makers not only cling to these ideas but push them with enthusiasm (along with an idea of 'progress'). It has been assumed, given 'natural' human rationality, that we will be able to correct past human mistakes and leave behind nature 'red in tooth and claw'; conflict-prone, hierarchical, selfish and thoughtless animal nature can be overcome by reason, learning, language, science and opportunities to cooperate for progress. Such a view has led many reformers to feel that they were/are the moral educators and guardians of humanity with a mission to banish all traces of cavemen mentality. In the process, particular utopian ideologies and/or moral codes have been presented as self-evident truths and opposing views as ignorance and superstitions.

This approach has viciously and cruelly destroyed many, many more humans than any beliefs that human life is a bit unpredictable, not always smooth, but that we should make the best of it while in the meantime enjoying: the foibles of children, sex, friendship, gossiping, drinking, sports, music, watching television, playing computer games, 'getting ahead' and just plain living. In modern times, utopian/moral approaches have often sought to enlist 'science' in their efforts to save us from our animal heritage, to eliminate superstitions and bring perfection if not utopia. Sometimes scientific claims from idealistic perspectives are more mythical than real – as in the conceptualization of 'the noble savage', who had lived altruistically and harmoniously and was, therefore, proof that human nature is benign (and could be a model for how we should behave).

More dangerous was the 'scientific' idea of the Aryan race as a 'pure race' best suited to rule the world. The Nazi regime in Germany, for example, set out to systematically exterminate at least six million Jews, and other 'lesser' humans, to clear the way for this race to inherit the world. Modern technology, science and rational bureaucratic organizational principles were employed to kill people efficiently. The Old Soviet Union and modern China have both significantly outdone the Nazis in terms of total numbers systematically eliminated – perhaps as many as 70–100 million people (Baumeister, 1997; see also, Dimont, 1962; Conquest, 1971; White, 2005; Chang and Halliday, 2005; Gray, 2007) – in the process of attempting to 'scientifically' reprogram human nature and to create the perfect society based on rational, non-exploitative principles. The leader of the Chinese Revolution, Mao Tse-tung (Mao Zedong) had pointed out that the new 'system' being introduced was, ' the most complete, progressive, revolutionary and rational system in human history' (Mao, 1966, p. 23).

In the old Soviet Union, under Stalin, a policy was undertaken to create the new rational social system through the use of 'Scientific Socialism' – of which Joseph Stalin was the sole professor (Nove, 1972, 1975). Scientific Socialism – based on tenets of Marxist 'scientific' theory, the aspirations of utopian socialism and a sense of moral superiority based on these – set out to replace tradition, superstition, religious beliefs and 'false' ideologies with a focus on the 'real' *material* conditions in which humans lived. These conditions were to be scientifically 'understood' so as to develop principles upon which to base the new classless, non-exploitative society. The full weight of the science of psychology was employed to create a 'New Soviet Man' (Bauer, 1952) fit for this new utopia. The science of psychiatry was turned over to the secret police (KGB), who, from their centre at the infamous Serbsky Institute, began to diagnose obstructionists as suffering from a whole number of new 'medically discovered' forms of insanity (Mee,

1971; Medvedev and Medvedev, 1971; Block and Reddaway, 1977). These included: 'Reformers Frenzy' (Joravsky, 1971), 'creeping schizophrenia' (Block and Reddaway, p. 131), 'sluggish schizophrenia' (Nekarasov, 1975) and 'paranoid development of the personality with the presence of reformist ideas' (Mee, 1971, pp. 6–10). At least 20 million people died, either directly by execution or in labour camps and mental institutions, in what turned out to be a failed attempt to get rid of selfishness, self-interest and politics and to implement a 'rational' political system.

So, even modern so-called scientific approaches have been used to destroy multitudes of fellow humans in a search to perfect them. An important thing to note is that the above examples of perversion of science – for the good of humanity – do not come from pre-enlightenment mysticism, religious dogmatism, tribal traditions or hillbilly farmers. They come from modern, technologically sophisticated, industrial societies with schools and universities and secular beliefs. However, the 'use' of science in the above mentioned utopian projects has no resemblance to Galileo's or Darwin's notion of science based on continuous observation, experimentation, theorizing and an on-going refining of conclusions (void of moral preaching); and if the results overwhelmingly indicate that the Earth goes around the Sun, and we are just animals, we will have to live and work with these discoveries. The teleological notions of inevitable direction, purpose, morality or the belief in the perfectibility of humans or of human societies, will have to be set aside. If there is a God we have to find him (and his methods) through observation and experimentation, not through postulation of a prescribed or required way of life backed up by creationist teachings, moral crusades, inquisitions, holocausts, labour camps or intransigent belief.

Science is neither a difficult nor a mystifying concept if left to its basics. As the Nobel Prize winning physicist Richard Feynman argued, science is about inspired guessing, finding things out, finding out how much we do not know; it is about how to handle doubt and uncertainty; it is, he argued, about developing rules of evidence, trial and error. Science includes a strong sense of invention, and it can get away with this because

> It does not make any difference how beautiful your guess is. It does not make any difference how smart you are, who made the guess, or what his name is – if it disagrees with experiment it is wrong (cited in, Gribbin and Gribbin, 1997, p. 179; see also, Medawar, 1969; Jones, 2013).

In the social sciences we most often would have to substitute 'observation', 'experience' and 'comparison' for experiment (all also highly regarded by Feynman), but the idea is the same. In so far as science has succeeded it has done so by continually checking for real evidence without postulating a world that 'should be' and then seeking to find it (Freud, 1961; Medawar, 1969; Vannelli, 2001, chap. 1).

This would mean, for example, that lust, anger, envy and jealousy may well be as important as thinking, caring and loving in explaining human nature. We will have to explain bickering as well as brotherly love, hate as well as devotion, war as well as peace – after all history suggests that we get much more excited and interested when talking and reading about wars than about peace. That people are apparently more interested in tobacco, alcohol, football, flirting, sex and joking than they are in philosophical contemplation or self-improvement is something to be explained, not judged. Divorce could be as normal as marriage and family. Hero worship is likely to be as much a product of human nature as 'rugged' individualism; seeking vengeance as natural as forgiving. Why are humans willing to die for causes? Why are they willing to kill for causes? Why do seemingly irrational desires and fears so often override *analytical* understanding, even when we know that this is happening and do not want it to be the case? In social life, why do messy politics constantly subvert elaborately established administrative procedures, legal systems or constitutions? And why do we so vigorously deny that the above behaviours are an intrinsic part of being human?

To argue for 'order in process' in human existence through the motivating power of species-typical desires and fears, an operational definition of desires and fears has to be made clear. Elsewhere I have argued (Vannelli, 2001) that species-typical bio-electric/ neural processes 'charge' (through the generation of 'mood states' – such as: manic excitement, heightened sensitivity and depression – pp. 112–118) evolutionarily derived emotional processes and potentials for cognition. Charged emotional processes take the form of: lust, love, jealousy, hate, envy, empathy, guilt, embarrassment, shame and depression. Cognition was seen less as problem-solving than as innate tendencies to: (1) make binary distinctions, (2) classify/evaluate, (3) stereotype, (4) make identifications, (5) think deterministically, (6) argue teleologically, (7) deceive and (8) self-deceive. I went on to suggest that species-typical desires and fears are a product of various mixtures of these that result in a relatively limited number of universal *propensities* for 'patterns of behaviour'.

It is to be noted that in this approach desires and fears are not excessively random because our propensities to emotionality and cognition are genetically anchored. Nor, however, do they become completely static entities, even in the life time of a given individual. Desires are based on powerful attractions and even addictions, and as a result they are almost never completely satisfiable – the obsession and pleasure is very often in the daydreams and fantasies that surround them rather than in their fulfilment, which, as a result, is often felt as disappointment. Jonathan Dollimore (1998) quotes Schopenhauer to the effect that desire is 'a striving that is bound to frustrate itself' (p. 173). This is because it is based on 'want and lack ... a condition of continual, restless, longing' (Dollimore, p. 173, see also, pp. 312–327). As a result of disappointment and longing, the objects and behaviours of desires have a tendency to escalate well beyond functionality; we want designer clothes even when they are less useful than much cheaper alternatives; we want more money even when we have more than sufficient for all survival or reproductive purposes; we want more excitement even when our lives are relatively fulfilled; we want more exciting sex regardless of what our sex life is like.

Fears too are volatile. Fears are based on perceptions of danger and avoidance, potentialities that existed well before humans evolved. 'The system that detects danger is the fundamental mechanism of fear' (LeDoux, 1998, p. 18, see also, pp. 127–180; Brown, 2003). However, 'When ... danger inducing stimuli – from external and internal sources – are dreamed, daydreamed and thought about they become more than activators of bio-electric precursors to flight, freeze or panic reactions', they are becoming human fears, a process completed when these feelings are to some extent at least cognitively ordered and these orderings are bio-electrically charged (Vannelli, 2001, p. 148). From here it is easy to exaggerate fears in our imaginations so that they seem much worse than an objective analysis would suggest. We turn dark places into the home of evil; we turn strangers into cut-throats; we see in heretics the end of civilization. Additionally, both desires and fears are easily imagined through time, forward and backward, so they become a glorious or dreadful 'future' to come, or lost opportunities brooded over and fretted about, or myths of a golden age of past happiness and pleasures.

Desires and fears represent the meshing of our inherited drives, instincts, passions, reasoning and cognitive tendencies into emotionally 'charged' motivational forces (however conflictual and/or ambivalent their impact often is). Desires and fears interface between the unconscious/conscious mind and behaviour, resulting in motivations for species-typical human behaviour, social patterns and culture. The goal of this work will be to seek to discover species-typical desires and fears, and to reflect on some of the species-typical patterns of the 'order in process' of human social and political behaviour that might arise from these. The idea is to do so without postulating design, direction, purpose or the supremacy of any moral order or of rationality.

'Reason' in this approach is a search for general explanations and meanings concerning our environments and our lives. As such, reason is not given a special power of understanding above and beyond normal emotional-cognitive processing. Practical reason can help us make a tool for a specific use, or to link reoccurring events, or to practice science; but it can also lead to hierarchically evaluating, classifying, deceiving, stereotyping and making up fantasy explanations about the universe, humans and self – including teleological explanations which assign meaning and purpose where there is none. When considering human nature as underpinning socio-political behaviour reason is to a very large extent the slave of passions, as Hume suggested (1958; see also, Freud, 1961).

Searching for desires and fears

First, I will look at human evolution with the emphasis on sexual selection. The task will be to consider what primate behaviours were most likely under selection pressure during human evolution in light of the observable changes in behaviours, especially in sexual/reproductive behaviour, we observe today (Chapters 2–5). The search will be for the sexual, developmental and psychological processes that could have been involved in the transformation of general primate sexual instincts, drives and reproductive behaviour into human sexual emotions, desires, fears and reproductive behaviour. Together with additional evidence from hunting and gathering societies, and from group behaviour, I will consider how these motivators could have been sublimated into a number of human social and political desires and fears.

Following this evolutionary line of reasoning, special attention will be given to the possibility of the evolution of human politics with a small p. The emphasis will not only be to seek out characteristics of a Machiavellian intelligence in which humans evolved skills for deceit, spotting and anticipating deceit and being able to calculate social advantage (Cf., Byrne and Whiten, 1988; Whiten and Byrne, 1997; Leslie, 2011), but also a number of other behaviours that deserve to be considered as political, and which, arguably, had a great influence on reproductive success (and consequently on human behavioural motivators). The personal politics that result from lust, love, jealousy, envy, hate, empathy and from the general presentation of self, for example, are just as important to our understanding of reproductive success as Machiavellian abilities for personal/social cost benefit analysis. In fact, it can be argued that politics represents a conflicting, confused, muddled, ambivalent human nature *at a social level*. So, the dynamics of personal politics may be one of the most significant environmental conditions in which sexual and natural selection took place as hominins became modern humans.

Discovering fears presents additional problems to those encountered in ascertaining desires because humans tend to not only deny the existence of fear feelings even more than denying desires, but also to seek to hide them. This is probably because many major fears have been (partially at least) repressed by selection. Our psychology includes fears that make us avoid many dangers (of snakes, for example), but also includes means of repressing them so that we continue to 'cultivate our gardens'. We build and rebuild cities on earthquake fault lines; we fall in love over and over despite how many times love goes wrong; we marry and marry again when the previous one does not work out; we play another game after each loss; we continue to go to war despite the horror of previous wars.

Some fears remain dormant – unconscious – for large periods of time until some objects and/or events set them off suddenly; early separation anxiety resulting in a fear of abandonment, for example, might emerge much later in reaction to a possibility of rebuff or infidelity. We tend to forget or avoid fear-generating thoughts and memories, often editing, or even 'reconstructing', those dreams, daydreams and thoughts that activate them. '"I did that", says my memory. "I can't have done that", says my pride, and remains adamant. Finally, memory gives way' (Nietzsche, cited in Ash, 2002, p. 6). But fears of all of these potential dangers linger – barely consciously, sometimes unconsciously. We somehow know that failure and loss in many aspects of life are always possible. But this does not make such fears easier to discover. Reactions such as insomnia, despair, frustration and simmering anger, episodes of disgust, depression, envy, jealousy, hate and guilt, which we would suspect are caused by repressed fear, are not easy to evaluate because individuals who experience them regularly are often the least willing to talk about them. And even when they are, the accuracy of human memory is notoriously selective, and, again, can be hard to tie to specific fears.

So, to go deeper into fears than an evolutionary perspective by itself might provide, we must also look for some relatively reliable 'beneath the surface' dimensions of fears. In Chapter 6 I will look for objects and events that trigger automatic, or at least semi-automatic, flight, repulsion, disgust, terror, panic, weeping, trembling, outbursts of anger, feelings of guilt and strong avoidance reactions as indicating fundamental fears. In everyday life, for example, it is common to observe weeping, wailing, anger, despair and feelings of guilt in response to the loss of a loved one, and to notice that such reactions are often very hard for individuals to control or conceal. Over a slightly longer term, we can suspect that a sense or fear of loss in the form of jealousy and/or feelings of failure or humiliation, for instance, can lead to obsessive worry, visible anxiety, a sense of guilt, and eventually to prolonged stress and states of depression; 'loss lingers long after the event ... often at the fringe of desperation of getting on with life ... Loss is a hard legacy to pass on' (Welch, 1994, p. 285). While humans may desperately try to hide these manifestations, they are rarely capable of achieving such deception.

At a social level, we often set out to avoid extreme 'abnormality', and to defend 'good' against 'evil' as a projection of our own fears onto external objects or activities. Definitions of abnormality and of evil can thus suggest a number of human fears. Social injunctions ('thou-shalt-not' taboos/laws) devised to prevent specific 'social dangers', including 'evil', can also be considered as additional evidence of fear-generating conditions and of specific fears. A consideration of evil and its control will also be undertaken in Chapter 6.

In Chapter 7 the desires and fears, derived from this analysis, that seem species-typical will be outlined. This will be followed by a look at the interrelationships – the compatibilities and conflicts – of the desires and fears compiled, and at their impact on general human behaviour. Evidence of both randomness and of 'order in process' from the above analysis will be considered. A suggested outline for a social science based on this analysis will be followed by a consideration of its implications for a notion of justice based on desires and fears (and an innate sense of fairness) rather than one based on reason, rationality, moral imperatives or utopian blueprints. The argument is that we can have a meaningful social science and a notion of justice based on species-typical desires and fears without backing away from the full implications of Darwin's approach; a social science which avoids the moralistic, spiritual and ideological traps that have hitherto hindered those philosophies, social sciences and political theories looking to explain and perfect man 'the unique animal'.

2 Human evolution: the background

In the decades following Darwin there was a desperate search for the big-brained ape, or at least one about halfway between apes (with a braincase size of 400–500 cubic centimetres) and modern humans (1200–1350 ccs). The general view in the press, the educated imagination and among amateur fossil hunters was: 'how could something be considered even proto-human if it did not have a brain much bigger than an ape?' This was also the view of a number of world-ranking (and greatly influential) British palaeontologists, anatomists, zoologists and geologists (Johanson and Edey, 1981; Reader, 1981). Some simply concentrated on the growth and development of the brain itself as the driving force not only behind human evolution but also behind the evolution of mammals and primates generally. This was because of its assumed superior powers for adaptation whatever else might have existed in the particular history of a species (Cf., Grafton Elliot Smith, 1927). Others with more varied causal explanations, such as T.H. Huxley, Sir Arthur Keith and Sir Arthur Smith Woodward, nevertheless had definite ideas about how big a brain should be before anything could even be considered to be ancestral to humans – approximately 700–750 ccs (Johanson and Edey, 1981; Reader, 1981).

Scholars of evolution set out to discover changes in the physical environment at the time of suspected human differentiation in order to see what dangers or opportunities might have acted as selection forces for the evolution of a large brain. In this approach, scenarios of human evolution were constructed from a postulated 'master' behaviour for dealing with problematic environments and/or exploiting opportunities in them. Such behaviour, it was argued, acted as a generator of a feedback mechanism that resulted in the development of rational skills and, eventually, 'rational man'. The environmental problematic for evolving humans most usually took the form of a move to the open savannahs from the forest-like conditions in which apes now live. The original master behaviours included: *bipedal locomotion* (Dart, 1925; Keith, 1931; Washburn, 1963a, 1968), *tool making* (Engels, 1940a; Oakley, 1958, 1961; Washburn, 1960, 1978; Wynn, 1988; Gibson and Ingold, 1993), *hunting* (Washburn and Lancaster, 1968; Suzuki, 1975), *seed eating* (Jolly, 1970), *gathering* (Zihlman, 1981; Tanner, 1981; Dahlberg, 1981), *scavenging/foraging* (Milton, 1988), *food sharing* (Isaac, 1978; Leakey, 1981), *household economy* (Quiatt and Kelso, 1985), *defence against predators* (Ehrenreich, 1997) and *meat eating/cooking* (Aiello and Wheeler,1995; Wrangham, 2002, 2009).

Regardless of which of these (or which specific combination of these) were/are placed at the centre of an evolutionary scenario, it has been argued that they in turn selected for the tool-using skills, intellectual skills and language capacities that provided the basis for 'efficient' human social organizations and cultures for savannah conditions. In doing so, they have generally implied a human nature that includes considerable rationality, while greatly downplaying anything approximating instincts, drives and/or emotionality. Furthermore, as Donna Haraway (1989) has pointed out, these scenarios have generated not only 'rational man' but also a liberal, anti-racist, co-operative man. It is almost as if the outcome of modern 'Enlightenment Man' was the reason for evolution in the first place – or at least for human evolution. Certainly, in these scenarios there is a clear suggestion that humans have risen above their animal ancestry (and the processes of biological evolution) because of coming to possess unique forms of reason, rationality and culture.

Problems with traditional scenarios of human evolution

Unfortunately for such approaches, while several of the above behaviours undoubtedly played a part in human evolution, especially later in the hominini line, scenarios which arise from them are generally not only teleological but are based on very little empirical evidence. More importantly, however, the earliest fossil finds raised a degree of havoc with the above propositions concerning 'humanness', because they suggested creatures with certain human-like characteristics, especially regarding bipedal locomotion, but brains no bigger than those of an ape (approximately 400–450 ccs). And, as the fossil evidence began to grow, this 'problem' did not go away.

The earliest candidates for bipedal locomotion seemed to have brains no larger than modern apes (for example, *Ardipithecus kadabba* – 450 ccs – approximately 5 myr), and this small brain continued up through the much more established bipedal *Australopithecus* fossils (from *anamensis* to *boisei*, 4 to 1 myr – 450/530 ccs). With *Homo habilis* (2.4–1.5 myr) there was a slight increase in skull size (550–600 ccs), but it was only with the emergence of *Homo georgicus/erectus* (1.8 million to 300,000 years ago) that the brain began to grow more dramatically, rising from approximately 600 ccs at the start and reaching about 1200 ccs by about 300,000 years ago. From there it grew relatively steadily until it levelled out at 1350 ccs or so in modern humans (see Chapter 4). If bipedalism freed the hands for tools, selecting for human intelligence and 'rationality', it seems that there was a lot of hominini evolution to get through before that happened.

Additionally, besides the problem of 'small brained man', evidence has increasingly suggested that there does not seem, in fact, to have been any *specific* environmental traumas or special 'intellectual' challenges/opportunities during the early period of human differentiation that would necessarily have set any of the above postulated mental developments in motion. Indeed, the 'move to the savannah' hypothesis has increasingly come under serious question. Certainly, there was a major rifting process in east Africa, the suspected place of hominin origins, from about 20 million years ago. And there was the beginning of a major global cooling from about 12 million years ago (Isaac, 1976; Jones, Martin and Pilbeam, 1992, appendix II; Adams, 2005), with a reversal at the beginning of the Pliocene (5–2.5 myr) before it began to turn cold again (Tattersall, Delson and Van Couvering, 1988). But such long-term changes, however dramatic in the long run, cannot be used as *proximate causes* of the evolution of a specific species. While such changes might generate 'evolutionary opportunity' (Lewin, 1993, p. 12), eliminate potential competitors through extinction (Foley, 1994) and explain why a particular type of behaviour and its supporting anatomy might eventually survive, if not thrive, it does not explain why or how the species came into being in the first place.

In looking at modern evidence it appears that during 'the earlier stages of human evolution ... Hominids were subject to restless fluctuations in their habitats, but not complete ecological disruptions' (Isaac, 1976, p. 133; Foley, 1994; Wood, 2002; Adams, 2005; Smithsonian, 2013a). Certainly they do not seem to have suffered savannah conditions. *Sahelanthropus tchadensis* (6–7 myr), for example, thought by some to be a possible common ancestor of humans and chimps (Brunet, *et al.*, 2002; Wood, 2002, 2002a; Roach, 2002; Parsell, 2002), lived in semi-jungle conditions with lakes and forest-edged rivers during the upper Miocene (7–6 myr). The area of the fossil find suggests fluviatile deposits as a result of 'episodic flash floods', a large lake basin (such as the 'Ancient Lake Chad Basin') and lake shores, and generally 'Swampy, well-vegetated areas ... [with] ... large and permanent waters bodies' (such as 'Mega Lake Chad') (Vignaud, *et al.*, 2002, pp. 152–153, 155; Leblanc, 2005); high sediment yields suggest extensive rainfall, including 'high monsoon activity' (Griffin, 1999, p. 1). In that period, even such current desert areas as the Arabian Peninsula were serviced by a major river system that flowed through numerous small channels, making up a 'braided network tens to hundreds of meters wide' (Whybrow, 2004). This may have been part of a larger system that included the Tigris and Euphrates rivers. This area, it appears, formed a forest bridge between Africa and south-western Asia through which animals migrated (Whybrow, 2004; Van Covering and Van Covering, 1976). These are conditions not unlike those in which chimps, especially bonobos, live today (see 'Primates' section of this chapter).

The scant hominid/hominin fossils of *Orrorin tugenensis* (approximately 6 myr) suggest tree living to have still been very much practised (Wood, 1994, 2002a; White, *et al.*, 1994; Senut, *et al.*, 2001; Galik, *et al.*, 2004; papers in *Science*, 2009). And in the more firmly established ancestral claim of *Ardipithecus ramidus* of at least 4.4 myr it is clear that 'Ardi' lived in wooded areas which included hackberry, fig and palm trees with small patches of heavier forest. And while Ardi was functionally and anatomically bipedal in terms of pelvis and lower limbs, she was also relatively adept in the branches of trees in terms of forelimbs (White, *et al.*, 1994; White, *et al.*, 2009). As we move through the Plio-pleistocene (5–2 myr) the climate was several degrees warmer than currently, with the oceans approximately 30 metres higher than today. However, there was 'almost constant flux and oscillation in aridity and temperature' (Adams, 2005; see also, Isaac, 1976). Tropical forests and moist savannahs reached as far north as 21 degrees (running across the very top of current Africa, well above most of the current Sahara Desert). There may have been a continuous east–west rainforest across Africa, with a great range of habitats, along with considerable fluctuation (Van Covering and Van Covering, 1976; Whybrow, 2004; Adams, 2005).

The very early hominins (*Ardipithecus/Australopithecus*) from the Kanapoi and Allia Bay areas of northern Kenya (4.2–3.9 myr), for example, seemed to live on the shores of a lake with river systems and 'gallery forests', where, besides mammals, remains of fish and aquatic reptiles were common (Leakey, M.G., *et al.,* 1995; Leakey, M.G., and Walker, 1997; see also, Hay, 1976; Leakey and Lewin, 1979). The Pliocene bipedal hominins from the Hadar region of Ethiopia (early Australopithecines, 3.9–3 myr) lived in an environment which seems to have consisted of 'a freshwater lake surrounded by marshes and fed by rivers ... The local environment was far more humid and wooded than ... today ... [M]egafauna as hippopotamus and crocodile are also consistent with these paleoenvironmental reconstructions' (Tattersall, *et al.*, 1988, p. 240; WoldeGabriel, *et al.,* 1994). The hypothesis that Australopithecus may still have had a foot adapted to grasping, and thus to being able to live effectively in trees (Clarke and Tobias, 1995; Lewin, 1995), and that they ate forest foods, including bark, fruit and palm products (Henry, *et al.*, 2012), further reduces the notion that a dramatic change to savannah living led to human evolution (See also, Wood, 1992; Andrews, 1995), as does the recent discovery of a contemporary cousin to the early Australopithecine 'Lucy', which appears to have been quite well adapted to tree living (Haile-Selassie, *et al.*, 2012).

Fossils from the Turkana basin (Kenya) suggest that pre-*Homo erectus* hominins (2 myr) 'lived in a wet – and possibly even a marshy – environment', bordering even on being 'lush' (BJS, 2010; Braun, *et al.*, 2010). Recent discoveries of fossil fragments of *Homo habilis* (2.4–1.5 myr) and *Homo erectus* (1.8 myr) found in the same location suggest an environment similar to that of gorillas and chimpanzees today (Spoor, *et al.,* 2007). Tropical/semi-tropical conditions with aquatic fauna, it appears, continued for some considerable time after bipedalism, not only in Africa but elsewhere as well. Recent work on Dragon Bone Hill, slightly southwest of Beijing, suggests that 'Peking Man' (*Homo erectus*) was a scavenger rather than hunter and 'in essence a tropical animal, shunning the cold and with a tenuous grasp on fire' (Boaz and Ciochon, 2004, p. 32). The time span for these Chinese fossils is somewhere between about 670,000 and about 410,000 years ago. At approximately the same time what appear to be different cultures of *Homo heidelbergensis* were competing for space in Britain, which has been described as being much warmer than today with a land link to the continent. Even more recently, approximately 160,000 years ago, the Hetro people (*Homo sapiens*) living in what is now Ethiopia occupied a 'warm tropical plain near a shallow freshwater lake, teeming with hippos, crocodiles and catfish' (Randerson, 2003, p. 5).

There does not seem to have been a food resource problem that would have selected for increased anthropoid brain power. Most researchers speculate that the earliest hominins practised opportunistic, generalist foraging, hunting and perhaps also fishing modes of existence. This probably took place in relatively dense woodlands (Andrews, 1989, 1995; WoldeGabriel, *et al.*, 1994; Adams, 2005), or at least in 'mixed habitat', rather than predominantly savannah conditions (Isaac, 1976a; Isaac, *et al.*, 1976; papers in Harding and Teleki, 1981; Andrews, 1992; Leakey, M.G., *et al.*, 1995; Leakey, M.G. and Walker, 1997; Vignaud, *et al.*, 2002; papers in *Science* 2009). Indeed, as more fossils are found, paleontological analysis is suggesting a considerable aquatic fauna associated with the earliest hominins; this includes a large variety of fish species, amphibious mammals, mussels, catfish and turtles (Parsell, 2002; Roach, 2002; Vignaud, *et al.*, 2002; Whybrow, 2004), as well as insects, birds and small rodents and possibly various root foods common to subtropical forests and marshy edges of lakes (Wrangham, 2002; Whybrow, 2004). *Ardipithecus ramidus* (4.4 myr) seems to have been a 'ripe fruit specialist' along with being a consumer of insects, birds and small rodents (papers in *Science*, 2009; White, 2009). Recent evidence indicates that as late as 2 myr pre-*Homo erectus* hominins in the Turkana basin of Kenya fed on 'turtles, fish, crocodiles and antelopes', and even hippopotamus (BJS, 2010; Braun, *et al.*, 2010).

In so far as all these food sources were as abundant throughout the majority of the period of human evolution as the fossil record suggests, food would not have been a major problem for evolving hominins. There does not seem to have been excessive predation pressure that might have made humans that bit more cunning (papers in Isaac and McCown, 1976; Leakey and Lewin, 1979a, 1992; Johanson and Edey, 1981; Reader, 1981; Dawkins, 2004; but see, Brain, 1981; Ehrenreich, 1997; Smillie, 2002). Some have argued that perhaps predation from other human groups was important (Alexander, 1974; Alexander and Noonan, 1979; and see Ardrey, 1961); but this last idea is derived more from theoretical speculation based on the not infrequent wars *modern* humans have engaged in rather than on clear evidence of its existence at the point of human differentiation. Such reservations do not make it wrong, simply that more evidence is required before this can be considered to be a major factor in human evolution.

Whatever the case, it is now clear that humans were bipedal (the first major morphological change in, and the definition of, the hominin line) long before savannah living became common, if it ever did. As the geological and fossil evidence has piled up, the savannah hypothesis has progressively faded from the scene (Stanford, 2002; Wood, 2002); indeed, it is all but dead (Strier, 2002; Wood, 2002a; papers in *Science*, 2009). Furthermore, although the fossil evidence remains scant, what does exist (and it is increasing all the time) clearly indicates a fully bipedal hominin for something like three to four million years *before* there is evidence of: extensive hunting (especially of any type of large game), tool use (at least of the worked stone variety), significant forebrain enlargement or language. In fact, all of these seem to have developed during the last two million (at the very most) to seventy thousand years of human evolution (Washburn, 1960, 1978; Wynn, 1979, 1988; Johnson and Edey, 1981; Reader, 1981).

This is important for our understanding because open country savannah living has been, one way or another, highly implicated in the master behaviours of almost all scenarios of human evolution. It is usually postulated as the new environment with which proto-humans had to cope. It was this environment, it has been argued, which selected for human rationality, either directly because of the challenges of finding food on the hot savannahs in competition with ferocious savannah predators, or through the demands of hunting and associated tool use generally. The point about the lateness of brain enlargement is very significant in that the early scenarios of human evolution place a great deal of emphasis on brain development as being the centre of the biological feedback processes set in motion by the postulated rationality-selecting master behaviour(s). But, as we have seen, the evidence is against both savannah living and brain enlargement for something like the first 3–4 million years of human evolution.

There is another major problem with the above scenarios. Even if early hominins eventually did spend some time in savannah-like conditions, used tools not preserved in the paleontological record, applied emerging 'computer-like' cost-benefit-analysing minds to survival (and even reproductive) ends, and cooperated in 'rationally' organized hunting endeavours, it is not easy to see why many of the particular characteristics of the human mind which *did* evolve would have done so. Many observable human mental conditions would seem to have been a distinct disadvantage for the above activities. Consider, for example, tendencies to anxiety, obsession, and overpowering feelings of jealousy, guilt and depression. And would fears of engulfment by adolescents, to say nothing about the disruptive influences of an infant 'will to power' and sibling rivalry, plus conflicts generated by lust and hate, by love and envy, for example, have been the dream ingredients for the evolution of a 'rational mind' seeking to use tools and to socially cooperate more and more efficiently on the way to becoming nature's ultimate rational species?

If a wise committee of nature (or of gods) were to devise an efficient forager, hunter, rational actor, master of the savannahs, destined for greater and greater achievements, it is far from evident that it would have selected the human psychosexual, emotional-cognitive processes we observe. Nor is it likely that it would have been keen on the development of human self-awareness and death-awareness, not to mention the distracting human ability to forever postulate 'the good' and to identify, despise, persecute (and kill) evil doers. Certainly, numerous species have become successful savannah foragers, hunters, scavengers and reproducers without all of this 'excess baggage'. Moreover, it remains unclear how bipedalism might have been a trigger for the evolution of 'rationality'; it certainly does not, *of necessity*, lead us to scenarios of human evolution in which cooperation and continual human perfecting are the only options. Free hands can be used for tool-making, but they can also be used for stealing, fighting, raping, throwing rocks and swinging clubs – for hiding things, for display, deception, kidnapping, masturbation and murder (as, of course, can intelligence). And, anyway, it does seem that hands were free a very long time before tool use became a major part of human life.

With the evolution of human nature we are into the 'explain the peacock's tail' game. We are into explaining the evolution of not just reason but also the apparent irrationality of humans, the random, the chaotic; we are into explaining not only tranquillity but manic happiness, desperate sadness, love and hate, war and spirituality. Our task, then, is to explain human evolution in terms in which humans actually evolved *despite* these tendencies and/or *because of* them, not to write these tendencies out of our evolutionary scenarios. The standard approaches to human evolution have tended to ignore the effects that changes in such things as birth intervals, signals of sexual attractiveness, triggers of sexual receptivity, patterns of sexual desires and fears, mate competitions, sexual deceptions, infant care practices and sexual politics might have had in influencing the evolution of species-typical human nature, including its more advanced cognitive abilities.

This may have been because such an approach would have opened up a seamier side of humanness. It would have suggested, for example, that competition, conflict, envy, jealousy, deception and violence, for instance, were as much a part of being 'natural' humans as 'rational' tool-making and 'intelligent-surviving' are claimed to be. And if one looks at sexuality, marriage and family among modern humans there can be observed much joy, but also boredom and drudgery; we see incest, child pornography, child sex, rape, wife beating, divorce, adultery and murder. Might all this too be rooted in human phylogenetic history? Might Freud, for example (having studied human sexual and personal conflicts in detail), have as much to say about human evolution as those palaeontologists who, from the smallest of fragments of sometimes a single bone, created scenarios of evolution culminating in humans as the rational animal: a rational animal that hunted cooperatively, divided up labour between the sexes, ages and skills, lived in families linked to wonderfully supportive kinship networks and worked in *adaptive* harmony with nature (awaiting the mental breakthroughs that would inevitably lead to civilization)?

The omission of reproductive and emotional factors in considerations of human evolution is even stranger in light of the importance of reproductive processes in the Darwinian framework. It is also surprising in that, from what is known about genetics, embryology, physiology, gynaecology and brain sciences, relatively fewer mutations would be required to affect such things as growth patterns, birth intervals, neural network potentials and other brain mechanisms for developing emotional-cognitive motivators (Cf., work reported on by Goy and McEwen, 1980; see 'Genetic and embryological processes' section in this chapter and Chapter 3), than those required for the more gradual morphological changes that are basic to most traditional evolutionary scenarios – morphological changes such as, bipedalism, an opposable thumb, restructured face, jaws, teeth and skull. Indeed, a number of developments in modern genetics and evolutionary theory itself have come to suggest possibilities of relatively dramatic phenotypic and behavioural changes in reproductive behaviour as a *result* of relatively minor genetic alterations and of genetic and/or reproductive isolation as major factors in processes of natural and sexual selection. For example, a change in a HOX (developmental) gene which maintained child-like characteristics into early adulthood would, through resultant muscle/joint flexibility, make bipedal walking easier (and more common) for a young ape. Bipedal walking would have affected further muscle and bone development, altering the posture and the way the head is held during growth, affecting the final muscle and bone formation in the neck and skull regions (see Chapter 4).

If we cannot find a dramatic, almost catastrophic, environmental reason for the selection of these morphological changes we certainly should look at reproductive processes, because, after all, it is through these that the selection process generally operates (Cf., Darwin, 1872, 1874; Fisher, 1930; Lovejoy, 1980, 1981a; Miller, 2000; Hamilton, 2001; Jones, 2002). It is for this reason that a number of the theoretical propositions from modern sociobiology, evolutionary psychology, sexology and general human psychology can be extremely important as a means of speculating about areas of reproductive behaviour which might have been subject to selection pressures during human differentiation. Sociobiology (Cf., Trivers, 1972; Wilson, 1975, 1978; Hrdy, 1981, 2011; Dawkins, 1989, [1976]; Barash, 2001), for example, brought the study of evolution back to reproductive processes by making the concept of 'inclusive fitness' (total reproductive capacity measured in relative terms) a central theme. But sociobiologists appeared either unable, or unwilling, to bridge the gap between their 'rational' cost-benefit propositions about 'selfish genes' maximizing themselves and actual motivations to behaviour. For example, mathematical calculations of degrees of kinship/genetic relatedness and/or anticipated reciprocal altruism became the suggested bases of human cooperative social behaviour without an explanation of how these were translated from genes into actual behaviour. Evolutionary psychologists recognized this problem but, for many of them (as noted in Chapter 1), their answer was mind-as-computer, mind as a rationally calculating mechanism (cost-benefit analysis in another guise) for *adapting* humans to their environments.

It can, nevertheless, be argued that sociobiologists provided an advance in evolutionary thinking by bringing us back to Darwin's concerns with sexual selection. Additionally, some way forward for understanding human sexual nature has been suggested by those among them who saw that sexual conflict could not be ignored, nor could such things as anger and jealousy. A number of feminist sociobiologists, along with some evolutionary psychologists, primatologists and paleoanthropologists, emphasized the potential female role and/or the importance of not only sexual conflict in human evolution, but also (among some) the role of emotional motivators (Cf., Hrdy, 1981, 1999, 2011; Fisher, 1982; Wilson and Daly, 1992; Buss, 1992, 1994, 2000; Miller, 2000; Campbell, 2013; Gray and Garcia, 2013).

This approach, when taken to its logical conclusions, does not have to be just about conflict between males and females, males and males, females and females, parents and children, but also between siblings, between elders and young men, between matriarchs and daughters-in-law, between mothers and sex-bombs, homemakers and *femme fatales*, providers and philanderers and between fantasies and realities. From this approach reproductive success can be seen, and analysed, in terms that might be described as 'sexual politics' in which such things as devotion, subordination, domination, 'dangerous' lust, envy, the violation of trust, jealousy, depression and physical combat are a significant part of the process itself. This approach, it will be argued, can be also developed to include a consideration of social (or political) politics. Social/political politics are about domination and subordination in the social environment; they are about centralizing and/or decentralizing personal and group independence/sovereignty; they are about struggles for control/power, over self, others, resources and social environments. Social politics, while undoubtedly having originated from sexual politics within social species generally, once established in the primate line have taken on a life of their own, and as such have contributed greatly to human social life in its modern form – including modern forms of reproductive behaviour.

It can be seen, then, that human evolution is a complex business. It is not just about trying to find out how humans ended up with big brains and then assuming that these brains were used to construct human civilization and will, eventually, achieve a state of perfection of both persons and societies. Evolution is not a teleological process, one designed to create humans, the highest level of earthly existence. It is what it is: a mindless, somewhat random, process in which characteristics that successfully reproduce are manifest, at least for a time, on earth.

Evolutionary processes

In a search for desires and fears (or, indeed, any evolutionary outcome) free from attributed meaning, direction and purpose there are at least three major propositions from evolutionary thinking that have to be emphasized.

(1) Chance happenings (for example, mutations, environmental traumas, new ideas, accidents, inventions, indeed even planned policies) occur independently of eventual outcomes; all teleological – or Final Cause – explanations must be rejected. Evolution works from a given starting point to an unknown future.
(2) Chance happenings only have an effect in relationship to their starting point (the environmental, genetic, embryological, developmental, reproductive and historical constraints and opportunities in the context in which they occur).
(3) The third element of evolutionary thinking is a lot more controversial than the first two and might not be so fully agreed by all evolutionists. It is that from an evolutionary perspective we are not looking for the evolution of a particular form of *adaptation* that shows a synchronic unity and overall (wondrous) design of nature. In the case of humans we are not looking for a form of reason or 'rationality' that has lifted us above the laws of nature. Rather, in evolution, we are looking to find out what survived and successfully reproduced (and how they did so). Any 'order' that we find must be treated as a by-product of evolution, not as an 'end-point' cause of evolution.

To set the stage, then, for a consideration of human evolution in a search for species-typical desires and fears let us look more systematically at some of the issues I have so far only alluded to when discussing traditional scenarios of human evolution. First, we must acknowledge that the environment remains important. This is because it is the contextual starting point of human evolution from an ape ancestry. It represents the constraints and dangers in which humans evolved. Second, a consideration of a number of genetic and embryological processes has to be included in our analysis. This allows us to assess the patterns of evolution that might have been possible – even likely – in the context – physical and social – in which humans started to evolve. Third, there are a number of patterns common to sexual selection that have relevance for the evolution of any species, because if mutated individuals do not reproduce *successfully* those mutations die out very quickly. And fourth, following this, the history of the species (anatomical, behavioural and social) has to be – in so far as is possible – ascertained to see how reproductive successes may have been possible, indeed enhanced, from postulated genetic and anatomical changes. Elements within each of the above categories will often blend together as they affect species differentiation. Nevertheless, for the purposes of clarifying the important factors within each category for a specific consideration of human evolution in Chapter 3, each will be considered separately.

The physical environment and human evolution

It seems strange that creators of scenarios of human evolution have not felt a need to consider swamps, bogs, pools, fenlands, marshes, bayous, creeks, rivers, lakes and regular flooding episodes as potential environmental conditions during human differentiation. It is strange in that it is difficult to conceive of tropical forests, the home of ancestral primates, suddenly dumping some apes onto a dry savannah (especially since considerable tropical areas remain even today with relatively low primate densities). The idea of a transitional geographical stage from tropics to open country, somewhat like the above wetlands, seems rather more probable; and, as we have already seen in discussing the environmental conditions surrounding fossil finds of the earliest hominins, semi-aquatic conditions seemed most likely to have been the conditions of hominin evolution (see fossil references above).

And, indeed, wetland conditions are where the vast majorities of pre-agricultural people have lived, and still live, in anthropological and historical times. While we have a romantic image of hunters and gatherers living on the great plains of the American West, the vast majority of American native groups actually lived in more aquatic areas such as the Great Lakes, in relatively dense forests along the coastal regions of the west and south east, and along rivers. Not only did more groups live in these regions, but also their populations grew considerably denser than among the plains hunters (Fagan, 1986). More generally, we hear a lot about the San of the Kalahari Desert or the Aborigines of central Australia, but they made up a tiny proportion of the hunters and gatherers of Africa, Australia or South America; the vast majority lived in rainforests and/or close to rivers, lakes and coastlines (Cf., Service, 1963; Lea, and DeVore, 1968; Woodburn, 1982; Fagan, 1986), as was also the case in southeast Asia (Cf., Mead, 1942; Benedict, 1946). Indeed, even most of the Aborigines in Australia lived in coastal areas (Service, 1963). At the same time, the role of fishing in pre-agricultural societies has been greatly underestimated (Cf., Palsson, 1991), indeed, even the San relied on fish to a considerable extent (Fagan, 1986).

Such environmental conditions have been a common feature of the earth for at least the last 100 million years or so and remain extremely prevalent even in today's dam building, water channelling times. Until relatively recently, for example, all of the existing major rivers of the world were prone to vast seasonal and cyclical flooding causing enormous disruption to plant and animal, including human, life. Such environmental changes can be very rapid even in terms of 'real human time'; and this is not even to mention hurricanes, earthquakes, flash floods and droughts, which are very difficult to predict even with modern technology (Cf., Kaplan, 2002; Alden, 2013). Over a slightly longer time span (although nothing in human archaeological terms, or human evolutionary terms, let alone geological time) we can observe that rivers have greatly changed their courses, lakes have come and gone, swamps become mixed woodland, scrub becomes desert, deserts become woodland (Cf., USGS, 1998). One hundred and twenty years ago or so, the Powder River Basin in Wyoming was lush grassland, feeding great herds of cattle (as it had bison a few years before that); now most of it is semi-desert scrub land and the Powder River is 'a mile wide and an inch deep, too thin to plow and too thick to drink' (Kloor, 2002).

What are the advantages of considering environments that are water-based, semi-jungle and fluctuating in our scenarios of human evolution? In the *first* place, such conditions provide abundant possibilities (intentional and accidental) for geographical isolation, including for very small groups of individuals, a condition related to relatively rapid speciation (Cf., Jones, 1913). The evolution of the 'dwarf' *Homo erectus, Homo floresiensis*, with a significant reduction in brain size over a relatively short time (Adams, 2004; Foley, 2011), shows how important geographical isolation (in this case, an island) can be in hominin evolution. And the isolating capacity of waterways and swamp conditions can be very great. Throughout the Middle Ages people in the fens of East Anglia in England, for example, had to build and maintain causeways across a multitude of naturally occurring waterways in order to communicate, but these could also be used to maintain separation and provide defence, and to hide robbers and rebels (Pearce, 2003). The marshlands of the Marsh Arabs in Iraq (thought by some to have been the Garden of Eden), very rich in reeds, fish, birds and water buffalo, were thought to be sufficiently separating and impenetrable by authorities that the marshes were deliberately drained by the Saddam Hussein government in order to control the Marsh Arabs (Randerson, 2003a). The might of the US military found it impossible to completely overcome the Seminole Indians in the Okefenokee and other swamps and woodlands of Florida because of the inhabitants' ability to use guerrilla tactics and fade away in the myriad islands and secret routes through the dangerous waters and sands (Seminole, 2013).

In the *second* place, geographical isolation can quickly become social isolation if individuals from specific groups avoid other groups for whatever reason. The pygmies of the Ituri Forest in Africa (Cf., Turnbull, 1965, 1976) most likely evolved their stature as a result of considerable jungle-based isolation from other peoples, as have the many, pygmy and non-pygmy, native tribes in the Amazon basin where similar conditions prevail (Raffaele, 2005). But it is likely that this soon led to social/sexual separation. Any form of isolation can soon result in reproductive isolation through heterozygotic incompatibility – different mutations accumulating in separated groups that prevent zygote formation if the groups come together again and try to mate (Roberts, 1980), leading to the evolution of different patterns of sexually attracting and activating features (Ridley, 1993; Miller, 2000). So, even when geographical isolation ends, different branches of what once were one species are either no longer able or are not motivated (or both) to interbreed, continuing the process of accumulating different mutations to the point where more clearly defined separate species exist.

Furthermore, when a particularly useful mutation, such as one that provides immunity to a particular disease or debilitating parasites, for example, occurs, it spreads very rapidly through a small isolated group (natural selection), and is then protected against dilution by both geographical and sexual isolation (sexual selection). This is not an insignificant example because it is most likely that a physical environment such as described above for human evolution would have included a multitude of parasites, such as flies, mosquitoes, body fungi, lice, ticks and a wide variety of socially and sexually transmitted bacteria and viruses. One can only wonder what the overall genetic effects of the apparent immunity of some individuals to AIDS would have had on the evolution of humans in a pre-scientific age, especially if breeding groups were relatively isolated geographically and/or sexually. And, of course, any non-harmful genetic changes linked to that mutation would have been selected (as riders), so a number of phenotypical changes in a particular population could show up relatively quickly.

In the *third* place, the idea of semi-tropical watery conditions suggests the possibility of relatively abundant food even richer than the fruit and berries on which ancestral apes survived. As we have seen, from the evidence of the Miocene, Pliocene and Pleistocene, food from the general area of early hominini fossils would have included a wide variety of fish, mussels, amphibians, snakes, water rodents, berries, fruits and small woodland mammals. These are relatively high value foods compared to the diet of common apes (Hamburg and McCown, 1979; de Waal, 2002a), even that of bonobos, who have a relatively good protein diet (de Waal and Lanting, 1998), suggesting the potential for survival and reproductive prosperity. Among the most prosperous Native Americans were the Kwakiutl of Vancouver Island who fished and hunted among the small islands and fiords of that area; they were so wealthy that conspicuous consumption in the form of massive potlatches – in which chiefs competed to give each other gifts – was the order of the day (Benedict, 1946; Harris, 1977). Today, marine foods remain the prime source of animal protein for a billion people on the planet, and a major source for many of the rest (MSC, 2014; UN, 2014).

Access to isolated sources of food, such as the aquatic food described above, would have avoided the dangers of intra-species competition for food generally and certainly would have been more feasible than trying to hunt game on the open plains. The idea of an Australopithecus trying to keep up with smaller game like an antelope in a chase hardly seems feasible. And picking-and-choosing among aquatic food possibilities would also have been very much safer than venturing out onto the savannah. As Jared Diamond (1991) observed: 'Along with drinking a strychnine cocktail, poking an adult rhinoceros or Cape buffalo with a spear ranks as one of the most effective means of suicide I know' (p. 39).

In the *fourth* place, such conditions could well have provided strong selection pressures for bipedalism, a capacity and tendency that already existed in most apes as a product of 'orthograde feeding' as described by Susannah Thorpe (2006). She suggests that in relatively tropical conditions orangs generally (see also, Isler and Thorpe, 2003), and chimps in at least one study, swing through trees as their major means of locomotion, but stand on the ground when feeding, using their arms to reach for nuts, berries and fruits in a pattern of 'terminal branch feeding'. If we add the possibility that some of these apes were living in watery or swampy conditions, the importance of orthograde feeding, and being able to reach high and look all around relatively easily seems to give even more importance to an effective bipedal posture, one that was comfortable for a sustained period of time.

Indeed, there are numerous reasons as to why once bipedalism became relatively common the environmental conditions described above would have provided additional selection forces for it. Creating and keeping clear pathways through tropical and semi tropical forests, as well as through marsh, swamp and wet terrain, is more or less a full-time job that almost requires a bipedal stance; moreover, it requires a bipedal stance that is stable and comfortable over a significant period of time. To clear effectively requires reaching a certain height to break, or cut, branches, stems, vines, brambles, reeds and bamboo, for example. These do not always snap easily – considerable twisting, pulling and jerking is often required. Keeping pathways clear necessitates constant attention and labour, or the way is soon lost. It might have been that this was done while also gathering fruit, nuts, berries and even the tubers and roots of the abundant growth that is found in semi-tropical conditions. So, the motivation to clear paths and openings may have been very similar to the motivation to eat. At the same time, paths to dry sleeping quarters, and to the sleeping nests of friends and family, to regular sources of food, and around settlements of enemies, may all have been important to the reproductive and survival success of particular groups of 'apes'.

Once 'work' on the environment had become established, and an advantage, building causeways, small dams and shelters on stilts would all have selected for bipedal abilities, as would standing up to see over the reeds or to see better while partially submerged in water, carrying babies or pulling them out of the water, carrying food and drinkable water back to the nest and signalling to one's mate and offspring over the bushes, reeds and abundant foliage. Swatting flies and mosquitoes, building and installing crude insect nets, picking ticks off mates and offspring (and friends), and making safe beds, may have been as important in selecting for tool-making abilities (and thus for free hands and, as a corollary, bipedalism) as were gathering, hunting and fishing.

In the *fifth* place, all of this requires a degree of intelligence. Existing apes have considerable intelligence as mammals go, and these conditions would probably have selected for an even more *varied* type of intelligence. In these conditions individuals would have to remember and mark their way through the equivalent of the Okefenokee Swamp, remember where mates and offspring were left, remember where all the paths led, and why, and when they required clearing and when was a good time to take a particular path. And it would be important to remember where the fish get stranded in easy picking pools and where the relatively consistently safe ground was to be found, and to learn and remember how to make safe shelters against ever changing (often unpredictable) weather and climatic conditions – conditions which included floods, monsoons, droughts and changing patterns of pests.

This kind of intelligence is not too much beyond what spiders building elaborate webs, ants playing their part in an elaborate division of labour, migrating/nest building birds and dam building beavers do through instinctive intelligence. However, apes start well in advance of all these when it comes to flexibility of intelligence, and so a more variable intelligence (without a much larger actual brain) would have been available for later selection to work on. And this form of intelligence is, of course, in addition to any social/sexual/emotional intelligence that might have been selected for (see Chapters 3 and 4). These kinds of evolution would require relatively minor genetic changes, without a sudden jump from one phenotype to another. But isolation would have provided for divergent accumulations of genetic changes, which, through embryological processes, could have generated noticeable phenotypical changes within particular breeding groups, in a relatively short time.

Genetic and embryological processes

First of all, it is important to emphasize that genes do not *determine* behaviour. A few genes, or simply parts of genes, or specific DNA (deoxyribonucleic acid) sequences, or scattered DNA bits (SNPs – single nucleotide polymorphisms) provide the 'instructions' that create the proteins which form the tissues, glands, organs, hormones, neuro-transmitters and their general morphological and inter-connection patterns that result in metabolism, movement, sensitivities, perceptions, feelings and memories. These, in turn, result in *propensities* for particular species-typical *ranges* of behaviour. Environments have effects all along the way. These include the biochemical environments in which genetic materials originally send out their instructions, the biophysical contexts in which embryological and developmental processes take place, and the external environments later encountered by an individual organism. Nevertheless, despite environmental influences, the blueprint established by the genetic instructions is usually sufficient to maintain a range of behaviour that can be spoken of as being 'species-typical' in so far as that behaviour continually results in the replication of the genes and DNA configurations that set the behaviour in motion in the first place.

Genes with relatively direct causal sequences – even for protein synthesis – are usually a minority of genes within the genome of a species. Many other genes are regulatory genes, or genes for switching on or off the activities of other genes. Some are what has come to be called 'junk DNA'. These are genes and DNA sequences that seem to have no effect during the development of an individual. However, junk genes and junk DNA sequences may represent a number of accumulated mutations that have been shut off by a controlling gene. As a result, a mutation in a controlling gene can suddenly release a relatively large number of genetic influences. This can have somewhat dramatic effects from one generation to the next in terms of phenotypic outcomes. Most of these dramatic results are detrimental to the recipient, and so neither survival nor reproduction takes place and the mutation that released them disappears in one generation. But this is not always the case. The new outcome might be better able to survive than the original, or to reproduce at an accelerated rate, and so it slowly spreads in a population.

Junk DNA may also have the evolutionary effect of preventing a return to cross breeding between members of emerging sub-species. This is because accumulated differences in junk mutations within each breeding group, over a relatively short time, can prevent zygote formation if the sub-groups attempt to 'come together' again (very often manifest as spontaneous abortion). Besides the accumulation of junk DNA, informational bits of a protein can change without affecting the activity of the protein, but leave material for later selection to work on (Goodman, 1963; Roberts, 1980; Ridley, 1999; Phillips, 2003; Silver, 2007; Rutherford, 2013). And it is not to be assumed that any of this works as a harmonious synchronized design process. W.D. Hamilton had an image of a genome as 'a parliament of genes . . . [or] . . . a company boardroom, a theatre for a power struggle of egoists and factions' (cited in Kohn, 2004).

Very significant among the genes related to evolution are the limited number of HOX genes that lay down 'developmental pathways' (Ridley, 1999; Lodish, *et al.*, 2003; Pomiankowski, 2003; Bürglin, 2005), which generate the symmetry and general body plans found all across the animal world (Brady, 2000; Wille, 2012), ranging from fruit flies through fish, fowl and mice to humans. HOX genes are largely identical across species but generate slightly varying patterns when regulator genes cascade from one point of development to the next, switching HOX genes on and off along the way (Rancourt, 1998; McNamara, 1999; Brady, 2000; Pomiankowski, 2003; Myers, 2004). Through the cascading processes, changes in the early expression of HOX genes can have quite dramatic down-the-line phenotypical effects.

For example, in hominin evolution – given the general principle that muscle deployment affects bone development – genetic changes that advantaged human bipedal walking through altered hip *muscular* deployment during growth would have had embryological and developmental effects on resulting hip *bone* structure – see Chapter 4. Similarly, as the vertebrate column migrated towards the centre of the human skull with bipedal walking, along with a sort of 'stretching' (and weakening) of the powerful neck and jaw muscles of apes, there was an alteration of both muscle and bone structures in the neck, skull and face, and, as a consequence, a somewhat new facial appearance. Overall there was a reduction in the sagittal and nuchal crests of apes and an emergence of a rounder, thinner brain case, a reduction of the jaw size (almost eliminating canine-like fangs) and a flattening of the human face.

All of this could have taken place largely during embryological and post-natal development without requiring numerous mutations for all or even most of them. HOX genes, for example, are part of a larger group of genes called homeobox genes, which regulate the timing and ultimate expression of developmental results, such as those that cause a switch from embryonic development to the onset of birth, and later to patterns of adult development and so on. Slight changes in regulatory actions in these genes can result in changing the forms of specialized organs, even changing development pathways from male to female in some cases (Ridley, 1999; Lodish, *et al.*, 2003; Bürglin, 2005). For example, a number of species have emerged from the processes of 'heterochrony', or different timing of development processes, which can include a return to 'child shape' known as 'paedomorphosis' (Schultz, 1963; Gould, 1977; McKinney and McNamara, 1990; Brin, 1995; McNamara, 1999). This process is thought by some to have been involved in the origin of a number of insect and vertebrate species, including among amphibians, birds, snakes and dinosaurs, for example. One version of paedomorphosis is called 'neoteny' in which embryological and childhood characteristics are retained into adulthood (Schultz, 1963, 1978; Lorenz, 1971; Gould, 1977; Brin, 1995; Fortey, 2001; Penin, Berge, and Baylac, 2002; Dawkins, 2004a). This is thought to have played a part in the evolution of, for example, the wide variety of dogs in existence today (Brin, 1995) and creatures such as the 'Big baby chicken!' known as the ostrich (Dawkins, 2004a), or the trilobite species (*Olenellus armatus*) which almost cries out: 'I'm an overgrown baby!' (of *Olenellus lapworthi*) (Fortey, 2001, p. 171).

Another process for potential speciation is an acceleration of sexual maturity called 'progenesis' (Dawkins, 2004a, p. 265). Progenesis is the attainment of sexual maturity while still in a larval or juvenile stage with the individual never reaching the later developmental stages achieved by their evolutionary ancestors. Yet another process of interest is known as 'peramorphosis' in which an existing shape or structure is elaborated, with corollary effects that have significant consequences for outcomes, as when individuals mature past normal adulthood and take on new traits in the process. Humans, for example, eventually got much bigger than their hominin ancestors, and developed faster and more extensively in some areas (for example, brain and skull growth) and slower in others (for example, skull shape, jaw growth and eruption of first molar tooth). So, although humans and chimps share approximately 99 per cent of basic constructing genes and related genetic material, they look very different, 'all that stretching and shrinking has made an enormous difference to the final result' (McNamara, 1999; see also, Carter, 1980; Penin, Berge, and Baylac, 2002). One major evolutionary 'advantage' of processes like heterochrony, progenesis or peramorphosis is that mutations which initiate such changes are less likely to initiate a mother's immune rejection of the foetus than those that set in motion relatively new processes of protein synthesis.

Another factor in the evolutionary process is that recessive genes often have some effect, perhaps only activated in particular circumstances. This may be the result of histone tails linked to genes that are especially sensitive to environmental conditions (Phillips, 2003). Or it could depend on gene location and gene movement during recombination, which also affects the outcomes of genes (Cf., Holmes, et al., 2003; Rutherford, 2013). This flexibility may be an evolutionary advantage in that genetic expression sometimes switches as a result of environmental triggers within rapidly changing conditions. Very significant for our understanding of human evolution – it will be argued – is that in some cases the heterozygous condition (a combination of a dominant and recessive gene pair) itself produces a mixed result, which has an advantage over either homozygous condition, as in the famous case of the heterozygous condition preventing both malaria and sickle-celled anaemia while the homozygous conditions are vulnerable to, on the one hand, malaria and on the other hand, debilitating anaemia (Harris, 1975, for other examples see, Wilson, 1975; McKnight, 1997). However, the result of heterozygous advantage is that through normal random genetic recombination we will always have some individuals born in a homozygous condition (Ayala, 1978; McKnight, 1997).

Furthermore, there is always the existence of many more alleles (different expressions from the same gene locus on a chromosome) for a number of characteristics than just the two used in the example above. So we might have blue, green, brown or black eyes, for example, emanating from the same place on a chromosome. Such 'polymorphic loci' are in fact relatively common (Roberts, 1980; McKnight, 1997). Or we might have a number of genes or DNA bits that are located on different chromosomes acting relatively independently upon the same characteristic, giving us a vast number of possible combinations (Silver, 2007) for a particular characteristic. All this means that most phenotypic characteristics appear on a continuum running from, for example, relatively dark skin colour to relatively light, or from tall to short, and so on, in the same breeding population. This applies to almost every characteristic we can think of, including those which act as propensities to behaviours and feelings: for example, 'A polygenetic model of sexual attraction permits a complex combination of genes and the expressing of degrees of both dominant and submissive forms of the trait' (Kauth, 2000, p. 187).

But we must not get carried away; entropy has not been given a completely free hand. Although mutations are relatively common – it is suggested, for example, that 'each new embryo has about 100 mutations its parents did not have' (Leroi, 2004, cited in Hawley, 2004, p. 9; estimates, in fact, vary between 60 and 180, Nachman and Crowell, 2000;

Jones, 2013) – breeding populations (species) can remain relatively unchanged for a considerable amount of time. This is because the vast majority of mutations are either ineffective or harmful. They are overcome by their non-mutant dominant allele (Roberts, 1980; Jones, 2002), shut off, consigned to junk or they destroy (or prevent the breeding of) their host. Genomes contain within their proteins a considerable capacity to funnel mutations away from sensitive areas (make them junk), and also to repair mutations and to cause cells with harmful mutations to 'commit suicide' (Cohen, 2002; Ainsworth, 2003; Rutherford, 2013). And in humans when an embryo seems to contain something that 'is not right', or something goes wrong with the parent, spontaneous abortion is the usual outcome, estimated to be between 25 and 50 per cent of conceptions in seemingly normal pregnancies (Estronaut, 1999; Wilcox, Baird and Weinberg, 1999; Wang, *et al.*, 2003; Gray and Garcia, 2013). In the case of pregnancy with foetal chromosomal abnormalities, the spontaneous abortion rate is about 70 per cent (Rull, Nagirnaja, and Maris, 2012).

Selection thus has a range of behaviours (and genes) to act upon, and because minor genetic changes operating through a variety of processes can result in relatively dramatic phenotypic and behavioural changes in a very short evolutionary time, a completely smooth fossil gradation for every feature of an evolving creature will not be found because they do not exist. They never did exist because a change in one area had relatively dramatic embryological and developmental consequences on a number of other areas (Dawkins, 2004a; see also, Bateman and DiMichele, 2002).

So, how might what we take to be a new species emerge? Consider, for instance, a mutation capable of unlocking certain elements of junk DNA, which, besides introducing a new characteristic(s), could also activate a cascade of embryological consequences, exacerbating the genetic effects in the final phenotype that demonstrated relatively dramatic changes. And if a female with the mutant gene gave birth to several carriers, and if these 'freaks' were shunned and so only mated with each other, there might be the start of a new breeding population to carry on the mutation. Such sexual isolation might be reinforced though geographical isolation if the freaks were driven out. Either isolation would allow for the build-up of sufficient mutations (active and junk) in both groups to prevent zygote formation if the freaks and the ancestral population wanted, for any reason, to breed together later on.

We have a relatively clear example of elements of this process in the case of 5-aR condition (5 alpha-reductase deficiency – type 2) that causes males to be unable to convert testosterone into a 'usable' male form during the foetal stage and so boys grow up looking like girls, including without descended testes or a clearly recognizable penis (McKnight, 1997; Dixson, 1998; Kauth, 2000; Marks, 2004). This was first discovered in isolated areas of the Dominican Republic and the condition was traced to specific families from one village in which cousin marriage was observed in a number of cases (Imperato-McGinley, *et al.*, 1974), although it has since been observed elsewhere in the world (Boudon, *et al.*, 1995; McKnight, 1997; Dixson, 1998; Bahceci, *et al.*, 2005; Gray and Garcia, 2013). It is now known to be a result of mutation at one specific gene site that is prone to mutation (Boudon, *et al.*, 1995; Genetic Home Reference, 2008). In the Dominican Republic, testosterone conversion kicked in at puberty and the girl-males became, despite having been largely reared as girls, almost 'normal' males (including exhibiting macho behaviour). They still looked a bit more feminine than non-5-aR males and some were always known as 'balls at twelve' among those who knew of the condition (Dixson, 1998, p. 289), also translated as 'penises at twelve' (Imperato-McGinley, *et al.*, 1974).

The importance here is that it shows how sexuality is very vulnerable to evolutionary processes. Consider, for example, that *if* the girl-males had developed a working penis and became fertile at puberty, *but* had not gone on to generate male secondary sexual characteristics or a macho approach to life, and *if* some females then found them attractive, we could have been on the way to a new species of 'feminized girl-males'. There are a few ifs here, but given that in every 2,000 births in the USA it is estimated that there are 1–3 anatomies that are sufficiently ambiguous so that it is difficult to know what sex to assign them – some people estimate it to be as frequent as 1 in 100 live births – and that in the 1996 Olympics one in every 400 female contestants were sufficiently ambiguous not to be given a 'femininity certificate' and had to drop out (Kauth, 2000; Jones, 2002), this possibility is not out of the question. Research increasingly makes it clear that: 'Sex is not an absolute, but depends on a web of hormones, of enzymes and of controls within the cell that can be torn in many ways' (Jones, 2002, p. 66).

The possibility of evolutionary change being related to sexual development and behaviour is further supported by evidence that suggests that reproductive 'systems' are not synchronic wholes but rather a number of 'sub-systems' that evolve somewhat independently of each other. For example, there is reason to believe that in primates sexual (1) attractivity (signals which attract members of the opposite sex), (2) proceptivity ('playing around', flirting, courting, convincing) and 3) receptivity (willingness to copulate) are based on different genetic and neural capacities (Beach, 1956, 1976; Baum, *et.al.*, 1977; McKnight, 1997; Dixson, 1998; Gray and Garcia, 2013). And we could add to these semi-independent sub-systems 4) consummatory behaviour (the physical and psychological reactions/rewards/fears in response to copulation). Changes in any one of these areas can have repercussions on all the others, even though the genetic changes involved in modifying just one may have been relatively small and completely independent from genetic changes in the others.

For example, in the hypothetical scenario deriving from the 5-aR case, the old form of male sexual attractivity – masculine appearance, macho behaviour – would have been 'lost' to the post-pubescent 'girl-males'. Some may then have become much more flirtatious and generous (proceptivity changes) to attract females, certain of whom gave in to feminized girl-male attractivity rather than be without a mate at all (selecting for a change in attractivity and receptivity genetic propensities). And if girl-males provided more emotional closeness during copulation and post copulation there might have developed additional, or at least different, pleasure desired by females from copulation (possibly selecting for a change in consummatory behaviour).

These changes in sexual behaviour would have provided a selection force for changes in both emotional and cognitive tendencies and capacities. Furthermore, from what we know about the processes of evolution generally, underpinnings of personality characteristics and motivations of behaviour (including patterns of learning, moods, emotions, desires and fears) are a product of human brain processes that operate on continua with several genetic, embryological and developmental processes involved, and so would be extremely vulnerable to changing genetic distributions and developmental consequences, and, as a result, to forces of both natural and sexual selection. The suggestion here is that in human evolution sexual selection was very important in the evolution of human consciousness. So, let us consider some of the general principles of sexual selection that might be of help because, as noted, if genetic changes are not replicated through successful reproduction they do not end up creating a new species (and humans are no exception).

Principles of sexual selection

Reproductive success (Darwinian fitness) among sexually reproducing organisms is correlated with those who (1) mate with like-phenotypes because like-phenotypes represent genotypic compatibility and thus the likelihood of successful zygote formation; (2) mate with phenotypes that provide offspring with genetic potential to later attract and make sexually receptive appropriate members of the opposite sex; (3) mate with phenotypes that derive from genetic capacities which are able to exploit congenial environments and to identify, evaluate, avoid and control dangers in those environments; and (4) mate with phenotypes that include genetic propensities for securing successful offspring survival. In terms of proximate causes particular genes create *mechanisms* – such as drives, instincts, desires and fears – that motivate the sexual, survival, social and 'political' behaviours that remain relatively successful in terms of the above criteria (in a particular environment).

In sexually reproducing organisms, behaviours that lead to reproductive success are somewhat different for females and males. This is especially the case with mammals that 'incubate and hatch' offspring within the body of females. Among mammal species females have vastly more time, energy and risk invested in each egg than males do in each sperm. (Each human male generates about two trillion miniscule sperm in a lifetime compared to about 400–500 relatively very large eggs that a female is born with.) Additionally, in mammals, a female not only has to provide food and energy for her embryo and then new born offspring, but she is taken out of reproductive action (if not partially incapacitated) during pregnancy, and in the case of primates, for the early years of an infant's life. In theory, then, females should evolve to be extremely choosy in terms of with which genes, at what times and in which places to become impregnated to give their very expensive and limited number of eggs every chance of success (both in terms of immediate survival and future replication potential). On the other hand, again theoretically, males should be much less choosy than females, trying to spread their extremely cheap sperm as widely as possible.

In practice, however, mating patterns can become quite complex. In general, female reproductive success is improved when females are able to (more or less) identify, attract and monopolize the genetic contribution and resources of strong, vigorous, healthy, attractive males. But in some species, such as primate species, this puts them in difficult competition with other females where the higher status, more dominant females are likely to win. Additionally, a problem arises if paternal care of offspring is a major factor in achieving reproductive success. The above type of alpha males will have little interest in protecting or providing for females or their offspring, because why should they bother – after all females tend to seek them out. Moreover, when males are displaced by a stronger, vigorous, more aggressive male it is likely that the new male will kill all of the offspring attached to that female, wasting all the reproductive time and effort she has already put in. In addition, such females render themselves vulnerable to being totally controlled (herded) by a dominant male, resulting in losing out on the benefits of sperm competition (see discussion of primate sexuality below in this chapter) and/or gifts and protection in exchange for sex from a larger pool of males.

Females can be extremely promiscuous, achieving reproductive success by setting up sperm competition inside their reproductive tracts, which have evolved to weed out all but the 'best' sperm. In primates, for example, 'the vagina constitutes a potentially hostile and vast environment for spermatozoa' (Dixson, 1998, p, 269; Fox, 2002; Jones, 2002; Gray and Garcia, 2013). It is estimated that among humans, for example, of the 40 million to possibly as many as 1800 million sperm deposited in a human female vagina during each ejaculation only 300 or so reach the site of fertilization (Zeh and Zeh, 1996, 1997 quoted by Pusey, 2002). During sperm competition fast, agile, tough sperm win out, regardless of the level of other fitness qualities (even including attractivity) of the male who contributed them. They may be sperm that are not only agile swimmers, but also best suit the female's immune system and so the chances of a successful, full-term pregnancy are increased.

Where sperm competition is common, males evolve big testicles (lots of sperm), a longish penis (gives sperm a head start), a lot of seminal fluid (to protect sperm in the female reproductive tract), a very fast ejaculation time (a more dominant, aggressive male may be moving in for his turn) and a relatively short recovery time for after copulation (to be ready for the next opportunity). Powerful males may attempt to dominate females to monopolize copulation, but this undermines females' possibilities for sperm competition, and also female choice and female abilities to maximize the use of their sexual attractiveness to attract a variety of males for more careful evaluation before reproductive sex is provided. Consequently, females often subvert strict male dominance. This is especially the case where aggressive males try to drive away more gift-giving and attention-providing males and kill all existing offspring of the female they have taken a shine to. Among higher primates, for example, (where sperm competition is relatively common) it has been observed in at least one study that females sexually presented most often to the least aggressive males that associated with them and their offspring. They did this in all stages of the female's reproductive cycle. Such males and females sometimes slept together and the males generally protected the females (Saayman, 1975; see also, Nishida, 1979).

So, while a female may be attracted to strong, vigorous, healthy, attractive males, she may coyly wait for the ones whom she can control, or at least manipulate. Or, it may be that selection favours females who attract and are made receptive by a variety of average, less aggressive males because it reduces the dangers of infanticide during male competitions for sexual monopolies, and because sperm competition is not diminished. But there is a potential reproductive disadvantage here even for apparently favoured males. They may, for example, be allowed relatively promiscuous sex by females who mother them a bit but are prevented from having sex by the same females, or by more dominant males, during the time of a female's oestrus. Such males may be enticed into bringing food in exchange for non-reproductive sex and so lose out both economically and genetically.

On the other hand, a clever female may fool a dominant male into thinking he has sole sexual access to her but in fact she spends most of her time sneaking off with younger, more attentive, males. These, in turn, may deceive her for sex through a presentation of 'come mother me', 'I am devoted to you', but secretly groom and give food gifts to another. In a context in which selection favours parental input it may be significantly better to attract and monopolize one average male's contribution (sperm and resources) on the grounds that one average male is worth more than an uninvolved alpha male on the one hand or an easy to control infertile youth on the other. Moreover Mr Average may be easier to control, and, indeed, to fool if she does occasionally stray with an alpha male.

Sexual selection, then, gives rise to an ever-changing dynamic. A particular basis of success (for example, dominance) can set in motion the very selection force that begins to undermine it. For instance, attempts by males at complete dominance of females puts males in fierce competition with each other and there is always a chance of many, if not most, males losing out completely to the few males that are most able to dominate all the females. But, even for dominant males, or would-be dominant males, fighting for sexual access can be energy consuming and physically dangerous. Even winners of fights often die of infections or other incapacities from their wounds, so complete dominance over any length of time is very rare. Sperm competition allows more males in with a chance, but there is always a tendency for high status, aggressive males to attempt to dominate, and for females to deceive lower status males into gift-giving and grooming while presenting to high status males during periods of their greatest fertility. Overall, there are complex patterns of deception and counter-deception, competitions for mates, patterns of domination and submission involved in sexual reproduction, and none of these patterns are completely stable.

Variations in sexual behaviour can be triggered by environmental factors. A change in the nature of feeding grounds, for example, can result in females either spreading out or coming closer together during feeding, making polygyny (mate guarding) harder or easier for males. And socio/political events, such as a change in the size and/or hierarchy of a breeding group can reduce or intensify competition for mates. Thus, even within a species there can be variations in both male and female competitions for sexual access and, as a result, variations for both sexes in social hierarchies, personal bonding and sexual groupings. As a result of all the contingencies possible, from the viewpoint of evolutionary success, there usually has been a tendency for 'genetic capture' (Miller, 2000) – that is, through selection, a focusing of male and female behaviours on just one or two of the relatively successful reproductive qualities listed above, which become species-typical in a particular environment.

Overall, the basic pattern of cheap sperm/expensive egg is modified in just about every reproductive situation imaginable. If a male simply attempts to spread his sperm far and wide, but none of them result in a reproducing offspring, his reproductive success is nil. Or, if a female is so choosy she never finds the right male, time or place, her reproductive success is also nil. At the level of proximate behaviour there is clearly more to a species' male and female sexuality than simply being aroused followed by copulation. Most males have to work very hard to be attractive to females, and then to be able to win them, and then to avoid aggressive competitors. And if a female constantly produces offspring from continuous sex but does not care for them or protect them it is usually not the most successful way to bring an expensive egg to fruition. Overall, a balance is usually reached in males between the evolution of showmanship (bright tails, big manes, large antlers, status displays), striving for dominance (in order to be highly evaluated), gift-giving and utilizing speedy sperm to best effect (generosity but with promiscuity when possible). This balance defines the reproductive behaviour of a male in a species, just as a balance among a female's propensities to be attracted to dominant males, to seek sperm competition and to seek safety and protection for herself and her offspring, defines the female reproductive behaviour of a species.

Protection of offspring brings us to the next complication. As can be inferred from the above discussion, reproductive success is very much influenced by the degrees and kinds of 'parental investment' in offspring. In sexually reproducing animals, when patterns of mating and parental investment are considered together, we can see two hypothetical extremes. At one extreme, males attempt to be widely promiscuous – possibly fighting for females or prime breeding locations or to defend egg deposits, and often performing extensive displays to attract females – while showing no other interest in females and definitely none in offspring care. In this scenario, females coyly try to attract (or be invited onto the breeding ground of) 'the best males', with very few brakes on extreme promiscuity as a fall-back if they do not gain access to 'the best'; they have a relatively large number of offspring but provide little, if any, care for them. Virtually no bonding of males with females or males with males or females with females takes place. The result of this pattern is that there can be a degree of competition for mates, and a lot of sperm competition, but a minimum of time, risk and energy on the part of both females and males invested in each offspring (r selection). Reproductive success is based on producing *quantity* rather than quality. These are patterns commonly found in insects and fish, for example.

As we move along a hypothetical continuum from r selection towards k selection (high parental investment in offspring), females increasingly extend a degree of care to fewer and fewer offspring born to them. In mammals (and many bird species) females generally produce relatively few offspring and often invest considerable time and energy in each (in the womb or in an egg) even before an offspring is born. And in some species of mammals (and birds) offspring require a significant degree of additional post-natal care. In herd or harem conditions, females spend their time in very close proximity to each other, with one, or only a few, males monopolizing sexual access to them – this is another case in which males evolve very large testes. Here females produce one to three offspring per year and do most of the child-care work, which usually is of relatively short duration, with some 'help and protection' from being in a herd/pride/pack/harem, or even from a dominant male that keeps predators at bay or infanticide-oriented males out.

Moving along the continuum towards k selection, we have multi-mating mixed-sex groups in which considerable promiscuity is the norm (although dominant adult males may have a majority of copulations) and in which male and female alliance groups within fission/fusion bands may be common. Females here usually do all of the child-care, which can be extensive and occurs over a relatively long time (2–4 years) – although there may be occasional aid from relatives and/or alliance mates. This is a pattern common among higher primates.

Nearing the high parental investment end of our continuum (full k selection), we have serial monogamy in which a male and female share child-care after a relatively long period of gestation and growth in the womb. Usually very few offspring are born at a time (or over a lifetime) and they are given considerable post-natal care, often by both parents. A few monkey species and humans are represented at this point. At the far end a male and a female bond and mate for life and share considerable child-care activities, although there may be relatively few offspring in the lifetime of the parents. This end of our continuum greatly reduces competition for mates (and the amount of sexual activity) and, as a result, allows for a maximum amount of time, risk and energy to be invested, by both females and males (perhaps in different ways), in each offspring. Here the emphasis is fully on *quality* rather than quantity. This is common in swans and gibbons, for example.

But for this end of the continuum to work, males must be 'tamed' and 'captured', made to commit, or they remain unreliable, if not dangerous. In humans, for example, one study in America found that unmarried pregnant women were three times more likely to suffer violence from boyfriends than married women from their husbands. This factor was a bigger predictor of violence than level of education, age, race or any other factor (Courtwright, 1998). With an aggressive boyfriend (the least 'tamed' in this sample), a female may well have been better advised to take the genetic potential from a dominant male and get on with rearing offspring on her own (or with the help of her relatives and friends).

But, in humans, most mating is done in pair-bonds – however transitory they might be. So, alpha males may have to be ignored, but even being able to 'capture' even a 'good-enough' male (Ridley, 1993; Miller, 2000) requires the evolution of certain female characteristics and abilities. For example, females who have reproductive success by attracting good providers and partners in offspring care will have evolved repressors of their wild lust for dominant and/or philandering males, thus forgoing benefits of dominant genes or of sperm competitions, for help in offspring care as a trade-off. Females will, for example, have lost showy means of advertising ovulation (or at least sexual receptivity), such as the 'high pink' of female chimpanzees that can be seen at great distances even in the darkness of relatively dense forests (Pusey, 2002). As a result of this pattern of female choice, males favoured by selection most likely will have repressed both aggressiveness and a blatant sexual drive/desire. They may, instead, get satisfaction through *helping* a female in exchange for sex, or in exchange for mothering, or in exchange for help against aggressive males and/or predators.

But selection rarely stabilizes a species at just one exact point along this hypothetical continuum. Some females near the high parental investment end may be best served by settling down with non-philandering second or third best males, but others may be better served by seducing a dominant male and then getting aid from a second or third-rated male. Others may be able to care for offspring on their own and enjoy a bit of sperm competition beforehand. For most females at this end, however, it is an advantage to not show sexual enthusiasm (oestrus) at all clearly, but to be able to coyly entice potential mates so they can be 'sized-up' for parenting potential. Some males will 'prove-up' nicely, but others will try to have their cake and eat it by acting like a dedicated father while secretly haemorrhaging resources, time and energy to a variety of enticing females.

Additionally, signals that attract and sexually activate the opposite sex can remain long after they cease to be valuable, indeed, even when they are almost harmful in a particular context (Fisher, 1930; Ridley, 1993; Miller, 2000). And they can, in fact, persist for a long time as occasional homozygous outcomes of heterozygous advantage. So, for example, a female may be attracted to philandering, non-protecting males long after a providing/protecting male would result in better reproductive success. At the same time another female may have inherited extreme lust-repressing propensities without interest in producing offspring of her own. As a result of a polygenetic balance between these propensities giving best reproductive success, both extremes can remain in a population through their occasional homozygous appearance during normal gene recombination.

So, within a species there will be a limited range of sexual and child-care approaches based on the phylogenetic history of that species in terms of how the issues of (1) expensive eggs versus cheap sperm and (2) r versus k selection have been dealt with in a particular set of environmental conditions. In many ways, an overall range of sexual signals and their results in terms of sexuality and parenting motivators can act as the definition of a species. This is because if members of a breeding group are not to some extent on the same wavelength, more or less correctly reading each other's sexual signals, they will not be aroused or otherwise enticed to breed with each other. The consequence is that as a breeding group becomes geographically (or socially/politically) split and slight alterations in patterns of sexual attractiveness separately accumulate, members of one sub-group can lose sexual interest in members of the other. So, even if the sub-groups come together again, individuals will not be sexually interested in each other – they are on their way to becoming separate species. Indeed, in a variety of different ways it is the case that: 'Divergence in sexual preferences has been splitting species apart for millions of years' (Miller, 2000, p. 431; see also, Darwin, 1874) or, in words of Ayala (1978), 'the question of how species arise is equivalent to that of how reproductive isolating mechanisms arise' (p. 12).

At this point another factor has to be considered: the Adam and Eve incest problem. If Adam and Eve were the first humans, did they and/or their offspring not have to commit incest if they were to establish the human race? It would seem so, and it also seems so in theoretical evolutionary scenarios in which sexual selection may have played a major part. For example, if a new mutation let loose some junk DNA, which created a creature that was significantly different from the norm in appearance and behaviour patterns, it is most likely that the only individuals that *might possibly* be motivated to breed with them would be their brothers and sisters carrying the same mutation. Or, if the brothers and sisters were driven out of a group as 'freaks', they would probably either go somewhere together or meet up later, and possibly breed together. Now there is no biological reason why this should not happen, indeed, the similarity of their genomes would make conception more likely. Certainly, in nature, where there are no alternatives, inbreeding takes place in a number of species; beavers, for example, mate for life, but if a wife dies, and no widow snaps up the male, he 'marries' his daughter (Morgan, 1985).

Incestuous breeding is not uncommon in primates (Pusey, 1979; Trinkaus and Howells, 1979; Roberts, 1980). Inbreeding, however, raises the dangers of vulnerability to parasites whose own evolution 'locks-in' on the genetic predictability that incest produces (Fisher, 1930; Maynard Smith, 1971, 1978; Ridley, 1993; Hamilton, 2001; Jones, 2013). But new mutations that affected chromosomal and genetic behaviour to any significant degree would not necessarily be immediately vulnerable to parasites, the parasites themselves always having to continually evolve to catch up with evolution in host species (Jones, 2013). And by the time the parasites caught up, selection would begin to favour new forms of incest avoidance. So, species could evolve through incest, but be subject to selection pressures to avoid incest at some point after.

Principles of sexual selection tell us that the instincts, drives, desires and fears that motivate sexual and parental behaviour are a key to an understanding of a great number of species. Variations in mating practices are evolutionary compromises between (often conflicting) tendencies inherent in male and female sexualities/sexual strategies, including modes of parental investment (as affected by particular environmental conditions). As we shall see, this is certainly the case with primates, including humans. But before looking specifically at the primate 'starting point' for humans, it is worth looking at one other issue involved in the evolutionary process. That is the issue of chance.

Putting the emphasis on genetic and embryological processes, in the context of a fluctuating physical environment, with sexual selection (including a range of breeding possibilities making up a single species) playing an extremely important part, we are taken away from a strict adaptationist approach. Emphasis on genetics, embryology and sexual selection introduces chance and randomness. This should not be taken to mean that everything about human evolution has been so random we cannot understand anything about it. Indeed not. I have been discussing some of the processes of evolution to argue just the opposite. However, what we must not forget is that the principles of evolution do not require an explanation that fits species perfectly into environments; and jumping ahead, they do not necessarily have to end up with 'humans, the rational creatures', 'the problem solving creatures', the ultimate end of an 'uplifting' process.

So then, how do humans fit into this picture? We have already considered some of the physical environmental conditions that likely existed at the time of hominin differentiation from a common primate ancestor, some of the genetic and developmental processes that might have been involved in human evolution, and some potentially relevant processes of sexual selection. Now let us look at what the sexual/social/political background of hominin evolution might have been in order to see what elements of primate behaviour may have been susceptible to the selection process.

The phylogenetic background to hominin evolution

I will first consider behaviour among primates in general then become more focused on our nearest cousins: chimpanzees and bonobos.

Primates

Primates appeared sometime in the Cretaceous (90–100 myr). However, the branches of tarsiers/anthropoid stock that we belong to emerged in the Palaeocene (55–66 myr) (Dixson, 1998). Members of this branch of primates are primarily adapted to tree living, even when a certain amount of time is spent on the ground. This means that regardless of differences among the 230–250 or so species, their morphologies, feeding patterns, means of escaping danger, sexual behaviour and offspring rearing are compatible with living in dense, semi-tropical forests. Primates are almost exclusively vegetarians with the only meat (occasionally consumed) being eggs, insects and small grass-living creatures – except for apes among whom other smaller primates and small antelope are sometimes eaten. Primates generally operate in a very small home range (compared to grazing animals or the human primate). There is extensive use of vocalization as a means of defining and defending territory.

Almost all primates are extremely social animals with a relatively strong social sense (they will stay quite near others), and they demonstrate social bonds that are above and beyond sexual bonds and mother-child bonds. This does not mean that social bonds are indiscriminate; primates have friends, relatives and allies. In general, however, males often act 'jealously' in their attempts to secure and monopolize females. Females demonstrate 'jealousy' of each other, and offspring are also 'jealous' regarding siblings and would be 'uncles'. In primate social interactions, facial recognition and recognition of facial expressions are very important. This capacity is demonstrably linked to brain processes in monkeys (Darwin, 1872; Dixson, 1998), suggesting a relatively long phylogenetic connection between brain processes and the potentiality for conscious communication, if not for eventual self-awareness. It has even been claimed, for example, that the faces of woolly monkeys suggest 'intentions and moods' (Morgan, 1985, p. 121). Apes are especially expressive via their faces. Among them, responses to a number of facial expressions elicit behaviours that appear to be sympathetic, empathetic and understanding (Parr, 2003) – although we may be reading some of these reactions into ape consciousness.

When we reach the level of the higher primates (monkeys, apes, humans) there is considerable behaviour that we might call 'primate politics' (papers in Hamburg and McCown, 1979; de Waal, 1982, 1988, 1989; Morgan, 1985; de Waal and Lanting, 1998; Dixson, 1998; Pusey, 2002). For example, coalitions and alliances are widespread among baboons (Hamburg and McCown, 1979; Harcourt and de Waal, 1992); among chimps an alpha male will protect females and children while attempting to gain female support for controlling rivals, developing alliances and keeping order; being a 'loser supporter' is a common tactic in chimp politics (de Waal, 1982; de Waal and Luttrell,

1988; de Waal and Lanting, 1998); males among Barbary macaques and olive baboons have been observed to employ the political ploy of seeming to be attracted to infants that are not their own (according to DNA testing) in order to be left alone by other males, as males are normally repressed from aggression around infants (Deag, 1974; Beck, 1975; Dixson, 1998). One can guess that the mother of the infant may not be too displeased with the male's caring attention to her offspring and there may be rewards of another sort for him (Cf., Smuts and Gubernick, 1992); humans kiss babies to win electoral approval (or so it is said), and are deeply engaged in the politics of alliances, patronage and factional conflicts.

Primates demonstrate considerable fission–fusion social behaviour (Dixson, 1989). This is a pattern in which group memberships and size vary over time, but in which processes of splitting and fusing regularly occur. Fission–fusion within larger groups is the norm with new world woolly spider monkeys, patas monkeys, hamadryas baboons, chimps, bonobos (and human foragers – see Chapters 3 and 4). Other species are less regular in this behaviour but can be characterized by fission–fusion behaviour in certain circumstances (Dixson, 1998; Wrangham, 2002). During alliance formations, new friendships and altered sub-groups are a constant feature. This is especially the case as we get to the higher primates where there is, in fact, a considerable variety and flexibility of group sizes to be observed. Take hamadryas baboons, for example, where a troop can range from being 100 to 700 individuals. Troops of baboons end up a certain size for social political reasons, not because of the availability or scarcity of food (Washburn, 1961, 1968a); in all cases troops are broken into bands and then clans, ending up with small groups of anything between one and ten individuals (Washburn and DeVore, 1961; Dixson, 1998). Sometimes baboon infants and juveniles break into sub groups for play. Troop identity among baboons is strong and real; members are attracted to each other, although the dominant male is attractive to many if not most of the troop; the mother/offspring bond is very strong but of very short duration; in large troops with lots of females, specific males and females sometimes 'hang out' together (Washburn and DeVore, 1961). Most primates do not allow complete strangers, especially males, to enter their group easily (DeVore, 1963; Wrangham, 2002). Nevertheless, in primates (except for chimps and bonobos, and possibly humans) males out-migrate for the purposes of mating.

There is extensive sociosexual activity among primates (Rowell, 1972, 1988; Bygott, 1979; de Waal and Lanting, 1998; Dixson, 1998; de Waal, 2002a). These include a great deal of grooming, sexual rubbing, mutual genital play, erect penile displays and non-reproductive sex; genital against genital (G-G) rubbing (including occasional 'penis fencing') in bonobos, for example, is common. In sociosexual activities, right across the various species, there is often little regard for the sex or age with whom an individual indulges. These activities act as means of communication and greeting, and for expressions of dominance, submission, reconciliation and friendship. The same motor patterns and brain areas are used as are involved with more direct sex. Many male mammals, it seems, are biologically compelled to have a degree of sexual release: bulls, goats, sheep, horses, stags, camels and elephants have been reported to masturbate. Considerably more documented, however, is its clear existence in old world monkeys, apes and humans, including child humans (Dixson, 1998, p. 139). For example, Kinsey (1948) reported that 50 per cent of 3–4 year old humans are capable of orgasm from genital play. They certainly spend a significant amount of time playing with their genitals.

Individuals have a variety of active methods of soliciting copulation. Males in general often display an erect penis but more subtle methods are also employed. Mutual grooming, a behaviour that is almost constant, can lead to regular sex, as can all the sociosexual behaviour related to general primate sociability mentioned above, including frequent sociosexual mounting, or 'Copulatory courtship' – thrusting if not intromission over a period following copulation (from William Eberhard cited in Miller, 2000, pp. 234–235; see also, de Waal and Lanting, 1998). This tends to increase the chances of a male being allowed to stay around and have a number of copulations, and it may also cause sperm to be 'pushed further along' by the muscular contractions involved in sociosexual mounting and genital play.

In many species females demonstrate various degrees of signs of ovulation, which usually activate male sex drives. And in these species most instances of copulation take place during the period of a female's ovulation. In a number of species females during oestrus are quite promiscuous and not very subtle when they themselves are interested, often showing a colourful swelling to a male and backing up to him. And when motivated to seek sex, those females with no dramatically visible signs of ovulation (or none at all) try to entice a male (or males) in an equally unsubtle manner, like playing with his penis and testicles. In all species there is considerable licking and smelling of the genitals of a desired sexual partner in order to ascertain the degree to which they might be ready for sex, or to 'evaluate' their potential fertility.

Females, however, are rarely totally indiscriminate. They have favourites, and they have friendship and economic interests to consider. So, young female orangutans, for example, have been observed to spend quite a bit of time swinging from the trees with their genitals in the face of a reluctant older, well established male (in a very good feeding territory) before he can be enticed to 'swing into action' – literally because orangs often copulate hanging from tree branches (Galdikas, 1979; see also, 1995). However, it can be noted that only in gorillas do females actually initiate the majority of mountings (Dixson, 1998). On the whole it is males who have to spend considerable time trying to impress females through physical prowess, dominance, 'creative subordination', generosity and/or being non-dangerous to females to get them to want to be around them and to play with their penises and testicles. It is for this reason, among others, that follower-tactics, or tactics of apparent subordination, often result in as much, if not more, reproductive success than take-over tactics (Dixson, 1998).

For a number of species, then, there are a lot of activities that can be considered to be proceptive and receptive (and even consummatory) behaviours. This is significant because it shows that mating is not just an instinctive reaction to very specific physical triggers (visual, auditory, tactile or olfactory), but rather is a complex process of sociosexual behaviour. This provides a prime basis upon which sexual selection can work, and for sexual isolation to develop, which, as noted in the discussion of sexual selection, may be a major cause of speciation. The degree to which primate sexual discrimination is a major factor in primate evolution is testified to by the fact that none of the 230–250 or so species of primates have the same facial hair colour and style (Miller, 2000, he counts 300 species). Indeed, in terms of fine details specific species' mating patterns tend to be identifiably varied among primates.

Mating patterns include solitary individuals who come together only to mate, as in orangutans, mouse lemurs and galagos prosimians, for example; individuals who stay together somewhat longer in a serially or 'loosely' monogamous pattern, for example, among mature marmosets (who often start out as polyandrous, one female being inseminated by several males but then she pairs up for a time with one male), tamarin monkeys, some titi monkeys and some humans. A greater permanency of mating bond is found in the more strictly monogamous species, such as gibbons, siamangs, some titi monkeys, owl monkeys, indris lemurs, prosimian tarsiers and some humans, for example. Polygyny (where one male monopolizes copulation with several females) such as among patas monkeys, black and white colobus monkeys, proboscis monkeys, gelada baboons, gorillas and (occasionally) humans, is relatively common among primate species And then we have the multi-male/multi-female groups in which promiscuity is the norm; prominent examples include ringtail lemurs, rhesus monkeys, chimps and bonobos.

Although monogamy is relatively common in primates, as is polygyny, adultery is a regular feature in both, and monogamy is often not very stable. On the other hand, in multi-mate promiscuous groups, when females are in oestrus at the same time, the dominant females tend to monopolize copulations with dominant males, and dominant males with females in general. In most species, males tend to display and try to aggressively control females while females exercise a significant degree of choice of acceptance/rejection and of staying (or not) with given males. And it can be noted that a given species can roughly demonstrate more than one of the patterns above, often depending on the 'economic', and other environmental, conditions. For polygyny to work, as with dominance generally, for example, males have to be able to offer protection to females (Washburn and DeVore, 1961; de Waal, 1982, 2002a; Ridley, 1993).

For some this is rather instinctive; dominant male baboons, for instance, place themselves between danger and the rest of the group (Hall, 1963), but for chimps a considerable amount of political activity is involved before a male can lay any believable claim to being a protector. And there are never any real sexual guarantees in polygyny, even for the big and brave; one half of offspring in one gorilla study were not the product of the dominant male, and even then he had to fight a lot for his half (Jones, 2002; see also, Smuts and Gubernick, 1992). Polyandry (one wife, many husbands) is extremely rare and episodic rather than species-typical of any species. In multi-male/multi-female promiscuous groups, 'consortships' have variously been observed to exist, lasting from 1 hour to 3 weeks (3 hours to 28 days in chimps to several days in bonobos). However, only fully mature males seem able to form consortships, although they can be of relatively low rank, and there is a degree of uncertainty as to how much reproductive success derives from them.

There is a lot of variation in methods of copulation itself, with most being front to rear (ventral/dorsal) but front to front (ventral/ventral) is not unknown and is a very common method among orangs, bonobos and humans, for example. Ejaculation times for males vary. For example, common marmoset males ejaculate in approximately five seconds, chimps in seven to twenty seconds, bonobos take about 12 seconds and humans 4 minutes; woolly spider monkeys last up to 8 minutes but woolly monkeys and orangs can go on for 14 minutes (de Waal and Lanting, 1998; Dixson, 1998; Jones, 2002). A high degree of sperm competition (common in multi-mate primates) is evidenced by the first three of these species with their short ejaculatory times, and much less sperm competition is observed among those in which mating is a more private, and longer lasting, affair, as among the polygynous and sometimes solitary woolly monkeys, the often solitary/secretive orangutans and also among humans.

Sperm competition can also be seen in the very high frequency of copulation for males among most multi-mating species, such as in the case of one stumptail macaque male, which was observed to ejaculate 38 times in one day, or mandrill males copulating and ejaculating 2–3 times per hour; bonobos are not too far behind with sex every hour and a half as a more or less constant feature of life (with other males, females, adolescents and even juveniles being involved) (de Waal and Lanting, 1998). Chimps are not quite in this league but nevertheless are far from being slackers with their six times per day during a female's cycle (de Waal, 1982, 2002a). Most primate males are 'multiple ejaculators' in that they are capable of more than one per hour (Dixson, 1998, p. 117). In general, the greater the promiscuity the less rest time in between copulations required for males (Dixson, 1998). On the other hand, the more monogamous the species the more time is devoted to other activities in-between mating. For example, in humans, depending on age of course, several hours, if not a couple of days, is required for a subsequent copulation.

This relationship between multi-mating and sperm competition is also supported by a consideration of penis size. In multi-mate mating species, male penises are decidedly longer than in monogamous or polygynous species. This is clearly seen in apes where the polygynous gorilla males have penises on average of two inches while the multi-mating bonobo males have an average of almost seven inches. Although a size correlation does not seem to exist for female vaginal length it does for female swelling; females with extensive swelling mid-cycle have deeper vaginae (Dixson, 1998). This may be because it provides a greater store for the vast amount of sperm gathered before the onset of maximum swelling, while nevertheless having sufficient capacity so that the 'good genes' of dominant males copulating later during peak swelling do not spill out. The existence of sperm competition is also evidenced by the existence of big testicles and an abundance of seminal fluid in multi-mate species (Dixson, 1998).

Multi-mate species require many more copulations per live birth than polygynous or monogamous species. In polygynous herd mammals, one quick penetration usually does the job. Semi-polygynous gorilla females indulge in approximately 10–20 matings per infant (Ridley, 1993; Wrangham, 2002). Among orangutans regular copulation takes place during a several day consortship and this seems sufficient for pregnancy to occur (Galdikas, 1979; Dixson, 1998). Human females appear to require regular copulation for about three months to produce one offspring (Miller, 2000) – this is, on average, about 50 matings (Jones, 2002; Pusey, 2002). However, a more extended period of mating with the same male seems to be a special advantage for a human female because her immune system tends to reduce its attack on the 'invading' sperm, and then on any resulting foetus

(Fox, 2002). On the other hand, multi-mate chimpanzee females engage in between 500 and 1000 matings with a variety of males, involving at least 135 ejaculations, in order for a female to become pregnant; and among bonobos (also multi-mate) it is more like 3,000 assorted matings (Ridley, 1993; Dixson, 1998; Jones, 2002; Pusey, 2002).

Even with the apparent free-for-all of multi mate copulation, frequency of copulation for female primates in general usually varies to some degree with their period of ovulation. Chimp females, for example, have very clear signs of ovulation and mate extremely often during their period of swelling, but then may not mate for two or three years after giving birth. In semi-monogamous marmosets, copulation only takes place during oestrus and then just one to three times per day (about one to three hours' worth for a bonobo just about every day); in the somewhat more monogamous owl monkey copulation also seems restricted to oestrus and 19 attempts at copulation were made during 278 hours of observation (Dixson, 1998). Bonobo females have much less swelling than chimps, but they copulate when not in swelling, starting again about a month after having given birth and indulging regularly and frequently throughout their lives. Humans are not restricted to oestrus and average four times per week (Jones, 2002; Pusey, 2002).

In fact, a number of primates copulate right through the female reproductive cycle. This is especially the case where visible and other signals of ovulation are absent (or at least not easily discernible by human observers), such as in stumptail macaques, vervet monkeys, orangutans, bonobos and humans. Non-cyclical copulations are also the case where signs of ovulation still exist but are greatly reduced, such as in most new world monkeys, chacma baboons and gorillas. It is noteworthy that visible signs of ovulation are absent in a number of species without pair-bonding, while some new world monkeys have pair-bonds with visible oestrus (Benshoof and Thornhill, 1979; Dixson, 1998), questioning the idea that hidden ovulation is the basis of pair bonding in primates, including humans.

Noticeable vocal and facial expressions are common during copulation; females look back and grab the males from behind, pulling them forward; males bite the neck of females (Galdikas, 1979; Dixson, 1998). Noise, however, does not seem to be related to the speed or efficiency of sperm transfer (Dixson, 1998). Therefore, noise and 'grabbing' seem to be psychological and bonding phenomena more than physical aids to reproduction. However, noise can bring unwanted attention. Indeed, attempts by offspring and other males and females to disrupt copulation are not unheard of, with a variety of effects. For example, among chimps, when a female is harassed during copulation subordinate males usually thrust right through it; but if he is relatively dominant it is another story – the harassers have to run for safety (Dixson, 1998).

Usually, however, much of primate sex leads to the evolution of tactics for sexual deception whereby a male or female's intentions are disguised as general sociability, such as in grooming, hugging, subordination, genital play, attention to offspring and frolicking. Most sexual encounters, in fact, are not observed, occurring at night, often when low ranking males try to entice females into the bush (Dixson, 1998). The young do a lot of sexual playing and, indeed, engage in sexual acts, but this deceptively appears to be 'playing' not reproductive behaviour. And it is unlikely that the sex play among the young results in reproductive success. Nevertheless, looking young may have the reproductive advantage that one does not get run off by dominant males. Success, in such cases, depends on retaining an adolescent look, perhaps while acting as a retainer for a father and inheriting some of his younger 'wives' when he dies. Looking young may also make it easier for out-migrating individuals to worm their way into a new society.

So, it can be seen that excessive sexual encounters do not necessarily represent reproductive success, and fewer encounters do not mean a lack of it. In humans, for example, the average peak fertility of a female is only about 12 hours per month and so timing is everything (McKnight, 1997), yet humans have had extraordinarily high levels of reproductive success. In all cases copulation seems to be a rather pleasurable experience for most primates. Compared to most mammals, primates are extremely sexual creatures, and whether it is quantity or quality that gives most reproductive success, primates are highly motivated to partake in sex and sociosexual activities on a very regular basis, and it seems clear that these behaviours are closely related to primate social/political behaviour generally. In the majority of primate species intensive care and protection of offspring is largely a female concern, although in many cases certain males play with infants and juveniles and protect their group (especially its females) against intruders and other dangers (Cf., Rowell, 1972, 1988; Hewlett, 1992; de Waal and Lanting, 1998; Dixson, 1998; Huck and Fernandez-Duque, 2012).

In primates generally (Dixson, 1998) child-care on the part of females is usually very intense, emotional and effective, albeit sometimes of a relatively short duration in the lower primates. Infanticide is not common, existing mainly in hanuman langur monkeys and humans, although it does occur in some other species of monkeys and in chimps at certain times. When it does occur there seems to be no distinction among polygynous, monogamous and multi-male/multi-female species. In all primates there is evidence of incest avoidance based on proximity of caretakers and kin during the early years of an infant's life – odour may be a factor; sometimes close relatives copulate but DNA analyses show few viable results. The 'Coolidge Effect' (Symons, 1979), or *'familiar partner phenomenon'* (Dixson, 1998, p. 405), has been observed in rhesus monkeys and humans. This is a phenomenon in which a sexual partner, usually male, is no longer sexually stimulated by a long-term partner, but is aroused by a stranger. It is thought that in rhesus monkeys this may be an anti-incest device.

More is increasingly known about the definite effects of pre- and post-natal hormones as they affect such things as rates of embryological development, post-natal growth patterns, the development of secondary sexual characteristics, sexual maturity, penis development and testicular descent in primates generally (Cf., Dixson, 1998; Gray and Garcia, 2013). It is now known that most primate females menstruate. A period of adolescent sterility exists in monkeys, apes and humans (it is longest in humans, although shorter in industrial societies than in hunting and gathering societies). In rats, there is a clear neonatal critical period for the development of sexual preference and behaviour. This also seems to be the case in most primates although the exact linkage has not been so well established experimentally. Testosterone metabolized in the brain, as well as in target areas such as the testes, creates a variety of steroid molecules, giving the neocortex and frontal lobes an effect on sexual activity. It is interesting, for example, that castration in adults does not end sexual activity in most primates; it slows it down. This fact suggests that a pattern of brain sexuality, with the power of causation, is established early on in the development of an individual.

Other evidence of sex being a 'psychosocial' event based on more than pure instinct is seen in a consideration of its relationship to stress. While tension and emotional excitement among certain monkeys cause low status members to eagerly partake in sexual behaviour (Carpenter, noted in Hall, 1963), it is thought that these are not likely to lead to reproductive outcomes. In marmosets the presence of dominant females dramatically represses subordinate females' sexual capacity, and there is evidence that stress causes sexual repression in subordinate females in cotton-top tamarins, saddleback tamarins, yellow baboons, talapoin monkeys and humans (Dixson, 1998). Similar effects, though suspected, are less well documented in a number of other monkey species and apes. It is clear that in most primate species both males and females have evolved not just to react to mechanical triggers, but also to be able to 'read' a variety of very subtle signs in order to know when sex might be on the cards relatively safely and/or productively.

We can speculate that as the primate sexual brain evolved different aspects of sexuality developed, in different ways and in different combinations. Furthermore, there is no guarantee that all aspects will be in complete sync. So, for example, an individual might be mentally aroused without the gonadal capacity to do anything about it (because of stress or just having had sex or because of age, for example); or individuals could find themselves physically aroused (a morning erection, for example) but not be able to identify an object/cause of that arousal; or, in the case of bonobos, for example, males sometimes develop an erection at the sight of food (de Waal, 2002a). Sexual selection usually edits these capacities sufficiently so that enough successful mating of (relatively) like with like takes place for us to speak of the existence of different species. It is the job of the evolutionist to see what combination makes up the species they are interested in.

So, in order to narrow down our analysis, it will be useful to focus more directly on the sexuality of chimpanzees and bonobos as they are our nearest phylogenetic relatives. The ancestral line leading to modern humans looks something like Fig. 2.1. In Fig. 2.2, we see the relationship between our nearest cousins and ourselves.

Chimpanzees

Chimpanzees live both on the ground and in trees in semi-tropical forests; they are currently confined to tropical Africa, largely north of the Zaire River. Chimps are supreme climbers, can swing from branch to branch (brachiation – reach their arms behind their backs) and can operate very effectively in trees, building nests of interwoven branches to sleep in safety (sometimes, however, building nests on the ground – Koops, McGrew, Matsuzawa and Knapp, 2012). When on the ground, however, they very often operate from a bipedal stance, such as when walking through wet grass, displaying (sometimes by swinging a stick or branch) or holding out an open palm to beg for food or using their hands for bodily contact or for support. They are

The phylogenetic background to hominin evolution 57

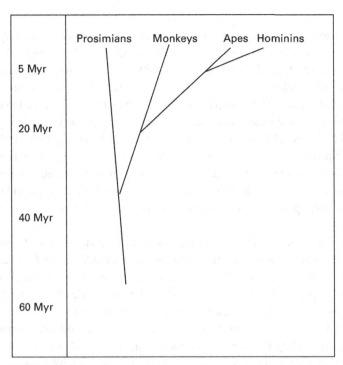

Fig. 2.1 Primate evolution

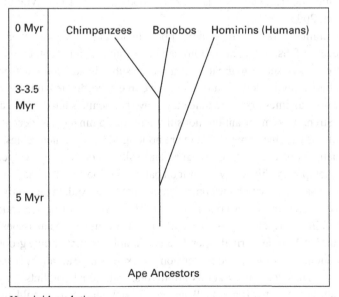

Fig. 2.2 Hominid evolution

largely vegetarian, eating mainly roots, tubers, fruits and nuts, but they do eat small animals and insects. Very occasionally they have been observed to hunt in groups (especially male alliance groups). Their targets are small game, such as small monkeys and antelope.

Fission–fusion is very common within chimpanzee communities, which vary between 20 and 120 individuals (Lawick-Goodall, 1971; Nishida, 1979; Pusey, 2002; Strier, 2002), with a mean group size of 41 (Dixson, 1998, p. 40). Sub-groups are made up of about seven individuals and are usually same sex groups, with male groups relatively strongly bonded. 'Males settle down as groups of friends that . . . are based more on age and status than on kinship and whose members build their lives on mutual grooming and shared meals' (Jones, 2002, p. 212; see also, Bauer, 1979; Fossey, 1979, 1983; Pusey, 2002). Kinship ties, other than mother and child, do not last long, and only consist of brothers at best. 'It appears that rather than staying with relatives, adult males are opportunistic in their relationships, making and breaking alliances for individual advantage as the relative power of each male waxes and wanes' (Pusey, 2002, p. 24; Strier, 2002).

In certain circumstances alliance politics (with a constant process of forming and reforming alliances through a number of agonistic encounters) can be quite intense among chimps, including the taking of vengeance on specific individuals (Lawick-Goodall, 1971; papers in Hamburg and McCown, 1979; de Waal, 1982; Dixson, 1998). In this process three-member 'power coalitions' and other forms of small alliances can last for a relatively long period of time. Male alliance groups tend to protect their community (especially its females) against intruders and other dangers. They also travel together over relatively large areas. And we have examples of male alliances trying to exterminate other groups and, more commonly, kill isolated members of other communities (Lawick-Goodall, 1977; Goodall, *et al.*, 1979; Wrangham, 1999; World Science, 2005).

Within a community rank is important for both males and, to a lesser extent, females. Numerous means of displaying rank, domination and submission exist. Dominant males attempt to stretch themselves to greater height when subordinates come with submissive greetings; at that point the dominant one steps over the subordinate who covers his head after bowing several times first. A female may shove her genitals into his face as a sign of submission. Strength is important but not sufficient for dominance, as deception, bluffing and coalitions can overcome it. There are no long-term fixed hierarchies, but rather fluctuating patterns of coalition, dominance and submission. Generally speaking, the more stable a temporary 'hierarchy' – built on balances of power at any given time – the more peace and harmony within the community at that time. And, in fact, only about one in a hundred conflicts ends up in a real fight; instead intrigue, bluffing, deception and the strength of coalitions (including with females) sort out most 'social issues' (Bygott, 1979; de Waal, 1982). Indeed, rivals spend a considerable amount of time grooming each other and reconciling in order to reduce tensions and keep future alliance prospects open. Non-swollen females sometimes act as arbitrators. Individual conflicts (physical and political) often come out differently than the 'formal' hierarchy would predict, but anyway the formal hierarchy changes on a regular basis.

Female groupings are based on friendship and in so far as rank exists among females it is based on age and personality more than on any other observable factors. This is probably because females individually migrate out from their birth-group to other groups and so do not have ties based on parental, early friendship or kinship bonds in their groups of adult residence. Nevertheless, powerful females can lord it over less dominant males. One basis of female power is that, in general, sons maintain a relatively high degree of 'emotional dependency' on their mothers until quite late in life. Females often group together to protect infants from males, who sometimes become irritated with the young ones.

Dominant males tend to monopolize copulations with females during the females' peak period of oestrus when females show very clear signs of ovulation, presenting large swellings and very colourful (red/pink) behinds. However, close allies are also allowed (or are able to sneak) copulations, and even lower status or younger males enjoy numerous copulations at the margins of the period of oestrus (often done out of sight of the dominant males). Subordinate males try to hide their erections when a dominant male comes into view as an erection is used to solicit copulation with oestrus females. During oestrus, females – although somewhat discriminating and generally preferring high status males – are extremely receptive to almost any male. The period of oestrus is approximately 14 days in a 35 day cycle; if she has been rearing an offspring the cycle only resumes after three or four years. Males in one observed group had sex once every five hours with a female in oestrus, but because there were four males the female mated six times in an eight-hour day (her approximate rate for fourteen days). Adolescent females in oestrus engage in sex even more often than sexually mature adults (de Waal and Lanting, 1998; Pusey, 2002), but, a given female has to mate hundreds of times before becoming pregnant for the first time. Out of oestrus females are usually not interested in sex and at that time males pay little sexual attention to them, although females do continue to copulate for a time after they have become pregnant.

Males, as one would expect from such sperm competition, have large testicles and a relatively longish penis as primates go (the penis averages approximately three inches long, so not as long as bonobos and humans). Males also possess a glutinous plug that can block out other sperm once the male has mated. Chimp males looking for sex have had their behaviour pattern described as: 'shaking, bipedal swagger, piloerection, seated position with penis erect, stretching one or both arms towards female' (Dixson, 1998, p. 107; see also, Nishida, 2003; de Waal, 1982). Chimpanzees may experience something similar to the orgasm experienced by humans (Goodall, et al., 1979; Dixson, 1998). Copulation is ventral/dorsal, except on a few occasions. Generally speaking, then, chimpanzees are a promiscuous species, but sexuality is clearly regulated by relatively clear triggers from female sexual presentations. However, from time to time individual males, of all ranks, go off – away from the group – with individual females, of all ranks, on 'consortships'. In these a considerable amount of female choice is exercised. Consortships usually last less than a day, but can last up to a couple of months. Copulation, however, takes place during this period, possibly with a high rate of pregnancies deriving from them (Tutin, 1975, 1979; McGinnis, 1979; Nishida, 1979; McGrew, 1981). However, there is some controversy as to just how much reproductive success is actually derived from consortships (Dixson, 1998; Pusey, 2002). Moreover, consortships seem to have very little effect on soliciting child-care support from males.

Gestation takes from seven to seven and a half months. Infants are born in a state of high dependency, but not for the same length of time that human infants are dependent.

Chimpanzee mothers are extremely devoted mothers and, as noted, come to each other's aid when an infant is in danger (often from males). Nevertheless, infant mortality rates seem to be relatively high. In chimps the cycle of birth is quite long compared to humans, approximately six years between offspring, and females do not come into swelling or copulate for about three to four years after giving birth. As with primate males in general, males spend considerably less time than human males aiding in child rearing, but they do regularly play with children. An offspring may stay with a mother for about eight years; aunt behaviour is very common.

Chimps demonstrate a lot of emotions (and sly deceptions) in their communications, as seen in friendships, male alliance politics and sexual politics (Bauer, 1979; de Waal, 1982, 1989; Savage-Rumbaugh and McDonald, 1988; Hallberg, Nelson, and Boysen, 2003; Parr, 2003). Chimpanzee communication includes a high degree of sociosexual play and comforting behaviour. For example, chimp males hold each other's penises and play with their testicles when frightened. Non-sexual mounting is common, starting at the ages of 4–5 for males. As with almost all primates, grooming is ubiquitous. Masturbation is common in chimps, at least in captivity. Chimpanzees have a very high level of intelligence as primates go and can make and use simple tools (Goodall, 1964, 1968; Lawick-Goodall, 1971; Nishida, 1973; McGrew, 1979, 1992) and, when pushed, can master simple communication with humans via learning various forms of sign language (Permack and Permack, 1972, 1983; Rumbaugh, *et al.*, 1975; Gallup, *et al.*, 1977; Savage-Rumbaugh and McDonald, 1988).

Although there is considerable debate about the degree to which chimpanzees have self-awareness to the extent of having a sense of self as a 'being' (Cf., above references and also Patterson and Linden, 1981; de Waal, 1982), there can be no doubt that they have a relatively well developed social/political intelligence (Lawick-Goodall, 1971; Mason, 1976; Goodall, *et al.*, 1979; de Waal, 1982; Permack and Permack, 1983) which allows them to manipulate each other, form alliances and be aware of some of the consequences of the social relationships they form, including both the pleasures and dangers that these might entail.

Given their evolutionary closeness to chimps, bonobos demonstrate a number of significant differences, especially in terms of sexuality and 'personalities'.

Bonobos

Bonobos live in environments that range from swampy, humid, muddy, relatively dense rain forests, to primary forests with sparse undergrowth because of the tall closed canopy, to more open secondary forest areas. All of these are south of the Zaire River, remote and isolated from chimpanzee and human populations. Previously known as pigmy chimpanzees, bonobos are now considered to be a species in their own right, not least because their behaviour is quite distinct from that of chimpanzees. They are shy animals and do not often come into contact with humans. They eat mainly fruits and 'terrestrial herbaceous vegetation' that has high protein content and tend to spread out considerably during feeding to find sufficient food. Animal products are extremely rare in their diets, perhaps about 1 per cent. Monkeys are treated as 'toys' rather than enemies or food. Overall, the food supply is good and relatively well concentrated (de Waal and Lanting, 1998, p. 65; de Waal, 2002a).

Bonobos spend a considerable amount of their time in a bipedal position. They do this for a variety of functions: securing food, display behaviour, caring for offspring and gesticulating when communicating. Yet they are well adapted to climbing and swinging in trees. Nevertheless, in terms of bipedal posture, the physical structure of bonobos is much more like the human body than that of chimpanzees; for example, the female vulva is well forward, and also their weight distribution is much more human-like than chimpanzee-like. This suggests that bipedalism co-existed with arboreal capabilities for a relatively long time in our primate history. In fact, De Waal (2002a) goes so far as to suggest that the bonobo face structure is not unlike that of Australopithecus and that bonobos possess a human-like distribution of neurons in the frontal cortex.

Fission–fusion is the social pattern; the only consistency of grouping is mothers and sons. Community sizes range from about 25–75 individuals, but in some circumstances some groups may comprise up to as many as 120 individuals. Sub-groups with eleven or more individuals are on average larger than among chimps (about seven). When smaller groups come together there can be a lot of shouting and drumming, followed by an exchange of sex among all varieties of individuals (but rarely between males). Bonobos have more mixed sex groups than chimps and male alliances are very short term and relatively weak when they do exist. Certainly, there is no evidence of male alliance behaviour with the kind of aggressive intent as has been observed in chimpanzees. But not all is rosy. Bonobos can be aggressive and males and females do fight, sometimes quite viciously. Nevertheless, 'reconciliation' behaviour is very common when fights do occur, and fighting itself is not common (de Waal, 2002a, p. 48).

Females form sub-groups and these have longer-term persistence than found among males. Female groups are based on mutual sexual attractions, sociosexual behaviours and friendship rather than on kinship. This may be because, as in chimps, female bonobos migrate out from their birth group to a variety of other groups, and so kinship or childhood playmates are not available for adult female group formation. A new female has to try to 'make friends' among existing female members (and she has to copulate with a lot of the males) in order to become integrated. Nevertheless, female/female bonds/friendships are sufficiently strong in bonobos so that females are able to dominate males through group action. They do seem to have a good 'psychological' head start in this because of the powerful bond mothers have with their sons; indeed, a son's status and ability to form friendships/weak alliances depends on the position of his mother. Whatever the case, any signs of male dominance have more or less disappeared in the evolution of the bonobos. Both males and females recognize rank, but it is considerably less clear and seemingly less important than among chimpanzees, and has less to do with political, 'economic', social or sexual domination than in chimpanzees. It is, again, based as much on personality, friendship and sexual attraction as anything else. Female choice in sex and friendship is especially evident in bonobos. Males recognize a dominant male among themselves but after that rank is considerably less clear or seemingly important.

Sociosexual activities are ubiquitous, with mouth-kissing, sexual rubbing and mutual genital play common. The latter includes G-G rubbing, oral sex, and among males, mutual scrotum rubbing. Males are not hesitant in presenting erections to both males and females, and, as noted earlier, engage in occasional 'penis fencing'. In sociosexual activities there is often little regard for the sex or age of the individual with whom a bonobo indulges. There is considerable fellatio among juveniles and occurrences of males masturbating younger ones; in fact, both males and females masturbate a lot. With regard to copulation itself, bonobos copulate more or less all the time. Copulation is used for greetings, reassurance, expressing friendship, dominance and submission, and for soliciting acceptance into a group by a migrating female (de Waal and Lanting, 1998; Jones, 2002).

It is, of course, also used for soliciting favours by a variety of individuals, especially females, and by younger bonobos of both sexes. It is indulged in between males and females, males and males, females and females, among juveniles and between juveniles and elders. From the age of two, males begin to seek sex and: 'Swollen females often accommodate the desires of these little Don Juans' (de Waal and Lanting, 1998, p. 117), but these are pushed out by dominant males at the point of the 'little Don Juan's' adolescence. But some of them, nevertheless, sneak a quick one with swollen females. An average bonobo initiates sex every 1½ hours (chimps every 5–7 hours when a female in oestrus is available). However, it is to be noted that many copulations do not last until ejaculation, being brief, 13–14 second encounters. Bonobos seem to experience orgasm.

Approximately 26 per cent of observed copulations in one study were ventral-ventral (face-to-face), and the rest dorsal-ventral; there may, however, be more pregnancies from dorsal-ventral copulations between males and females (Dixson, 1998). This may be because sub-adults engaged in face-to-face copulations more than mature adults. Females, nevertheless, attempt to solicit copulations by lying on their backs. Females do show signs of ovulation with a longer swelling phase than in most primates and a less abrupt detumescence of swelling during the second stage. However, a female is only fertile during about 4–5 days at the high point of her swelling. Copulation, however, takes place throughout the cycle (which is not the case with chimps). Females have recognizable external breasts, sometimes quite large and prominent. Consortships can last for several days and are usually undertaken by more mature adults.

Males have large testicles and relatively long penises, longer than humans but thinner and more spike-like (de Waal and Lanting, 1998; Dixson, 1998). Their ejaculation time is approximately 15 seconds; resting phase between copulations for males is approximately 30 minutes. Birth intervals are approximately four and one-half years apart, but females begin to copulate one month after giving birth. Child-care is exclusively a female affair, and they are devoted mothers, although males often play with babies and youngsters and have sex with juveniles. Babies are dependent on mothers and have to be carried until they can cling on at the front or back while she walks upright. Females often help each other with child-care, each female being part of a sub-group of regularly interacting females. Infanticide does not seem to happen among bonobos. The young take about 13 years to mature and females have their first offspring at about that time.

Bonobos, like chimps, demonstrate a considerable number of emotional expressions, both vocally and facially. This is clearly the case when they play and tickle each other, as well as when they engage in psycho-socio-sexual behaviour. These expressions and behaviours seem to communicate information about the moods of individuals, as well as about intentions and desires. Bonobos, like chimpanzees, are highly intelligent and inquisitive, with, it appears, a significant degree of self-awareness. Bonobo behaviour especially, however, implies a capacity for empathy, if not sympathy. It is suggested that they 'read each others' eyes' and can stimulate others without being stimulated themselves. Bonobo intelligence is related to social and relationship knowledge much more than to tool use, as they use almost no tools in the wild. However, they do establish '*Taboo nests*' for daytime playing and permission is required to enter (de Waal and Lanting, 1998, pp. 152–154, 158, their emphasis; de Waal, 2005).

So, bonobos, as well as chimps, seem to have a relatively strong capacity for complex social behaviour, including an ability to form coalitions and alliances and to recognize rank if not more permanent status positions. In both species there is a considerable degree of individuality, friendship and flexibility in group formation and inter-group behaviour, and an ability to 'read' each other's intentions in a number of sexual and social areas. However, politics and alliance formation seem to play a lesser role among bonobos than in chimps, but sex plays an even greater role. Like chimpanzees, bonobos have a capacity for using deception in order to achieve personal, sexual and social ends. Males compete with each other for access to females, and dominant males tend to win out, especially during the female's oestrus in both species, although non-dominant males have considerable sexual access, including a degree of reproductive success, especially among bonobos. In neither species do long-term pair bonds exist, nor do males help with child rearing.

It has been argued in this chapter that attempts to explain human evolution through the notion of a rational brain evolving to cope with the difficulties (and opportunities) of savannah living are found wanting by the evidence. In the first place, the physical evidence clearly does not support savannah living at the time of human differentiation. It suggests the real possibility of a continuation of at least semi-tropical conditions in forest land which contained a lot of swamps, lakes, rivers, creeks, small islands and fen-type areas. In the second place, a bipedal hominin existed long before a large brain evolved or there was extensive use of tools or the development of language. In the third place, the 'rational human' hypothesis leaves out way too much of what is observable human emotionality. It is further argued that if we take a close look at a number of processes of evolution, from genetics to sexual selection, a real possibility of understanding evolution in terms other than a teleological/adaptationist or rational/perfectionist mode presents itself. The importance of considering the evolution of human sexual and reproductive life gives us a window into the emotional, social and political human nature that we observe around us every day. A look at the background to human evolution suggests that a great deal of our reproductive, emotional and social life was already there in early primates, and continues to be shared with chimps and bonobo? So, we can ask, what have we abandoned, repressed and/or added? Answers to these questions will be our quest in Chapters 3, 4 and 5.

3 The evolution of human species-typical desires and fears

Humans are fully bipedal primates, with no opposable toe; we are ground living and almost useless in trees, but we do use material from trees for making tools and shelters. Humans are largely hairless with a layer of subcutaneous fat under their skin; both features are relatively unique to humans among primates. Humans have a much greater home range than primates in general. This was likely the case from very early on; Australopithecus, for example, probably moved more than any other previous primate (Washburn, 1960). Although humans have come to be able to live in a wide variety of environmental conditions, including in extreme cold and heat, they prefer what has come to be called a Mediterranean climate that ranges from mild to hot with seasonal rains. When such conditions do not exist, humans try to create them through the use of clothing, shelters, fire, the provision of shady areas, fountains and, eventually, irrigation and air conditioning. We like to live relatively close to significant bodies or sources of water. Humans eat fruits, nuts, vegetables, cereals, fish and meat. In fact, humans have learned to grow and rear (and cook) all of these in order to provide a relatively secure source of food. This also requires the development of storage facilities.

Human mating patterns

Humans mostly live in family nests or households based around a monogamous mating pair (Murdock, 1957, 1960; Dixson, 1998; Gray and Garcia, 2013). However, the pair may change over time (by choice or one partner might die and the other take up with a new partner, for example). Mating pairs usually, but not always, end up with offspring who will share their nest for some considerable time. Additionally, however, widowed in-laws and nephews and nieces may be accommodated in a household in particular circumstances; occasionally young strangers are taken in, as are servants and/or slaves. At the other end of the age range, one or more of the parents of a mating pair may be taken into the nest (usually near the end of the lives of these parents). And there are always a number of widows, widowers and otherwise older single people living alone. It is not uncommon, in modern times at least, for both females and males to live on their own for a while as single individuals after leaving a parental home, or for a few same sex individuals to get together and share nests for a time before seeking a sexual and/or marriage mate. Sometimes females live on their own with their offspring. So, there is a considerable variation in the exact make-up of human 'households'.

Nevertheless, actual mating is largely monogamous, not polygynous (except in selected cases, and then usually only among the rich and powerful) nor polyandrous (except in very rare cases, which usually do not last long). Although monogamous pairs do change, and we might even speak of the human form of mating as being serial monogamy, the multi-mate promiscuous pattern of many primates, including chimpanzees and bonobos, clearly does not exist (except, perhaps, as a form of recreation for a very few people). Mating pairs, and households, are based on some mixture of sexual/ love bonds, parent/child bonds and friendship bonds, while 'single-hood' is based on particular circumstances and/or a strong desire for personal independence.

Besides mating pairs and parent/offspring bonds, young males often form strong friendships, if not alliances, with a few other males. Members of such groups variously play, drink, fight, defend territories, participate in sports and delinquency, punish anti-social individuals, count coup and seek vengeance (as a team), and, above all, respect the mating claims of each other (and defend these claims against foreign males). In no other primate species (or mammals for that matter) do males live in such close proximity as human males and still maintain a high level of monogamy. When they do pair up with females, human males contribute to child-care (and to juvenile and adolescent care) and provide economic support for selected females and often several offspring over a relatively much longer time than found in any other primate. Indeed, in a large number of primate species effective male offspring care is more or less absent, including among the higher apes (Cf., Hamburg and McCown, 1979; Hewlett, 1992; Dixson, 1998; de Waal and Lanting, 1998; Huck and Fernandez-Duque, 2012).

Females form much weaker friendship groups that aid each other in: child rearing, supporting males in protecting territories, gathering, growing food and running households. Families rub shoulders with other families and sometimes form bonds of friendship with them, perhaps through kinship links and/or the mating of their offspring. In hunting and gathering societies, for example, groups of thus related families, and odd associated hangers-on, form very weak larger groups ('bands') that range in size from about seven to twenty-five individuals. When groups get much bigger than twenty-five there is a tendency to fission, and when they get near to seven there is a tendency to fusion with (or incorporation or submergence into) another group. This process seems to have only a weak relationship to food supply or environmental conditions (Birdsell, 1968; Steward, 1968; see also, Binford and Binford, 1966, 1968a; Dixson, 1998; see Chapter 4). Not only are families relatively changeable within themselves, but families also break off connections with other families and form new connections, or just go their own way. These weakly related groups tend to come together from time to time in a sort of tribal gathering, for fun and for trading, for gossip and for seeking mates for the adolescents that are coming into sexual maturity. In post-hunting and gathering times, very similar sized group linkages have been based more on friendship and colleagueship than on marriage linkages, kinship or geographical propinquity.

According to Y chromosome and mitochondria evidence, human females out-migrate – as they do among chimps and bonobos – while males stay put. It is not as clear and distinct as in these two ape species because there are times and places where men have clearly been the migrants. But female out-migration seems to have predominated in earlier stages of human evolution. When males do migrate genetic markers suggest that whereas females spread out males tend to migrate and settle as groups (Jones, 2002); this is a pattern seen in almost all patrilineal societies studied by anthropologists, for example (Pusey, 2002). It would seem, however, that females do not just set off, as they do in chimpanzees or bonobos, but rather wait until a number of groups come together and there is an eligible young man who might ask her to go with his family and be with him. His family may well be behind him, which gives the female a sense of security in terms of the future of her children; on the other hand, she is at the mercy of relative strangers, and this might not be so congenial.

In humans a powerful male or a powerful male's family may try to dominate a number of other individuals or families by accumulating several wives for their male kinsmen. Perhaps the father has already taken a second wife, possibly his widowed sister-in-law, and perhaps one of his nieces not cared for by an absent father becomes wife number three. Multiple wives come at a cost, but they provide labour and they bring security into the future through providing many offspring to care for the father, and they bring alliances. The sex provided may be good, certainly it is varied, the status may be exhilarating, but, as noted for primates generally, for polygyny to work males have to offer protection and security for females and their offspring. For humans, as for chimps, this is not always easy in emotional or political terms. The patriarch not only has to have enough power and wealth to keep all wives happy (and from harming each other) but also rivals (including perhaps his brothers) at bay, and must also have the ability to keep his jealous sons in order. Most likely, some of these sons have different (jealous) mothers who would not mind some vengeance for the fact that Sir took in a new young wife. And the sons may well hate each other because they think the other is the favourite of the patriarch. It is as a result of these problems that polygyny only exists in certain circumstances, and then only among small sections of a population; as noted, the most common form of human mating is monogamy.

Human sexuality

So, what are some possible characteristics, drives, emotions, desires and fears that underpin the above behaviour? To start with, human females show no signs of oestrus. Therefore sexual attraction operates through a number of more subtle physical and behavioural characteristics. Human males are attracted to females on the basis of: sizable external breasts (even larger than those of bonobos, the only other primate with significant external breasts); a slight enlargement of buttocks and thighs; a waist/hip ratio of about .7 (Singh, 1993; Buss, 1994; Singh and Luis, 1995; Dixson, 1998; Miller, 2000; Singh and Young, 2001); smooth skin; child-like features; and, above all, a face that communicates youthfulness, safety, kindness and welcome through characteristics such as a small lower jaw, full lips and large eyes (Morgan, 1985; Buss, 1994; Jones, D., 1995; Dixson, 1998; Miller, 2000; Jones, 2002). Young human females are often very concerned about these features, including about not having large, firm breasts; throughout history some have padded the insides of their clothing, and in modern times have had surgical implants to try to create larger breasts. Above all, human females have been concerned to remain young looking as they begin to pass through and beyond adolescent years. In this quest they have used a variety of creams, salves, herbal remedies and medicines to attempt to maintain infant-like smoothness of skin and freshness of face.

Females like strong, healthy, athletic-type males who are taller than themselves. But females add a number of psycho-social characteristics to their list. They like men who appear, in terms of both facial expressions and behaviour, to be kind, sincere, trustworthy, likely to be successful and to have a sense of humour. Nevertheless, males who possess wealth and status have a better chance with females than those who do not (Symons, 1979; Buss, 1989, 1994; Ridley, 1993; Miller, 2000, 2010). These parameters make both males and females quite choosy compared to other primates. It has been suggested that men and women are about equal in choosiness (Miller, 2000), but whatever the exact ratio, there is no doubt that humans of both sexes are supreme *evaluating* creatures. Even for temporary sexual encounters, most human males are surprisingly selective in terms of seeking the above desirable characteristics when compared to male chimpanzees or bonobos when these spot a female in swelling (or one simply passing by in the case of bonobos). And females are generally even more selective with only the most handsome, wealthy or famous males having much of a chance for brief temporary encounters. A female chimp during swelling will accept almost any male, and a female bonobo will accept almost any male, at almost any time, several times a day.

We cannot overestimate the importance of the *face* as the major sexually attracting feature in humans. A female can have what is considered to be a 'very nice body, indeed' but be generally considered 'ugly' and have very little chance of sexual success. But a relatively overweight female, without noticeable breasts and with slightly fat legs, *but* with a very pretty face, has every chance. A male with either a 'negative' face or body has little chance until he builds up a base of status and/or wealth. The type of human face that seems most attractive appears to be what might be called a 'normal face', that is, a sort of average of evolved human facial features (Morgan, 1985; Thornhill and Gangestad, 1993; Miller, 2000). Average faces, it seems, are more attractive than perfectly symmetrical faces, a condition that was once thought to be the basis of human beauty and handsomeness; it appears that symmetrical faces do not show emotional expressions very well (Dixson, 1998).

It has been suggested that the power of the human face in terms of human attractiveness may be its evolutionary role as a general 'fitness indicator' (Miller, 2000). Certainly a deformity of face is especially off-putting for humans, and cross-culturally the least tolerated of all disabilities (Cf., Richardson, *et al.,* 1961; Safilios-Roschild, 1970); when relatively severe it is much more damaging to initial sexual access than a deformity of any other part of the body. But it is hard to see why the face should be a fitness indicator any more than any other feature of the human body. In terms of strength, endurance, fertility, general health and intelligence, for example, we can think of many more possible indicators which would be more reliable. However, the use of the face in emotional communication is widespread in the primate background of humans; infant faces are attractive across a wide variety of primate species, and they tend to elicit care;

among primate adults facial expressions communicate aggression, submission, threat and friendliness. And it is these factors that are deeply involved in primate reproductive behaviour. So, it would be very surprising if a social species that evolved out of this phylogenetic background did not also communicate through facial expressions; it would be doubly surprising if a species that came to depend on caring relationships did not experience a significant enhancement in this facility, including its spread to all sorts of human interactions, from courtship, cooperative, status and political behaviours.

Whatever the exact details, it is arguable that the communication of emotional expressions via the face is an essential part of human mating and general social behaviour. Additionally, there is evidence that both female and male faces that tend to the 'feminine', or child-like, are even more attractive than average faces (Perrett, et al., 1998). This may be derived from the fact that human infants are born in a high state of dependency and require considerable support over a relatively long time while growing to sexual maturity. In this evolution it is not surprising that selection favoured infant faces that caused considerable *pleasure* (Cf., Robinson, Lockard and Adams, 1979; Thornhill and Gangestad, 1993; Brin 1995). In humans, infant faces, and their expressions such as smiling and laughing, not only generate pleasure, but also a powerful sense of wanting to hold, cuddle and protect infants. And when the baby cries and screams it only intensifies the desire to protect it, to make it smile and laugh again. These triggers are sufficiently strong so that humans generally will come to the aid of a distressed infant even if the infant is completely unknown to them (Robinson, Lockard and Adams, 1979; Brin, 1995; Hrdy, 2011); they will do so by first looking into the face of the infant and trying to calm them by speaking into that face using baby-talk.

It is not a long way from here to postulate that a number of these responses began to apply to adult human interactions given the degree of neoteny of humans (maintaining of childlike features, especially of females, into adulthood). The evolutionary result would have been the selection of emotional mechanisms that attracted both males and females to offspring; males to females; and females to males, for more longer-term mutual caring relationships than is observed among most primates – certainly than is observed in the sexual free-for-all among chimpanzees and bonobos. Additionally, it may well be that facial communication does more than signal caring potential; it may also signal likeness in terms of genomes, temperaments, personalities, and desire and fear propensities.

Other things being equal, mating with such similarities would provide reproductive success, from the level of facilitating zygote formation to the enhancement of long-term monogamy. It is certainly the case that humans are attracted to *certain* faces – not all faces, not even to faces that would generally be objectively defined as beautiful or handsome. Indeed, some beautiful/handsome faces appear emotionally cold, they shriek danger. They say, 'why would I bother with the likes of you, I can do so much better? And if I do bother with you for the moment, don't expect me to give you any sweet (baby) talk or sympathy or to hang around for very long'. Some faces, on the other hand, are compelling to certain individuals; they inspire confidence in those individuals, making it easy to approach and talk to the owner of that face; it is easy to like them well before much other information has been gathered.

To be sure, the human face is not the only indicator of trustworthiness and compatibility. Humans do a lot of things in order to be noticed, to advertise their reproductive potential and niceness: they dance, play sports, make jokes, indulge in conspicuous consumption, dress themselves up, paint themselves, exaggerate, sing, gossip (to suggest they know a lot of secrets and interesting people), write poetry, talk fancy and quite a few other things in their sexual advertising. Humans use 'psychological warfare' through 'conversation, charm and wit' (Miller, 2000, p. 13; see also, Ridley, 1993) in trying to attract members of the opposite sex. And when a longer-term commitment is desired, quite a number of other factors come into play; indeed the attractiveness of a face may come to be downgraded somewhat. A vast cross-cultural study (Buss, 1989) has shown that for longer-term relationships women worldwide like high status, high income, intelligence and energy in the man they would be attracted to, but above all they are looking for kindness. Men prefer kindness, intelligence, beauty and youth. For both males and females a sense of humour is considered very important (Ridley, 1993; Buss, 1994; Miller, 2000; Eagleman, 2011).

So, both males and females have turned sexual attraction into attractions for friendship and compatibility desired to last for a considerable period of time. As a result, humans enter into complex patterns of courtship to find out about their potential partners. Sex may or may not play a part in courtships. However, a long courtship without sex provides a male with a degree of security against being cuckolded (Trivers, 1972), so it is not surprising that sex during courtship is frowned on, if not made taboo, in a number of cultures. Also, humans do not engage in a vast amount of sociosexual behaviour in public as is so common among primates in general. Although infants indulge in sexual play (with their own genitals and with each other's genitals – often referred to as 'playing doctor') this is largely done out of sight of adults and is often condemned by adults. Certainly, adult sex in humans is practised in secret, but so too is sex-play, masturbation and sex-talk. Sociosexual behaviour in humans is largely done inside the head of individuals in the form of sexual fantasies, and sometimes in joking and/or teasing among close friends. Its most public expression is its depiction in mythology, art, dance, drama, comedy, poetry, limericks, advertisements and pornography, rather than in explicit, on the street, personal behaviour.

Even then, in most societies there are those who would ban these forms of depiction as well, and their creators most often have to be very careful in just how far they go, using euphemisms, symbols and innuendos much of the time. And these rarely suggest that sex should be anything other than a human secret. Storytellers and artists, with very few exceptions, show us considerably more heroics and romance than raw sex; they present heroes and heroines as 'all round' achievers and as potentially very good mating partners rather than as great sexual athletes. It is interesting that even most pornography portrays the participants as having the very types of faces and personalities which humans find attractive and generally trustworthy (as described above). The faces of mainstream porn stars, male or female, do not suggest frenzied sex maniacs, continually putting their genitals in the face of the opposite sex as inducement; nor do they solicit sex from every passing stranger. Instead, they show seductive faces that communicate 'sex is fun, and my body is beautiful. While we indulge I will hold you tight and get great pleasure from you and give you great pleasure, and I will not tell on you or humiliate you; what we do together will be our secret'.

The nearest thing to overt sociosexual behaviour in humans is found in sex play among children. And indeed, human proceptive behaviour generally, and 'foreplay' preceding copulation, has many of the characteristics of children at play, and not just sex play but play in general. There are sly glances, funny faces and coy looks, joking, teasing, biting, giggling and laughing. And in 'foreplay' there are many characteristics of a parent tickling a child to make them laugh, or soothing and lulling a child into complete relaxation. Sexual partners too rub each other gently. Kissing is common, often mouth-to-mouth, and licking is not unknown. When these reach the level of rubbing and kissing the genitals of the partner it has gone beyond child's play and begins to approximate the sociosexual behaviour of our ape cousins; however, in humans, again, it is done in private and confined to the mating pair. And the pair is said to be 'sleeping together', suggesting that there is more to it than just copulating together; it suggests that they stay snuggled together afterward, at least for a while.

Human mating pairs copulate about four times per week on average. But average is a very misleading notion; individuals vary a great deal and age, length of relationship, states of mind and numerous other factors affect individual frequencies. And it is to be noted that we are talking about mating pairs here, not simply the males or females of a troop or band as might be seen in most other primates. Except for the elite of extremely sexually attractive individuals, film or sports stars, or very wealthy males, or a few highly promiscuous females, female camp followers and/or prostitutes, non-paired humans copulate very infrequently.

Humans mostly copulate face-to-face, lying down with females on their backs. But other positions are possible and also engaged in. Human males have relatively large penises (5–6 inches on average); significantly longer than chimpanzee males but slightly shorter, although much fatter (and non-spike-like), than bonobo males (Dixson, 1998; de Waal and Lanting, 1998; Jones, 2002; Lynn, 2012). This suggests a degree of promiscuity and sperm competition. However, human males do not have enough seminal fluid for continuous copulation to have reproductive success via sperm competition (Dixson, 1989). The larger than predicted size of the male penis may be due to bipedalism and ventral-ventral copulation; a shorter penis would mean that sperm would be ejaculated some way short of the cervix and thus run a great deal of risk getting there (Miller, 2000). Additionally, human males have lost the small penal bone found in old world apes (but not found in a lot of mammals or new world monkeys). As a result, human males have to be aroused to participate, which may account for a larger penis full of nerve endings which enhances arousal potential from manual stimulation (by self or a partner). Certainly, human males get a great deal of pleasure from their penises, and this pleasure is linked through fantasies and daydreaming to an ever-present desire for sexual release. There is a general male tendency to wish their penises were bigger.

Human testicles are small, predicting monogamy or polygyny (Dixson, 1998). Monogamy is also suggested by the relatively small degree of sexual dimorphism (difference in size and strength between males and females) in humans. Once aroused and participating, males almost always achieve orgasm with ejaculation.

Human males get excited to ejaculation during sex, not only because of the stimulation of the rubbing of their penis inside a vagina, and the flow of their fantasies, but also by females making noises. This may be because males make noise, hear noises – at least in their own heads – when thrusting and tend to project their own desires and reactions onto females, therefore noise from a female triggers male excitement and suggests the male's ability to 'keep' the female happy and willing to have sex with them (and perhaps want to stay with them).

Although many females get a great deal of pleasure from the rubbing of a penis inside their vagina, they most often require additional stimulation for orgasm to occur (Symons, 1979; Dixson, 1998; Leland, 2000), and even then it is not guaranteed. A recent survey suggests that even in today's scientific, supposedly modern, feminist age, only 14 per cent of females regularly achieve orgasm during sex, with 32 per cent less than a quarter of the time and 16 per cent never. Although 34 per cent could reach orgasm every time through masturbation, 14 per cent could not (Hooper, 2005). These figures have not changed much over the past 50 years (Kinsey, 1953; Leland, 2000; Roberts, 2005; Furuhashi, 2005; Lloyd, 2005). However, there is very little relationship, if any, of female orgasm to sperm transfer (fertility) or reproductive success (Dixson, 1998; Furuhashi, 2005; Lloyd, 2005). Female satisfaction in most cases comes from the attention paid to them, the closeness of the encounter and the fact that a male seems fully dependent on them for his satisfaction (Cf., Maccoby and Jacklin, 1974; Symons, 1979; Unger, 1979; Duncombe and Marsden, 1993, 1994). Achieving longer-term relationships through these desires and regular sexual encounters with the same partner certainly does give humans reproductive success.

As noted, it can be argued that the greatest non-copulatory sexual expression in humans is not *socio*sexual – even in its storytelling and artistic forms – rather it is what might be called *psycho*sexual. It is in the realm of daydreaming and sexual fantasies. Daydreaming and fantasizing can be romantic or lust-filled, or some combination of both. It is this process, I suggest, that turns our phylogenetically inherited unconscious sexuality into sexual desires and fears, but this is usually done as privately as sex itself. And it is the reduction in overt sexuality and the role of daydreaming and fantasizing that makes humans so relatively good at evaluating potential sexual partners and longer-term mates, and also so choosy – certainly as far as primates go – all of which has given humans considerable reproductive success.

And most members of the human species contribute to this success, including a very high percentage of males. As Darwin (1879) observed regarding sexual selection generally, when the numbers of two sexes are equal, and polygyny is not practised, even the 'worst-endowed males will ... ultimately find females, and leave as many offspring, as well-fitted for their general habits of life, as the best-endowed males' (Darwin, 1879 [2004], pp. 247–248). This is largely due to the fact that while a given female might want the alpha male, thinking nothing less will do, it is nonetheless more likely, as in many species, that 'she may accept, as appearance would sometimes lead us to believe, not the male which is the most attractive to her, but the one which is the least distasteful' (Darwin, 1879, p. 257; see also, Ridley, 1993; Miller, 2000). In humans, her reward will likely be a loyal husband and a house full of offspring.

There are often a number of offspring hanging around in a condition of dependency (probably becoming increasingly demanding because they become increasingly jealous of each other). This is because human females are capable of giving birth every one or two years. This is an important factor in light of the fact that human offspring are born in a state of very high dependency as far as primates go. They have to be fed for several years; they cannot walk very well until about two years of age and cannot walk any distance until about four or five. They are born naked and need to be covered up in order to stay warm and dry. They are prone to a number of childhood illnesses; they throw-up and do not have control over defecation until about two or three years of age; they raise havoc within the family nest. They cry often for the first few years. They become extremely demanding; they want things now, they do not understand about deferred gratification, waiting is not in their nature.

Children interfere with parental sex and parents feel obliged to react to the children's distress and greatly curtail their sexual activities. In other words, children require not just care but an incredible amount of indulgence and patience. And they get it; they get it from both parents. So much so that from a 'rational' point of view one could ask why any human should want to have offspring at all? And it is doubly surprising, considering that the evolution of the bipedal stance in humans has made it both difficult and very painful for females to give birth; indeed it can be life-threatening. And humans became aware of this, perhaps very early on in the trajectory from ape ancestor to human. But they still give birth, and they do so with great enthusiasm, and they care for their offspring with the same enthusiasm.

When comparing human reproduction with primates, especially chimpanzees and bonobos, we can note that:

— human infants at birth are more dependent, for a much longer period of time;
— the human birth cycle is shorter (1–3 years) than in most primates, especially higher primates such as apes generally (but with the possible exception of bonobos);
— human females do not show phenotypical signals of ovulation;
— humans indulge in very little overt physical sociosexual behaviour, but a considerable amount of psychosexual behaviour in the form of fantasies and daydreams;
— human copulation takes place more regularly throughout a female's cycle than in many species of primate, but less regularly than among bonobos, and less frequently than among chimpanzees when chimp females exhibit sexual swelling;
— human copulation is done in secret; monogamous pairings are much more common and very much longer term, and are where most sex takes place;
— there is considerably more human male investment in offspring provision and rearing than found among other primates;
— a higher percentage of human males probably have some reproductive success;
— humans joke and gossip about sex but also turn it into romance and love; and
— above all, when compared to primates – especially primates in oestrus – human sexuality is very repressed.

Repressed sexuality

Human copulation takes place in relatively restricted conditions and with a very limited number of acquaintances in a lifetime. In terms of cultural development, sexual restrictions have variously come to include a number of sexual taboos. Sex is not to take place with strangers or children (nor in front of children), between close relatives, with friends (or even friends of friends), with violence, when a female is menstruating, as an economic bargain, when becoming old, sometimes when a female is nursing and, in some places, not on special spiritual days. Compared to primates, relatively long and sophisticated periods of seduction are entered into before copulation takes place. These can range from several hours to several years (about 5 seconds for a bonobo). Humans usually only copulate three or four times a week, and only this regularly when in a pair bond, doing so in secret, sometimes in the dark so that even the partners do not see each other.

It has been argued that one of the unique features of humans is that females are 'continuously receptive', and that this has been very significant in the development of human male/female pair bonding. It would be much more apt to say that human females are continuously 'non-receptive' and continuously 'selective' (Alexander and Noonan, 1979, pp. 449–450). For one thing, clear indicators of sexual receptiveness and triggers for, and/or cycles of, hyperactive sex have disappeared in females, and secondary sexual characteristics that might trigger sexual excitement in males are generally hidden, at least in public.

Males too are expected to keep any active sign of their sex drive or interest – such as an erection for example – concealed. In fact, even in cultures where a minimum of clothing is worn, clothing regularly covers genitals of both males and females (Symons, 1979). And if a human purposefully displays their genitals in public (as is common practice in a number of primate species, including the higher apes), they are shunned, if not expelled or incarcerated. Sometimes humans cover much more of the body than the genitals, breasts or naked flesh for the purposes of modesty; this can even include the face. In fact, this makes a degree of sense in so far as arguably the face with its emotional messages is used more as an indicator of sexual possibilities among humans than any signals from sex-related organs. The repressing emotions of shame, embarrassment and guilt are also often expressed, and exposed, though facial expressions. The apparent automatic climax-like physical response observed in a number of primate species has been diminished, if not almost eliminated, in a majority of human females (Symons, 1979; Dixson, 1998). At a very practical level, for females especially, there is always a possibility of soreness or thrush or cystitis acting as a potential brake on promiscuity, even with a regular partner.

Humans generally consider that sex should be confined within pair bonds, and that it is meant to solidify the bonds and produce children. It is the pair bond and the children that are held culturally important, not the sex – however pleasurable and culturally acceptable the sex might be if done discreetly. Kinship is often recognized and turned into a 'higher essence' – responsibilities for one's family and loyalty to it are considered to be moral obligations and to represent a 'higher calling'.

Some of the moral codes that surround sex can go so far as to suggest that sex is almost sinful, if not evil and dangerous. At this extreme there have been those throughout history who have argued that abstinence is among the highest human virtues because it allows an individual to achieve more worthy, lofty things, such as the purifying and perfecting of self, serving others in a quest to help humankind make progress and build civilization. And many who would not take it to such extremes still consider that philosophical contemplation, visiting museums, yoga and attending self-improvement lectures are certainly more important than sex. It is generally felt that a person who thinks too much about sex is somehow deficient as a human being; such an obsession is often treated as a dangerous addiction or illness. In developed societies (and in many others before that) children have come to be generally considered pure when it comes to sex; they certainly are not to be treated as sexual creatures. Infants should be isolated from any innocent encounter with sex, and as they develop, sex play and masturbation are strongly advised against. At adolescence, in most cultures, 'sex education' is given to try to convince them of the dangers of promiscuity and of sex too early in life.

While repressing sexuality may have worked relatively well for 'civilizing' the human primate, it has never succeeded completely. As both Freud (1973) and Kinsey (1948) observed, very young children are capable of sexual feelings and sexual responses (see also, Dixson, 1998; Gray and Garcia, 2013). As noted in Chapter 2, Kinsey, for example, suggested that 50 per cent of 3–4 year olds are capable of achieving an orgasm. Mothers enjoy the sucking of their infants on their breasts, and infants enjoy it too, and infants do not always like to be weaned, especially when a new one comes to the breast. Adolescents, especially males, masturbate in secret, regularly and many with gusto (Kinsey, 1948, 1953) and they fill their daydreams and fantasies with all sorts of ideas about wild and/or romantic sex, including with Mrs 'next-door-neighbour'. Incest, although relatively rare, does occur in humans, usually between brothers and sisters and fathers and daughters, but occasionally between mothers and sons (Cf., Armstrong, 1978; Forward and Buck, 1981; Summit, 1982; Renvoize, 1982; Nelson, S., 1987; Gonyo, 2006).

Among many adolescents and adults, social interactions are peppered with sexual joking and teasing. Pornography and sexually inspired art, mythology, literature, music, dance, advertising and gossip are common in human history, seemingly irrepressible from the point of view of those who would repress it. Nor have the 'higher callings' that have been deemed more important and more worthy than sex as outlets for human energy displaced sexual gossip, sexual scandals, sexually suggestive stories and sexually vivid visual performances. For many humans unauthorized sexual desire bubbles away just beneath the surface; sometimes it surfaces.

Adultery is not uncommon and in one survey in the West about 60 per cent of males and 40 per cent of females have at some point tried to lure mates away from others using their sexually attractive attributes; 70 per cent report having been approached (Schmitt, and Buss, 2001; Rew, 2004). A certain number of teenagers and adults set out to be as promiscuous as possible, flaunting their sexual energies as publicly as they can get away with. Some couples experiment in secret, trying out the kitchen table or the bathtub – but do not tell their friends. Clothing designed to be seductive is sold in shops as 'high fashion'. Sex aids can be bought in special shops; more than 50 per cent of women and almost 50 per cent of men in modern America have used a vibrator (Aitkenhead, 2012; see also, Herbenick, *et al.*, 2009; Reece, 2009). Eleven per cent of men and six per cent of women in North America have experienced 'fetishistic activities' regarding sexual release (Dixson, 1998, pp. 141, 144; see also, Gray and Garcia, 2013). Prostitution has been around for a very long time, and has existed in almost all known cultures (Cf., Winn, 1974; Roberts, 1992; McKeganey and Barnard, 1996; Gray and Garcia, 2013).

Sex sublimated into human civilization

So, although the cap on primate sexuality in humans generally works, it is never altogether secure. And however 'elevated' it might be, we follow our primate ancestors in having powerful desires to display because, just as for them, the better our display the better our sexual (and often reproductive) success. But, significantly, our displaying is more about ourselves as *individuals*, or about our *achievements* and status, than about our physical sexual prowess or desires for sex. This moves sexual desire into the realm of social display and communication. Indeed, it can be argued that in many ways civilization is a result of human desires to display. Powerful individuals display in order to tell challengers just how rich they are, just how much they can waste. And they get followers and reproductive partners as a result. But even less-dominant individuals display: they dress and paint their bodies, they paint and furnish their nests, they tell jokes, they sound witty and smart and they smile a lot. They play sports and act like they are good at them. And they gather things; they look to have each new toy to show how 'modern' they are, and how rich they are. They write books, read books and make movies – and they acquire sexual mates and friends as a result. And they deceive themselves into thinking that they do all this not to improve reproductive success but because it is *good* for them (and humankind): it makes them a better person, they have ambitions, they care, they have their animal-like lust (and other 'destructive' emotions) fully under control – they are moral beings, they are civilized.

So, sexual repression and sexual display work together, underpinning human civilization. We call it 'civilization' because we not only display and delude ourselves into thinking that we are doing *ourselves* a moral good, we see it as doing *humankind* good. We transform attempts at monopolizing sexual and feeding privileges by dominant males and creative subordination by less dominant males (and by females) into patronage systems loaded with codes of honour, obligations, rules and taboos. We turn seduction/courtship into 'love, romance and marriage', and a mating pair and a nest full of offspring into 'family'. We turn conflict-prone sibling competitions and generation gaps into kinship obligations – 'blood is thicker than water'. Unsuppressed youth rebellion becomes an indication of continual cultural vigour and renewal. And when all of this becomes elaborated (as in the grand display of royal court life for example); and marriage and family obligations become legally controlled; and the tools employed in display become more complex (or deadly: like spears, guns, tanks, airplanes and atomic bombs, for example); or when mass production makes display competitions more varied and widespread, we call it 'progress'.

This process is driven by unconscious drives and emotions, as well as by a number of semi-conscious desires and fears, and by a notion of 'rationality' that postulates progress and a spiritual (not necessarily religious) quest as the 'purpose' for human existence. We must seek an understanding of what motivates all of this if we are to start to comprehend humans in a scientific manner. In Chapter 2 and in the first section of this chapter I placed human sexuality and patterns of parental investment in a comparative primate social context in order to set the stage for considering some of the specific drives, emotions, desires, fears and advanced cognitive patterns which it seems likely evolved in the human line. It is now time to look more closely at potential scenarios of human evolution in order to become more specific in our quest for discovering such underpinnings of species-typical human behaviour. We can remind ourselves that any idea of an ever larger brain, using tools, out on a dangerous savannah, has to be abandoned. In its place we can consider the evolution of human reproductive patterns as forces for general human evolution.

Scenarios of human evolution

Three bio-physical characteristics extant in modern humans which could have set in motion the evolution of the emotional-cognitive processes which came to be significantly manifest at the end of the hominin line are: first, the increased birth rate (compared to other primates) of human females; second, the relative immaturity of human infants at birth (Cf., Lovejoy, 1981, 1981a; Mellen, 1981; Dixson, 1998; Hrdy, 2011); and, third, the retention of a number of infant/adolescent-like characteristics into adulthood. The potentiality for both an increased birth rate and extreme dependency of infants at birth was clearly in the ancestral line. Bonobos in captivity and when the offspring are removed soon after birth, for example, can produce offspring at the accelerated human rate of slightly more than one every two years (and at Wamba in the wild an almost similar gap to humans has, in a few cases, been observed). And bonobo babies are also quite tiny and helpless at birth (de Waal and Lanting, 1998).

The increased birth rate seen in modern humans could have been based on minor genetic changes. For instance, in a particular village in India consisting of 2000 families, about 220–300 have twins, 6 times the national average; this has been a very recent phenomenon (last 60–70 years) and seems to be increasing with time (*The Telegraph*, 2009; Russo, 2009; *Mail Online*, 2009). Though rare, this phenomenon has been observed elsewhere, such as in a town in south-western Nigeria and in a village in the south of Brazil, for example (Attah, 2014; *Mail Online*, 2014). If there is a genetic component involved (women who married out of the Indian village also produced twins at the accelerated rate), and provided this did not slow down the rate of successful pregnancies, consider the evolutionary effect of a similar phenomenon within a small isolated group of proto-humans, given that mathematically speaking an increased fertility rate can spread relatively quickly through a relatively isolated genetic pool (McKnight, 1997). This, of course, would have depended on the ability of such proto-humans to successfully nurture offspring being produced at an accelerated rate, which would especially have been a problem if the increased birth rate coincided with the greater immaturity at birth of human infants. The degree to which there is a genetic cause of immaturity at birth among primates is suggested by the fact that early or late births tend to run in families (Winkvist, Mogren and Hogberg, 1998; Bhattacharya, *et al.*, 2010).

The third possibility for human separation from ancestral apes, neoteny, is evidenced by the fact that a number of adult human characteristics resemble those of infant and juvenile apes (and monkeys), and that many human infant and adolescent characteristics are maintained well into adulthood in modern humans. While baby apes and baby humans look much more alike than the adults of the two species, it is humans that retain certain of these characteristics into adulthood, such as the flatter face, rounded skull, smallish teeth, greater hairlessness, and weaker physique. We can also observe that all of the fossils of bipedal hominins from about six to two million years ago suggest a smallish creature, very much weaker and smaller boned and muscled than modern apes. The recent discovery of *Homo floresiensis* (Foley, 2011), a sort of reduced *Homo erectus,* also shows that the process of miniaturization may well have played a role in hominin evolution. The fact that variation in these developmental processes can be observed when comparing primate species, and even when comparing individuals within a species (Carter, 1980; Dixson, 1998), further suggests their vulnerability to the evolutionary dynamic. At the same time, changes in any one of these areas not only bring change in reproductive potential, but also, through normal embryological processes, quite new appearances, morphological characteristics and even behaviours.

All three of the changes suggested above (an increased birth rate, the relative immaturity of human infants at birth and the retention of a number of infant/adolescent-like characteristics into adulthood) would only have required alterations in the functioning of, for example, the growth hormone HGH (Williams, 1968; Carter, 1980; Corballis, 2002) or other maturation mechanisms based on HOX and/or regulating genes, or even on various SNPs (single nucleotide polymorphisms). In all mammals, a number of genetically based hormonal effects influence foetal development; in humans this includes a rush of testosterone at seven weeks of gestation that sets in motion the development of male genitalia – a lack of this rush results in female development. The relatively slow development of modern humans in the post-natal period means that many of the crucial genetic and embryologically influenced effects which result in the fully mature breeding adult are stretched out over a period of 12 to 14 years. During this extended time, post-natal circulating hormones (especially various androgens) affect the general rate of growth (including various growth spurts), the development of secondary sexual characteristics (such as the growth of body hair, breasts and penis development), and the onset of reproductive maturity of each individual.

Consequently the end result for each individual appears somewhere on a continuum between 'extreme male' and 'extreme female'. This is because sexual characteristics continue to be influenced by subsequent developmental processes even after the original genetic messages for the differentiation of male and female genitalia and for a male or female trajectory has taken place. For example, a lack of 21-hydroxylase may 'masculinize' women, making them aggressive, career-oriented and not very interested in children (Dixson, 1998; White and Speiser, 2000). This is a major form of a condition known as 'congenital adrenal hyperplasia' in which the result is often the development of ambiguity of female genitalia, to the point that genetic females can be seen as being male, later such females can grow beards and act like men. This genetic condition affects both females and males, although the observable effects on males is not always so clear. In both cases an early onset of puberty (with the development of secondary sexual characteristics) can take place (Dixson, 1998; White and Speiser, 2000; Jones, 2002; Kenny and Knott, 2012). We have already seen the example of the delayed development of masculine features in some men in the Dominican Republic based on just one gene.

Along with a number of other primates, humans are capable of being sexually attracted to others, partaking in proceptivity behaviour and indulging in sexual intercourse after castration (Dixson, 1998; Jones, 2002). For males this is because there are a lot of neurotransmitters in the brain and neuro active peptides in the penis that are capable of generating sexual activity (and fantasies and daydreams) following castration. It is also because the effects of such things as dopamine, noradrenaline and serotonin – all involved in sexual responses (including during REM sleep) – are still operating in the brain. Testosterone is metabolized in the brain as well as in target areas, such as the penis, to create a variety of steroid molecules that give the neocortex and frontal lobes an effect on sexual activity (Dixson, 1998). Female sexuality is even less tied to specific hormones or to specific organs, so an ovasectomy has even less effect than male castration (which does not mean that it has no effect). In both males and females, hormone treatment can regenerate aspects of sexuality when natural production ceases.

Thus, from the point of view of genetics, embryology and developmental processes it is possible that a minor mutation could have freed up some previously repressed or previously inactive primate DNA that had relatively dramatic phenotypic effects in terms of heterochrony (different timing of developmental processes), including a maintenance of 'child shape' (paedomorphosis) and even incorporating the process of neoteny (the retention of childlike features into adulthood). Whatever the exact genetic, embryological and developmental sequences (and the feedback processes involved), maturation among primates does, in fact, differentiate more or less with the degree of phylogenetic separation (Daughaday, 1968; Roberts, 1980; Jones, 2002). In so far as changes in one area provide reproductive advantage, they become selection forces for the others. An increased birth rate, for example, selects for maternal (and possibly paternal) behaviour, and enhanced maternal behaviour selects for an increased birth rate. At any rate, whatever the exact factors that set the human line onto its own trajectory, we can assume, given their importance in modern humans, that the above processes were extremely significant in moulding its path to becoming *Homo sapiens* and, in the process, in generating the vast array of backup emotions, desires and fears that appear to be ubiquitous in modern human life.

The emergence of youth-apes

Let us postulate for a moment that a genetic change resulted in offspring that remained youthful in appearance throughout their lives (perhaps by blocking certain normal developmental pathways). Nevertheless, after about eight to twelve years females were able to give birth and males to impregnate (as chimps and bonobos can even before this age). Let us also postulate that this was accompanied by a bonobo's near capacity to give birth every couple of years or so. If she was especially fertile she might, as is the case with certain pre-industrial human females, have produced 10 to 12 (perhaps more) offspring in her lifetime – a possibility enhanced by a longer life span which tends to go with the above postulated changes. Through inheritance of the embryological and development processes involved, these offspring would also *look* quite different from their ape grandmother and grandfather; we would have some kind of fertile 'youth-apes'.

The problem with this as a means of speciation is that a small genetic change (say a point mutation or chromosomal aberration of this sort) would remain in only a small minority in a population if 'normal' genes overrode it during the breeding of the carriers. The mutations would remain extremely marginalized in a normal breeding population with a few 'odd' individuals appearing in a breeding population for a few generations until the mutation became extremely rare in the gene pool. If, however, these new 'youth-apes' were not a sexual 'turn-on' for dominant males or females, or even aspiring subordinate males or females, but they bred with each other, then the swamping of the mutation would be considerably less. It is worth noting, for example, that chimp males seem to prefer middle-aged to older females, and even those with rather 'wrinkled skin, ragged ears, irregular bald patches, and elongated nipples' to those with the characteristics of younger (although fertile) females (Muller, Thompson and Wrangham, 2006, p. 2236; see also, Tutin, 1979; Goodall, 1986). So, youth-apes might have been oddities – freaks – but not sexual competitors to 'normals'. However, as among bonobos, youth-ape brothers and sisters (and possibly cousins) might have engaged in a lot of sex play including grooming, pseudo-mounting, fellatio, genital licking, masturbation *and penetrative sex* just for the fun of it. This possibility is supported by the fact that considerable infant/child sexual play – playing doctor – remains in modern humans (Gray and Garcia, 2013).

Indeed, the existence of infant/juvenile sexuality has long been noted in humans. Infant males commonly get erections – sometimes when playing, sometimes when breast feeding and during dreaming (REM sleep). Genital play starts at about 6–11 months for both males and females, continuing up to at least 6 years of age. In one 1980 survey of 796 undergraduate students in America, 15 per cent of females and 10 per cent of males reported some form of sexual experience with a sibling (Finkelhor, 1980). In another survey, 17 per cent had a pre-adolescent experience with a sibling, and 45 per cent had a sexual encounter with a non-sibling (Greenwald and Leitenberg, 1989). No identifiable long-term effects on adult sexual life – certainly no dysfunctions – were identified in this study. In another longitudinal study, 48 per cent of children had engaged in interactive sex play by the age of six, with no observable significant effects on later sexuality (Okami, Olmstead and Abramson, 1997; see also, Gray and Garcia, 2013).

If, after about eight to twelve years, babies began to appear from the bodies of the youth-ape females, many of them would be youth-apes in their turn. If our original Eves/ Adams were relatively productive, they might have produced quite a number of youth-ape children, and eventually they in turn a significant number of grandchildren, as a result of interbreeding with each other, and without even 'knowing' – any more than a chimp or bonobo 'knows' – when having sex that it was reproductive behaviour they were engaged in. In so far as youth-apes remained children/juveniles as far as 'normals' were concerned (normal males being too busy looking for some proper swelling and females not being very interested until swelling commenced, and then largely only with dominant males) a degree of *sexual isolation* would have been introduced. And an element of *social* isolation may also have been at work in so far as 'normals' began to shun and be aggressive towards freakish looking youth-apes (who never seemed to grow up). The result of both would have been that youth-apes had only each other to play with.

If youth-ape freaks started to become too much of 'a pain' as far as the 'normals' were concerned, they would have been the first to be driven out (or have wanted to leave) when normal patterns of fission occurred – that is if they were lucky and were not killed. Certainly, many modern apes have a tendency to shun what appears to be 'strange' to them (Pusey, 2002), and modern humans generally view other humans with deformities with a degree of horror that often leads to their rejection and isolation, if not death (Cf., Safilios-Roschild, 1970; Fiedler, 1981; Parkin, 1985a; Davidson, 1991). Among chimps we have documented examples of individuals being driven from their band, sometimes being beaten to the point of great physical harm, if not death, and even an example of genocide by one group of chimps against another (Bygott, 1979; Goodall, et al., 1979; Pusey, 2002).

If social isolation led to geographical isolation, the tendency to 'friendship' playing, reproducing and caring would, of course, have been reinforced. We can recall the potential for geographical separation suggested by the type of environments in which humans first likely evolved (see Chapter 2) as a basis for arguing that youth-apes could have found safe havens 'to play in', and to hide-out in. These were environments, it was suggested, with a high availability of fish, berries, nuts and other relatively easy to secure high value foods, greatly reducing survival problems. A number of youth-apes, thus, could relatively easily have become an isolated breeding group on the basis of geographical, sexual and social criteria, leading to the accumulation of in-group mutations that would make reintegration with 'normals' very difficult, if not impossible.

This overall scenario (or something like it) can thus claim support from the fields of genetics, embryology, developmental biology, modern sexual behaviour and geography, but also from a consideration of human incest. Many of the chromosomal 'rearrangements in man today' that have become firmly established 'must have occurred in a small population in which incestuous unions were common' and, indeed, genetic studies in Yanomamo villages suggest 'that tribal societies were more inbred than current evidence suggests ... [so] ... westernisation must have resulted in a marked relaxation of inbreeding' (Roberts, 1980, pp. 35, 40; see also, McKnight, 1997; Dixson, 1998; Jones, 2013). In fact, cousin marriage has remained relatively common throughout human history, and brother/sister sexual experimentation has been noted in rural societies for some time (Fox, 1962) and, as noted above, has continued into modern times. In certain ancient societies, such as ancient Egypt and among the Hawaiians and Incas, members of royal families had to mate with brothers and sisters because no other would be pure enough (Steiner, 1967; Fagan, 1986). Additionally, a recent survey has found that when brothers and sisters have been separated at birth and then reunited as adults, about 50 per cent fall desperately in love with each other, almost at first sight (Hill, 2003; Gonyo, 2006). As noted above, in modern times even father/daughter and mother/son incest, although rare, is not unknown.

This approach, however, does suggest a number of problems for emerging youth-apes. Not least of these is child rearing; remember the mother is herself a youth-ape (some might say, child-ape). And while most youngsters love playing with babies, it is not at all clear how good they might be at keeping them alive. On the other hand, age 10–12 is the point that chimp and bonobo females give birth and they are good, very dedicated mothers. By about 12 years of age it is possible that a degree of child-ness would have disappeared and primate mothering capacity manifest itself. Nevertheless, gathering and fishing away from the nest could have been cumbersome and very difficult, even for the best mother (or grandmother or aunt) if an extremely dependent infant could not cling to the mother while the mother foraged or fished (Hrdy, 2011). Leaving infants alone, even in a nest of sorts, would have been precarious, especially if there was a lot of water around. In her description of the twentieth-century stilt-house, lagoon-living Manus of New Guinea, for example, Margaret Mead (1942) points out that a parent could never let a very young child out of sight and they had a strong belief that: 'Until a child has learned to handle his own body, he is not safe in the house, in a canoe, or on the small islands. His mother or aunt is a slave, unable to leave him for a minute, never free of watching his wandering steps' (p. 27). Even today in every rural area of the world irrigation canals and rivers are the nightmares of parents with toddlers.

So, moving about in response to changing environmental conditions, or in search of food, would not have been easy, especially if there were a number of infants to be moved and cared for at any one time. All in all, child-care would have become a rather full-time activity for our youth-ape mothers. The best bet would be to stay with offspring, or leave them with someone who could be *trusted*. But trust, like love and good sex, can be a scarce commodity. Original ape and youth-ape grandmothers and aunts may have played a major role as helpers, continuing with the 'aunt behaviour' observed among bonobos (and in chimps to a lesser extent). Females could also have attracted child-care aid from other females, such as cousins, daughters or best friends, as is also seen to some extent in chimps and bonobos.

And we cannot ignore the possibility of females attracting their sons into child-care activities, after all bonobos show us that the potential for a very strong mother/son bond existed in the ancestral line. Other male possibilities include brothers or male cousins, who may well have been the father of their offspring anyway. If – in emerging humans – sisters, cousins, friends, daughters, sons and brothers shared the freakishness of being youth-apes, and played together, emotional mechanisms for motivating mutual child-caring could already have been in place. This friendship behaviour could have continued into the period when youth-ape females were reproducing on a relatively regular basis. So, the possibility of sharing child-care may have had at least a partial solution relatively close at hand.

Another problem generated by this scenario of human separation, however, is that there is considerable evidence that incest eventually leads to definite biological disadvantages, if not disabilities (Cf., Ridley, 1993; Miller, 2000; Jones, 2002, 2013).

Primates appear to avoid incest on the basis of some form of instinctive mechanisms, including the out-migration of one sex or another for mating purposes (Bischoff, 1975; Dixson, 1998). As noted, however, this may not have been a major problem early on. Although selection pressures would eventually have been against incest, it would have taken some time for new parasites to evolve to attack the *specific* new genetic combination represented by youth-apes, or for deleterious mutations and/or genetic deformities from in-breeding to pile up sufficiently to become harmful to overall reproductive success among youth-apes.

Nevertheless, at some stage incest would have become sufficiently harmful to have been selected against. This would mean that the genetic propensities for offspring and siblings being attracted and enticed into sexual activities through infant/juvenile play would have been selected against. Indeed, in humans a high degree of 'anti-incest' seems to have been achieved by the evolution of emotional-cognitive mechanisms, which repress actual sexual intercourse during play and which, later on, repress lust being activated by sexual signals from those with whom an individual has intimately played during infancy (Shepher, 1971, 1979; Fox, 1980; Gray and Garcia, 2013). Reproductively successful youth-apes, at this point, would have been those motivated to find slightly 'strange' opposite-sex individuals sexually exciting, and to look for sexual partners from outside their immediate group.

The evolving underpinning mechanisms would have been based on mutations or genetic drift in the direction of a reactivation of primate child/juvenile/early adolescent sterility and a bonobo female's state of adolescent 'sexual inactivity' (Hashimoto and Furuichi, 1994). An evolution of maturing more slowly – through minor mutations or simply genetic drift – in the post-natal period (for something like the first 10–14 years, for example) would have been part of this process. In behavioural terms, maturing more slowly could have entailed being a baby/toddler longer and not so quickly being interested in play-like sex games. This would be a 'stretching out' of the early human life span rather than just an elimination of juvenile fertility – the baby and juvenile periods being expanded and fertility starting at about 12–15 rather than at 8–12 years of age. And there is no reason in these developments that as females became fertile only after a period of child/juvenile infertility that they would have lost the capacity to produce more immature births, at more regular intervals – the capacity postulated in our original scenario. Indeed, such stretching out, as noted, could have resulted in a longer reproductive span for emerging humans (as well as a longer overall life span).

Sexual activation itself would come to be based on attraction to more adult-like characteristics (in a way a reversion to certain ape-like characteristics); these would be what we call secondary sexual characteristics (in place of more child-like, play-like features). In so far as humans did not feel lust for those they played with as toddlers and juveniles, it would be these characteristics of *outside* males and females that became sexually attractive. On the part of females it is likely that this would have included becoming attracted to more dominant (possibly older) males. The result of this would be males becoming more aggressive as they increasingly competed for migrating females and for status to attract females; this would explain why males have evolved more ape-like features than females.

The emergence of puberty and the taming of sexuality

The emotional-cognitive motivators selecting for the above genetic changes and/or drift would have been something like the development of a capacity in human children to be extremely shy, embarrassed and anxious about their early sexual development, often accompanied by anxiety (Freud, 1962, 1973 Lecs. 21, 25), as observed in most modern human children (Cf., Klein, 1952; Erikson, 1965). It might have been that as human development became stretched, elements of 'baby-ness' remained into what became a toddler stage and as children moved from the toddler stage to the juvenile stage the previous forms of pseudo-aggressive overt sexual play as seen in juvenile apes became too rough and painful, and so individuals tried to avoid such play – they cried rather than laughed and giggled. Indeed, the human capacities for fear reactions and apprehension when near another seemingly rough or powerful individual certainly existed in the ape ancestry. Moreover, with more and more infants in a nest, sibling rivalry may have become almost violent, even among infants.

With the evolution of self-awareness (visible at about the age of 3 in modern humans) individuals may have become especially protective of their bodies – of the soft parts, the tender parts, parts that were so easily pinched, punched or grabbed by teasers. The evolving feelings which supported hiding and protecting these parts could have been embarrassment, shame and shyness about themselves (and their bodies), lasting through to adolescence and on into adulthood. This 'checked sexuality' and later embarrassment, shame and anxiety would approximate Freud's latency period between about 4 and 6 years of age and the later stage when pleasure transfers from various parts of the cuddled, caressed body to erogenous zones (1962, 1973, Lecs. 21, 25).

Whatever the exact sequence of developments, embarrassment, shame and shyness – especially regarding sexual parts and behaviours – tend to become quite strong (in most cases) and remain throughout the life of individuals (albeit often unconsciously motivated). This, of course, encourages parents to restrict their own children from presenting themselves sexually (after about the ages of 3–4), perpetuating, and even enhancing, the pattern of behaviour described above. Such emotional-cognitive motivators would have repressed infant through to young adolescent sexual behaviour, selecting for a later age of sexual indulgence.

However, the secrecy and parental injunctions in the above developments can lead to mystery, intrigue and, given the physical pleasure that can be derived from these 'hidden' parts of the body, an excitement about these 'sinful' areas, as well as to later fantasies and daydreams of both the pleasures and dangers they might bring. Human sexual parts became more than physical attributes, they became an essential element in the development of human consciousness, strongly related to the embarrassment, shame and guilt underpinning the development of a number of sexual taboos within human communities.

As child/early adolescent playful/sexual attractiveness lost its effect as a trigger for copulation and impregnation – in tandem with delayed sexual maturity – both females and males evolved a number of new means of advertising their delayed fertility and for stimulating opposite sex sexual interest at that later time. For females, these include the growth of external breasts, widened hips relative to their waists, firmly rounded buttocks, long shapely legs and retention of the baby face attractiveness and high pitched voices of their neoteny. At some point (perhaps as early as 3 myr ago, and very likely by at least 2 myr – Wheeler, 1984; Winter, et al., 2001; Rogers, et al., 2004) humans started to become generally hairless; this might have been for a number of selective reasons, but among them most probably was a sexual attraction to the baby-like nakedness itself, especially on the part of males (Cf., Giles, 1984; Jablonski, 2010). They did, however, develop long-growing head hair, eye lashes, underarm hair, and they retained (or developed) ample pubic hair, the first two and last at least becoming sexually attractive features.

These sexual/social attractions of nakedness provided an early opportunity for selection to work on as evolving humans seemed to have become naked well before humans started wearing clothing (based on evidence that the body lice that depend on clothing appear to have evolved well after humans had become naked, about 170,000–80,000 years ago for body lice – Toups, et al., 2010). Once hairless, nakedness rendered the human body a 'canvas for decorations' for the possibility of attracting attention, status and mates, and for declaring membership of a particular group or network (Jablonski, 2010). As post-puberty females looked for males somewhat older, somewhat 'strange' and exciting, males in competition for them grew slightly bigger and stronger than females (about 20 per cent). They maintained more body hair than females (on legs, arms, chest, face and sometimes backs), developed a deeper voice and experienced the growth of the penis in both length and thickness (the human male penis in adult repose is similar in shape to a baby bonobo penis, but unlike in bonobos sees an expansion of the same shape in erections – see picture in de Waal and Lanting, 1998).

The existence of pubic hair is not easy to explain other than in terms of sexual selection. As humans progressively became upright, and, as a result, the vulva of females became relatively well hidden from view (unless purposely exposed), it seems that pubic hair replaced previous visual indicators of the major area of female anatomy that can be sexually stimulated through contact and where penetration is to take place. Female (and to a lesser extent male) pubic hair could have acted as an attraction to grooming, a practice representing affection that often leads to sex in primates generally. Such grooming (and associated rubbing in the genital area) likely would have been especially exciting for both males and females, who, after a bit of reciprocating, had sex – certainly, it is not uncommon that tactile stimulation of the female vulva and male penis often leads to sex in modern humans. As such, pubic hair on females would have become a major signal for igniting sexual excitement among males because those males who persisted in wanting to find it and groom it and gently press it and kiss it would have been favoured by selection. If a female wished for a considerable amount of time to size-up potential males she could have worn a covering of some sort and only allowed males who had proved themselves in a variety of ways to have a touch, and be even more selective before allowing a viewing or prolonged grooming. Additionally, pubic hair may serve to retain sexually exciting smells.

Head hair is the only human body hair that can grow very long before it breaks (beards come second). Head hair may have evolved as something for back-riding infants to cling on to before back-strapped seats were perfected (Cf., Morgan, 1985; Dixson, 1998). However, that would encumber the mother (and probably be extremely painful), so I am more inclined to see it as part and parcel of the process of sexual selection that I have been describing. Even more than pubic hair, long head hair lends itself to grooming; it invites lice picking and being played with. It can be twisted and combed and shaped and polished and decorated. Long head hair can be spotted from some distance as a signal of 'look at what I have that you can play with'. Certainly, long head hair has often represented sexual and physical virility, masculinity and/or femininity in human cultures. It requires a lot of care to keep the lice out and the tangles under some control; the bearer seems to have time and energy and resources to spare. The human head of hair is the peacock's tail; its cost is high, but if an individual can afford it they must be a good bet. There is no doubt that in modern times a bald man is at a distinct disadvantage, especially in entertainment and political activities, as is a bald woman or a woman who never seems to look after her hair. Because there are very few bald women compared to men, it seems that through sexual selection long hair in females was more of an attraction than for head hair in males. With head and facial hair we see a major early link between nature and culture; in many human cultures hair and facial styles have come to represent a person's social standing as well as individual desirability.

The evolution of external breasts may be because they reminded males of the buttocks of females in the days when the buttocks were the site of both sexual advertising and sexual entry, and/or of the swelling that indicated sexual receptivity (Morris, 1968; Anitei, 2007); or it may have been that firm fully rounded buttocks were signs of good health and fertility, and that sizable external breasts came to share this role. It would have been very easy to become attracted to breasts because as children we get so much pleasure from sucking them, and many females derive pleasure from holding a baby to their naked breasts (Giles, 1984). The fact that, in general, adult human males like to suck female breasts, and females like this, as part of sexual play gives some credence to this. It has been suggested that the evolution of the attractiveness of the 'hour glass' waist to hip ratio of human females may have been because female fat reserves in their buttocks released crucial fertility hormones, or even acted as an energy reserve to be employed in rearing offspring successfully, or such fat could have been an indicator of optimum health. Another possibility is that the specific ratio that seems most attractive (approximately .7) appears to be the best for giving birth while still allowing a female to walk efficiently – given the problems bipedalism caused for both birth delivery and locomotion in humans. Still another possibility is that the waist/hip ratio is the best indicator that a female is not already pregnant (Cf., Ridley, 1993; Singh, 1993; Singh and Luis, 1995; Singh and Young, 1995; Marlowe, Apicella and Reed, 2005).

The above characteristics as triggers to sexual desires (and fears) most probably evolved in fits and starts, and relatively easily. This is because, as with maturational processes generally, they are based on genetic packages in which a wide variety of alleles already exist ready to be arranged and rearranged (ones that are also vulnerable to mutations). Further, they are also based on the effects of resulting neurotransmitter and hormonal changes that are subject to embryological and environmental influences, and so the whole process is especially vulnerable to selection. Take body hair, for example;

in humans, body hair seems to operate as several 'systems' (Morgan, 1985; Dixson, 1998; Dawkins, 2004a) and hair follicles are sufficiently present in embryos to produce a fully haired human, but are repressed by early hormonal actions so that only certain restricted patterns emerge.

The degree of vulnerability to genetic drift and minor genetic alterations of secondary sexual characteristics can be seen in the fact that most of the secondary sexual characteristics described above, for both males and females, show considerable variation even within a breeding population. And, indeed, there can be an overlap of males and females. There are, for example, women who are bigger or stronger or more hairy than some men, but on average this is not the case. Both men and women vary considerably in terms of the amount and location of body and facial hair, growth and shape of external breasts, development of genitals, and depth of voice, and so on, but in terms of the mean there is in all societies an identifiable gender difference in these.

And it is to be noted that even with the evolution of sexual attraction to secondary sexual characteristics, the power of being sexually attracted to (and to be pacified by) childlike characteristics remains relatively strong in humans (providing support for our original scenario). For example, we maintain a variety of cultural images that portray adults with childlike characteristics, especially females, as being attractive. Most cultures throughout history have advocated a very young, innocent looking age as being 'the best' for marriage for females (Hajnal, 1965). The increasing use of girls as young as 12 years old as fashion models in modern times follows in this tradition. The attraction of pubic shaved females in a number of historical paintings, in a great deal of pornography, on certain internet sites and among skimpy bikini wearers, suggests, again, that the characteristics of modern infants and youth have long been, and remain, characteristics which act as powerful sexual attractions for a significant number of humans.

But there is also ambiguity in our sexual natures. At the same time as we are sexually excited by childlike sexual attractiveness, this attractiveness also motivates us to want to care for our young, to coo at them, to hold and protect them, to guard them with our lives. So we also idealize children as (usually) being innocent victims of the dangerous world into which they are born. These dangers often include specific adult misbehaviours and 'evil' adult intentions concerning sex with them. The conflict between childlike characteristics being sexually attractive and children as 'spiritual' entities to be loved and cared for at all costs, then, has a long evolutionary history (Brin, 1995), one that has not been edited by selection to everyone's satisfaction, one which still seems to be in process. This may be because the sexual attractiveness of children continues to elicit care; it attracts care from mothers (and fathers) directly, and it attracts child-oriented males to females and to their existing offspring, while attraction to secondary sexual characteristics tends to avoid incest and to lead to mating between more mature, more competent individuals.

Whatever the nature of specific parental relationships with infants and adolescents, there is often conflict between parents and offspring when puberty-driven sexuality hits full stride and an offspring 'will to power' kicks in full-time, with the desire to break away from parental control. This can generate feelings of guilt in offspring; after all it is difficult to forget the loving care given to them by parents throughout childhood – the unconscious does not forget even if the conscious rationalizes away its significance. Embarrassment and shame have led offspring to hide their emerging sexual characteristics from parents, and now they try to keep their sexual experimentations secret from parents. Sex becomes a generally tabooed area for humans. A young person hides their sexual parts, to be later shown only to that special 'friend' with whom a variety of other, non-sexual intimate secrets might be shared – like: 'I have a mole on my behind' or 'I am terrified of ghosts'. And when they have an infant of their own they are drawn to the infant with the power that sexually drew them to each other in the first place, that is, the infant-like presentation of a smiling /'I am safe'/'be my friend'/'come care for me' face. But now this does not generate a desire to have sex with these infants; rather the urge is to protect them at all costs – and later to encourage them to hide their sexuality from others.

At the stage when females were becoming attracted to secondary sexual characteristics of strangers, selection had a new set of problematics to work on. Original youth-ape females may have become pregnant rather indiscriminately but, for an out-breeding adolescent youth-ape female, choice became vulnerable to selection. Would a female get it right in her choice of mate? Would the stranger harm the female, and her offspring, especially any offspring she might already have from a previous consortship? Would he be attracted to the offspring for sexual rather than caring purposes? Thus, abilities to identify and to attract 'tame' child-oriented male strangers (but ones not interested in sex with the offspring) to aid in child rearing in exchange for sex and mothering seems to have been selected for. Trustworthy '*adult*' males would have increasingly been favoured by selection in so far as child-attractiveness was no longer a sufficient indication of fertility or safety.

Given the extreme dependency of human offspring at birth – and for several years thereafter – in both cost benefit and emotional terms, mating with a specific male who has been 'evaluated' in terms of child-orientation and a willingness to help rear children, but without sexual interest in children, would seem to have had considerable advantages over needing to attract several competitive, potentially aggressive, jealous, sexually-aroused males to provide the same total benefits. Anyway, if one of these competitors became dominant, he would most likely drive the others away, and the dominant one might well be the very worst at child-care and at bringing in or sharing resources. The evidence clearly suggests that among primates, including among humans, evolution does not stabilize at polyandry (Cf., Dixson, 1998; O'Neil, 2006).

Mutual playing brought youth-apes together in the first place. But now we are looking for the evolution of being able to select a new 'best (opposite sex) friend' to play with and go off with (or at least hang out with) and to have sex with and to rear children with. Without visible signs of ovulation females had enhanced possibilities for the use and control of their own sexual attractivity to these ends than most primate females. Without visible signs of oestrus females had opportunities to be coy and deceitful; for example, females could try to get genetic material from a powerful, dominant male and deceive a more domestic type (or types) into caring for her (and the dominant male's) offspring. This, of course, would have set up selection pressures which favoured males who themselves sought a degree of confidence in a female before committing themselves to child-care. And males could have been deceptive too; they might have communicated caring, loving intentions and then left after sex, or conversely, after sex have become extremely jealous and aggressive and dangerous to the female and her offspring. So it can be noted this evolution never settled at harmonious, or strict, monogamy.

The above evolutionary possibilities fit with the observation that humans mature at different rates (Eveleth and Tanner, 1976; Waber, 1976; Waber *et al.*, 1981; see also, Excerpts, 2013). They fit the fact of observed incest avoidance among those who play intimately together when very young, as well as incest avoidance between parents and children. They fit the fact that 'polymorphous perversion' among infants (Freud, 1973, Lec. 13, p. 246, 1962; Kinsey, 1948) disappears through shyness, embarrassment and anxiety, and through adult condemnation (Freud, 1973, Lec. 13). This scenario fits in with the observation that humans develop relatively powerful (albeit often somewhat ambiguous) sexual and personal identities around their sexual feelings. And the above speculations conform to the fact that more or less formal sexual avoidance behaviours and/or taboos exist in all known human societies (Fox, 1967, 1980). This scenario does not exclude sibling rivalry, the widespread existence of feelings of jealousy, envy, love, hate and anger in human relationships or the elaborate courting patterns found among modern humans. It does not contradict the idea that humans are largely a monogamous species (albeit often serial) in which both parents usually lavish an incredible amount of attention on offspring.

It is to be noted, however, that although the above possibilities bring together genetic, biological, psychological and social factors, it does not necessarily point to a completely *stable* (or even happy) personal or social outcome for humans. As selection pressures changed youth-apes from being emotionally geared to (and dependent on) seeking and receiving the attention of mothers, siblings, cousins and playmates, to desiring attention and playfulness from sexually exciting 'strangers' of the opposite sex, ambivalence and a degree of resentment among newly competing friends and relatives must have existed. There certainly is no special reason to think that female/female and male/male and male/female competitions, deceptions, manipulations and manoeuvrings (all relatively common among primates), and the potential animosity among the various individuals involved, would have disappeared with the evolution of youth-apes to humans. We would expect a continuation of a degree of envy and jealousy among humans, something already relatively strong in the primate line (and certainly something clearly observable in modern human children, adolescents and adults).

Some emerging reproductive patterns would have generated feelings of shame, embarrassment, anxiety and guilt as repressors of earlier primate sexuality, and would also have had repercussions in a number of other areas of human life – some perhaps good as useful 'lessons', others debilitating as stifling restrictions. Certain sexual avoidances and taboos would have led to painful emotional conflicts and ambiguities, and some to secret fun, such as the grooming of pubic hair, and to sex in secret. Sex in secret could have had noticeable evolutionary implications. For example, human males have retained an element of chimp and bonobo tendencies to relatively rapid ejaculation, especially with a new partner; but staying together, playing, for a time afterward, unhassled by others, could have resulted in a second (and later third) copulation of longer duration, increasing the play and holding and hugging time, and thus the potential for pleasure on the part of both males and females. This would have intensified identification of pleasure with a particular partner. It would also have increased the production time for oxytocin and vasopressin – hormones involved in the forming of bonds between mates (often activated during sexual play and sex) (Cf., Szalavitz, 2002; Lim and Young, 2006). This pattern could have fueled sexual exclusiveness, and shared secret sexual jokes and teasing and desires to be 'private' about secret sex with a partner. But secret sex, restricted sex, tabooed sex, can also lead to fantasies about sex generally, and about objects of desire. Shame and embarrassment may have led to the wearing of clothing to hide genitals, but then to fantasy visions of what a certain person might look like with no clothes on, and to clothing designed to titillate the imagination.

From here it is easy to consider the manipulation of hair, body shape and clothing becoming a major part of an individual's presentation of self, which could have become elaborated and exciting as an activity in and of itself. With this elaboration of self-presentation and potential for deception we would expect that selection favoured individuals who could 'read' the presentations of others to ascertain who might be the best bet to link up with for the successful undertaking of a variety of survival and reproductive tasks. Eventually in this evolutionary scenario individuals could gossip about 'personalities' and about the 'real intentions' of others, and what and whom to beware of; lust, love and romance would have become the mainstays of fantasies, gossip, stories and hopes (and fears). Those most able to survive the emotional minefields involved would have increasingly contributed to the evolution of human desires, fears and patterns of cognition.

If, as the hominin line moved through the incest problem, females, even after the evolution of later sexual maturity, nevertheless remained drawn to (and were attractive to) males on the basis of *some* of the playing and friendship characteristics that attracted youth-apes to each other in the first place, the solution to the problem of finding a suitable mate would, in part at least, have been in situ. This is because the males attracted would have been those whose aggression was more or less repressed by the female's infant-like characteristics, and who were attracted on the basis of an emotional propensity to play, friendship, and even a desire to bond with her, but now, in addition, who were sexually aroused by her *secondary sexual characteristics*.

This, however, would still not have eliminated all mating ambiguities, confusions and conflicts. Some males, for instance, might have ended up so dependent on the females that they were useless as helpers or partners; he had come to be mothered once again, just like his first mother had done; he liked the sex but afterwards always wanted to go play with his male friends. Initially she may not have minded loving and mothering him, just like she imagined she would her children; and she may have been happy to stay faithful, and he too; but he could have become quite demanding. And, she may have had her hands full with offspring, and he may have resented her mothering going to them, especially if his sex was curtailed as well. And she may have got fed up with his demands for attention and attendance, and remembered how her dreams and fantasies about a hero partner had not been fulfilled by him.

Nevertheless, variable and conflict prone as it sometimes might have been, a new pattern (and its motivational processes) for pair bonding between male and female primates was emerging. Previous phenotypical indicators of such things as: genetic fitness, fertility potential, parental capability, sexual attractiveness and sexual receptivity were no longer 'accurate' and selection within the emerging species was favouring a genetic potential for the ability to 'read' phenotypical signs of *emerging* behaviour *potentials*. Among bonobos a number of facial expressions, gestures and play-like behaviours successfully invite copulation, so we can assume that the possibility for directing these at very *specific* individuals was already in the ancestral genetic make-up; only now the invitation was to play for a while and see if copulation might be a good idea in the future.

While it is unlikely that all the characteristics that appear sexually attractive now are exactly the same as those at the time of human differentiation (where youthful play and youthful appearance were paramount in the developments presented here), the importance of play remains, and the 'helpless', innocent, baby-like face has, if anything, grown in importance as an invitation to friendship/mate-ship (rather than as a direct invitation to sex). It is the human face that communicates kindness and *trustworthiness*, the characteristics most essential to know about an individual in deciding whether or not to sneak off with them on a consortship, or to let play turn into long-term regular sex with them, or if they are someone to snuggle up to in the night after sex in anticipation of a shared breakfast, or to leave an infant with.

In modern humans reproductive success for both females and males can still be equated with a baby-like, smiling face that says: come play with me, protect me, like me, trust me, look after me (mother me). While it is true that the human face can be deceitful as well as inviting, it is surprising how difficult it is to fake facial messages over an extended period of time. A fake smile or laugh is detected right away. Trying to look interested in an uninteresting person is extremely hard work and never fully succeeds; a look of disgust is very difficult to conceal. On the other hand, the facial expressions of joy and happiness of newly courting couples simply at the sight of each other are instantaneous, spontaneous and 'silly'. Two individuals in the process of matching-up look intently at each other and giggle and squeal and laugh like a couple of youth-ape/human children in deep play.

At the same time, a face can look serious and be 'full of character'; we begin to read moods and personality characteristics from (or into) the faces of individuals as they mature away from the pure baby-face of an infant, adolescent or young human. To maintain a degree of desirability, however, a maturing face must 'read' as accumulated wisdom, kindness, trustworthiness, strength and success, not as threat, despair or weakness. And it never hurts to retain a hint of youth through disguising wrinkles and having a strong twinkle of vigour in the eyes, perhaps with a full head of hair dyed a bit; certainly a nicely styled, or well-trimmed, white beard with a smile never goes amiss.

Modern psycho-socio-sexual motivators, then, are largely derived from youth-apes becoming motivated to progressively mate out from their immediate group. It may seem unwise to generate scenarios based on the above kinds of speculations and to expect so much from them. But is it more foolish to speculate on the basis of *observable* general primate and human behavioural, psychological, reproductive and social characteristics than to create a mythical 'rational man' on the basis of very, very few fossil fragments?

In fact, until recently all the fossil fragments could have fitted into a small closet in an obscure laboratory somewhere. Some time ago S. L. Washburn (1973) noted the problem involved with this, adding the observation that much more than morphology must be used in understanding human evolution because, he reasoned: 'The evolutionary game played with molecular information, immunochemistry, functional anatomy, and behaviour is a very different game than one which is limited to fossils – especially when the record is as fragmentary as is the case of the primates' (p. 68). Earlier he had noted the importance of psychological and cognitive criteria in such an understanding (1963a), while later noting that bipedal walking came well before tool use and well before the large human brain (1978, p. 7). I have argued, as was put forward by Darwin (1879), that the processes of sexual selection will have played a very important part in human evolution.

It must be said that more complete bones, and even partial skeletons, have been found since Darwin, and even Washburn, made these observations. And the work of palaeontologists who seek to explain human evolution from archaeological evidence is crucial. Their painstaking reconstructions are often extremely impressive in terms of the science involved and the interdisciplinary methods increasingly being employed. But the truth is, the fossil evidence remains extremely fragmentary, and none of the new finds contradict the statements of Washburn or Darwin's suspicion about the importance of sexual selection or the role of behaviour and emotions in the evolutionary process. Indeed they strongly support the order of development of bipedalism, tool use and the increase in brain size. Tool use and brain development in fact came quite late in the evolution of humans – a bipedal ape/hominin still somewhat adapted to trees evolved long before a larger brained, tool-using, hominin arrived.

Furthermore, a look at the fossils themselves tells us precious little about the behaviour, social life or mental states of these ape/hominins. There is a tendency among evolutionists to see apes (and other primates) as being highly sexual, conflicting, sometimes violent, emotional creatures, at one end, and humans as being 'rational', problem-solving creatures (almost above emotions) at the other end, with all transitional species being almost robot-like creatures increasingly able to solve survival problems. We have in many such portrayals an ape/man or cave man out risking his life cooperatively hunting for mammoths while his 'wife' is in the cave keeping the fire burning and sewing some clothing for the winter, one kid is practicing with a small stick to become a skilled hunter and another is attempting to make a clay bowl to eat the supper out of. We do not seem to see lust, love, sexual attraction, obsessive friendship, sexual jealousy, adultery, sibling rivalry, envy, hyperactive offspring, ruckus play, cheating hunters, cowards, masturbating teenagers, sexual frustrations, a will to power, status striving, obsessions, personal conflicts, fighting, wars, prostitution, depression or suicide, for example, as having played any part in the transition from ape to human. Perhaps this is natural in that we only have bones (and a few camp sites) to look at rather than reports of hominin psychoanalysts; but this has not stopped many from creating rational, liberal, cooperative – non-emotional – man from these bones, even when such a man is hard to find even among modern humans.

By engaging in the types of reproductive speculations in this chapter we are at least able to ask questions about some possibilities of human evolution that are compatible with observed human reproductive, psychological and social behaviours; 'designing a plausible scenario of a species' evolution is a scientific endeavor ... its outcome is open to falsification ... [its assumptions] ... can be proved wrong' (de Waal and Lanting, 1998, p. 137). If palaeontological evidence (fossils, camp sites, tools, butcher sites, environments) comes in future to clearly contradict the types of speculations of this chapter, the speculations must be revised accordingly; if the palaeontological evidence suggests new speculations, all the better for the sciences of human evolution and human psychology. Overall, however, at this stage the advantages of considering the above possibilities relating to sexual selection are:

— that it is not necessary to discover or postulate a specific major, dramatic, sudden environmental change behind human evolution;
— it is not dependent on using a postulated end point of rationality as the teleological pulling force causing human evolution;
— nor does it depend on the idea of a big brain driving a 'rational, purposeful' evolution; and
— it might account for many of the sexual and social emotional-cognitive processes, and resultant patterns of behaviour, *which have been observed* to be extremely common throughout human history.

It, in other words, would account for a species-typical human nature (emotional-cognitive processes/consciousness) which underpins modern human life.

The evolution of human emotional-cognitive processes

Human consciousness, like any species' motivational package of instincts, drives and patterns of learning, is a blend of what has gone on before and what has emerged during a new evolutionary radiation. So, the question is, what specific drives, emotions, desires and fears emerged as human emotional-cognitive motivators? To explore this question it will be useful to focus specifically, as far as is possible, on some of the problematics in terms of reproductive success that faced emerging humans to get an idea as to the specific motivators that might have worked. I will do this by looking separately at offspring, females and then males.

Offspring: evolutionary 'problematics'

The issue of infant survival – especially when there increasingly were more than one at a time in a state of dependency – probably cannot be overemphasized. Human infants are extremely dependent and helpless at birth and the infant mortality rate during human differentiation may have been extremely high (Mellen, 1981; Babchuk, Hames and Thompson, 1985; Hrdy, 2011). Indeed, it is very high among gorillas (Fossey, 1979) and chimpanzees (Lawick-Goodall, 1971, 1977; Saayman, 1975), and remains so among humans in many developing countries (Schell, *et al.*, 2007; CIA World Factbook, 2013; Wikipedia, 2014). Infanticide rates also may have been relatively high during the period of human differentiation (Alexander, 1971, 1974; Babchuk, Hames and Thomson, 1985). Infanticide, though not common, certainly exists in some mammals, including monkeys, gorillas and chimps (Lawick-Goodall, 1971; Suzuki, 1975; Fossey, 1979; Goodall, *et al.*, 1979; de Waal and Lanting, 1998), and rates of it have been high in a number of human hunting and gathering societies and during certain historical periods (Alexander, 1974; Monter, 1976; Harris, 1977; Dickemann, 1979; Jones, 2002). Many suspect that it is still high in a number of societies.

Therefore, it does not seem likely that there was an evolution of an intensified drive for mothers to love offspring more than is usual among other primates just because offspring were more dependent at birth. It seems that there was an evolution in the opposite direction when compared to chimp and bonobo mothers who normally have a very powerful emotional attachment to their offspring (Cf., Hrdy, 2011). More accurately, it appears to have been an evolution of an emotional-cognitive system that motivated human mothers to invest in child-care when times were good and not to when times were bad. This would take the form of something like heightened sensitivity (elation) at the sight of a new born when there was general heightened sensitivity and/or tranquillity, and behavioural depression or disgust at the sight of a new-born when there was a general degree of agitation, anxiety, anger, hate and/or fear. In other words, for humans good times mean good emotional times just as much, if not more, than good economic or physical times. It would be difficult to ascertain the degree to which primates in general might suffer from post-natal depression, but it is observable that among modern humans it is far from uncommon. It is not clear as to what might set in motion post-natal depression in humans, but its existence alone certainly demonstrates that the whole process of giving birth, and then contemplating caring for that resultant infant, can activate moods that result in inaction with regard to the provision of child-care.

Humans can also perform infanticide for calculated economic and social reasons (or be forced to by elders or husbands, for example), so a 'rational' capacity for resistance to producing unwanted offspring also evolved. Indeed, with modern contraceptive methods and the availability of safe medical abortions in many parts of the world, controlling (reducing) birth rates and the timing of births in relation to external factors is very common. Modern humans (at least those who have the option, based on science, affluence and a degree of female choice) have decided that 2–3 children are just about the right amount 'to love and cherish'. This, significantly, is most prevalent at a time in history, and in geographical locations, where humans could most easily afford many more. Birth in the modern world is a controlled, pharmaceutical and medical event (for those with access) to provide two or three bundles of joy, not a gathering of excited playmates and friends attending a miracle of nature or a gift from the gods. We may never know what goes on in the mind of a chimpanzee or bonobo female, and they do very occasionally abandon infants, but it is hard to imagine them emotionally comprehending abortion, sexual abstinence, 'family planning', wanting only male children, wanting male aid in child rearing or seeking very special mates so their offspring could have a nice daddy. They intensely love and protect their offspring (*one at a time*) until the offspring more or less leave and get on with their own lives as best as they can.

There is a seeming paradox here in that an increased dependency of human infants at birth did not necessarily lead to a stronger bond between mothers and offspring, but rather to a weakening of it. The paradox is solved if we consider that a weakening of the maternal bond is compensated for by the fact that, except in the worst of times, more individuals, including fathers and families, and sometimes friends, were emotionally enticed into aiding in child-care – something a chimp mother, for example, would very rarely allow (Hrdy, 2011) – and in relatively good times many more offspring could be born in short order. It is as if some of the bonding power between a chimp/bonobo mother and her offspring has been diluted but compensated for by a human female's capacity to bond with significant others. We have seen that it is likely that females drew males into child-care with offers of mothering and sex. The two are not as unrelated as it might seem. Face-to-face copulation means that a human female receives her lover in the same (often naked) manner as she receives her child (Cf., Hockett and Ascher, 1964; Shepher, 1978); some of the intensity of a mother/child bond in primates generally (and the powerful mother-son bond in bonobos specifically) was transferred to the female-male sex partnership in humans. If a human female was to give total devotion to one offspring (as a bonobo or chimp mother does), she would have had little left over for a male mate or nest full of offspring.

The emotional processes involved in this evolutionary change may have been the development of attraction to general *baby-infant-ness* instead of to the odour or appearance of a specific offspring, making it easier for females to not bond during difficult emotional times, and for others to be drawn into child-care activities at all times. The extension of infantile physical expressions and characteristics among offspring for a considerable amount of time after 'babyhood' gave these attractions a longer effectiveness. Much baby 'cuteness' might disappear as they grow but toddlers and young children remain attractive; smiling, giggling, laughing, sounds of pleasure when tickled or physically played with, snuggling-up to adults, expressions of coy innocence and curiosity, thankfulness and appreciation draw attention, resources and help from parents for a considerable time.

Thus, the power of child-like attractions would not only have been extended over a longer period, but also have more easily been shared among a bunch of siblings, and have drawn support from adults in general, not just from a mother (Cf., Hrdy, 2011). In modern times, for example, it is virtually impossible for an adult human not to say 'ahh' in a low, deep, drawn-out voice and then go into baby talk when a very young baby smiles in their direction or is just held close to them, even if they are strangers. Although I have not seen reference to ape babies getting mixed up, there are cases of human babies being given to the wrong mother, and birth professionals go to great lengths to make sure that this does not accidently happen. Additionally, step-parents can become very protective of step-children, and people can adopt babies and love them, all suggesting that baby-ness is, in and of itself, a potent attractive force.

Baby-ness/child-ness can indeed be powerful. So much so that in good emotional times it can overcome all the inconveniences and unpleasantness of a number of the 'not-so-wonderful', and non-care-enticing, expressions and behaviours of infants. Infants vomit on parents and carers, they cannot control defecation/urination, they often mess on the parent, they do not always sleep through the night; they demand immediate attention right now, not later when it might be more convenient or seemingly reasonable. Toddlers get into everything, they break everything, and they have tantrums. Young adolescents hit their younger siblings, make them cry, demand every new toy, refuse to eat healthy foods. So, selection would have favoured parents who relatively easily become *addicted* to signals of 'child-ness' despite a great deal of unpleasantness and inconvenience.

But baby-ness can also lead to rejection. Quite apart from the unpleasantness babies can generate as described above, smiling, giggling, laughing, sounds of pleasure when tickled or physically played with may instead be crying, throwing food, tantrums in public, screaming, running away and violence towards parents and siblings. Expressions of coy innocence and curiosity, thankfulness and appreciation can feebly mask lying and deviousness. Rather than inducing an addiction, infant/child behaviour can lead to a lack of care, especially when parents are in a state of anxiety or stress. The addiction to child-ness is an addiction, then, that can, in particular circumstances, be 'cured' by the behaviour and expressions of the very objects that cause it. And, of course, the moods and background of the parents will also affect specific outcomes. And strong memories of sometimes very difficult births can act as a deterrent to more pregnancies, or at least to more birth experiences. So, besides birth control we also have infanticide and 'child abuse' and negligent child-care provision. Most parent/child experiences will be somewhere between the extremes suggested above; many will move back and forth somewhere along the middle of a continuum between the extremes. Love them as they might, after having a couple of healthy children many parents could have 'had enough' and been ready to embrace various means of birth control, abstinence, abortion and infanticide – and, if these methods were not available, to farm their young children out as servants to more prosperous individuals (in modern times, perhaps, to have them taken into care).

Indeed, the 'control' of an addiction to baby/child-ness starts even earlier, but, paradoxically, can go some way to preventing complete rejection. Discipline and 'reasoned restrictions' begin to replace undivided attention and unquestioning care as children pass their first year (although the exact timing does vary among cultures and individuals). It is interesting, for example, that the very time of one's life when almost total attention, care and unqualified love are abundant is the very time of life that no human can remember. Is it that offspring that remembered the amount of care, attention and love they received when first born, and for about the first year or two of their existence, and continued to demand it for the rest of their childhood, would end up being kicked out (or greatly abused) by their parents, shunned by playmates and in later life isolated as adults for being too demanding, just too 'self-centred'? Babies enjoy a degree of narcissism that few growing children or adults are allowed. In physiological terms this 'memory blackout' may be related to the fact that there is considerable pruning of neurons and neural connections between the first and second year of life (Bjarklund and Kipp, 2002).

Sibling jealousy undoubtedly existed when new arrivals made an appearance, but at this point the type of care demanded by a 2–4 year old is somewhat different from that of an infant. The 'terrible twos' may be a reaction to sibling rivalry or unconsciously trying to regain the intensity of baby care previously provided, or be a rebelling against restrictions, restrictions and boundaries that were selected for because they prevented complete narcissism. Children progressively want to explore the world and want help in this activity, or at least to be allowed to move freely. But by about three years of age they often have accepted certain restraints imposed for safety or nuisance reduction reasons. Bedtimes become regular and imposed, certain locations become off-limits, there are objects not to be touched, toilets to be used, standards of cleanliness are introduced, and so on (again, these are more or less accepted, and they vary in degree of implementation among cultures).

As they become juveniles and adolescents, such things as respectful mimicking of adults, facial expressions of curiosity and understanding, 'silly mistakes' born of earnest effort, verbal recognition of caregivers, evidence of self-awareness and naive explorations of self and the world come to be perceived by parents as endearing characteristics; elaborate praise and even material rewards are provided for following rules. Human parents become *proud* of their offspring (well mostly); they begin to generate fantasy expectations for them. Parents counsel 'deferred gratification' and the importance of duty, not lust and fornication. Parents often do not really mind if offspring stay around to help care for new siblings and even themselves in old age. All of this involves a number of emotional propensities and mental abilities not readily observed in other higher primates.

So, offspring survive but now there is a new threat to their reproductive interests in that parents might resist losing them to the adult world of reproduction. But puberty rears its head; testosterone overrides the fun of being a child; rebellion is the order of the day. Boys advertise their genetic worth with body decoration, swagger and bravado; and with booze, sports and fighting. And girls advertise their genetic worth through body decoration and fashion, with hints of sexy attributes and dancing skills. And the sexes come together over music and flirting, joking, walking together and dating. In this world, parents are the single biggest embarrassment a growing adolescent can imagine; their parents' music, clothing, sense of style and idea of fun sounds foreign and 'old-fashioned' and definitely not to be emulated. Both sexes begin to look for exciting strangers to romance and lust with.

Although the behaviours discussed here may have been crucial in the evolution of humans (given the degree and length of time of dependency of newborn infants, the increased fertility rate of females and the flexibility of offspring care these behaviours suggest) they, nevertheless, can carry a cost in emotional terms. Not all babies get the early attention, care and love described above. There are ample studies of the life-long emotional problems caused to individuals who did not receive sufficient attention, or received the wrong kind of attention, or received attention in extremely ambivalent and contradictory forms when they were infants (Cf., Klein, 1952; Bateson, *et al.*, 1956; Erikson, 1965; Laing, 1965, 1971; Winnicott, 1965; Bowlby, 1969, 1973, 1980). Even in the best of cases, a clean break from infant demands does not always take place; indeed, it is questionable whether humans do completely forget the loss of love, care and attention they received when very young. Some people, in fact, do end up as adult narcissists and wonder why nobody loves them. As Freud (1917, 1973, Lecs. 26, 27) observed, neuroses of narcissism can be the most difficult to deal with. More often, however, growing children may consciously forget the details of infancy, but do they unconsciously remember the 'fears' of abandonment that instinctively motivated care-enticing behaviour at that time? Do these remain as unconscious fear propensities, acting as a basis for later motivators for feelings of and reactions to rejection, perceived failure, jealousy, envy, anger, shame, embarrassment and guilt (for example) when puberty propels them into an angst-ridden world of sexual competitions, and later when they go into a full adult world of work and family?

Another potential problem for human offspring was competition from other offspring. It is a very different personal, social and political proposition to be born first rather than second. It is an even greater difference to be born fifth or eighth; and the amount of gap and the mixture of sexes all make a difference. Competition may have been for resources, attention and devotion from parents, or from older siblings, or for 'best friend' or for sexual partners among siblings and playmates; later in our evolutionary scenario it could have been for external mates. Given the high mortality rate that likely existed during human evolution it is very probable that when one parent died (or left) the other would take up with another lover. Such arrangements would often mean that two, genetically unrelated, families would be brought together. Here we would expect that the competition and potential for conflict would have greatly intensified (even though there are some very good ones, step mothers and fathers and half brothers and sisters do not usually feature as heroes and heroines in the stories humans have handed down through the ages).

However, the possibility of cooperation and sibling-love also clearly existed in family life. Older siblings can look after younger ones; siblings can gang-up on outsider children, or protect each other from outsiders. The primate background did not exactly prepare humans for several offspring being dependent on a parent (or parents) for an extended period of time, but it did provide patterns of child and juvenile play that could be quite congenial; and it provided the basis for rather ad hoc alliance behaviour. Both of these features among primates yielded a great deal of flexibility from which young human sociability could evolve. However, given the potential conflicts and ambiguities involved we would expect that for offspring to live together in any significant numbers (especially if they were only partially related biologically) selection would have favoured a number of political and diplomatic skills in the areas of both parenting and family living, skills that remain with us today.

What evolution has clearly done is provide infants and developing young adults with a considerable ability to read and react to the facial expressions, babbling and behaviour of adults, and of each other, in a fashion greatly elaborated over what our ape cousins can achieve. While in primates powerful attractions between parent and offspring (and in social species between and among certain other individuals) operate on the basis of instincts, drives and primitive emotional feelings, in humans these instincts have largely been expanded into emotional-cognitive feelings (human emotions), mood states, desires and fears. Additionally, however, as human mental capacities evolved, with tendencies towards emotionally charged, desire and fear-induced daydreams and fantasies, the result has been the evolution of an almost *spiritual* sense among parents for the act of parenting. Parents feel good about parenting, they feel elation, they consciously remember the smiles, laughing and other attracting characteristics of the child; they forget the vomit, or at least they joyfully tell others about it and laugh and joke about it; they love to be complimented on their parenting skills. Parents secretly condemn themselves for some of the anger and lack of attention they occasionally let slip into their spontaneous reactions to a child; they feel guilty. The disrupting behaviour of the child becomes their own fault; they swear to themselves never to be so quick to react again. They are ever ready to condemn others found wanting in the parental department. Parenthood has become a higher calling, a *moral* obligation.

The spiritualizing of parenting is part of a process that further helps to explain our paradox of humans restricting the number of offspring they produce. Above we have seen a number of emotional forces which might cause parents to feel that 'enough is enough' after two or three surviving, seemingly healthy, offspring, but this is aided by a human desire to display and to show off. Some of this is done through nest building, body decoration, successful hunting/farming, warfare, ritual performances, material accumulation and political achievements, all of which detract from full-time child-care. So, in societies in which health care is sufficient to keep most babies alive into adulthood, females have a degree of choice, physical survival is relatively secure and status opportunities generally are available, it is relatively easy to rationalize negative emotions concerning the difficulties of child-care into 'quality' parenting for a very few offspring, who will grow up relatively quickly, at the expense of a lifetime of quantity parenting of a multitude of offspring. Display activities that attracted mates to each other among evolving humans, thus, also contribute to reduced family sizes in modern developed societies, and, it can be argued, to the evolution of species-typical human social behaviour.

Parenting has introduced a decided social/political dimension into human existence. Over the duration of human evolution, some aspects of infant-like attractiveness (of both physical characteristics and non-verbal signals) became general stimuli for attracting not only care-giving towards infants but also lust, love, friendship and caring responses among adult humans (see also, Morris, 1968; Eibl-Eibesfeldt, 1971; Morgan, 1985; Miller, 2000). Indeed, it can be argued that these stimuli became the basis of many human personal and social interactions. For instance, teenage females can often get things from their fathers by acting the innocent young baby; adolescent males can get things from their mother by forever being a baby in their presence (basically by never spontaneously doing anything to look after themselves); females can get help from stranger males by acting a bit helpless, males can get things from females by acting as if in desperate need of some mothering.

The result of this evolution is that humans have acquired a mechanism to induce a flow of power and resources from the apparently more powerful to the apparently less powerful – that is, from the target of 'helplessness' signals to the sender of the signals. The recipients of such signals are usually potential providers/protectors (that is why they have the signals directed to them in the first place). The flow of power from them takes place because the more powerful come to depend for their own self-respect and happiness on, for example, smiles, close attention, love, personal adoration and general positive recognition from an apparently weaker person. Sometimes stronger/more wealthy individuals seek 'underprivileged' individuals to offer to them material help and social protection because of the emotional reward achieved by doing so. As a result of this process, the sender of dependency signals gains the ability to remove the signals in order to punish or manipulate receivers. Infants, for example, use such things as 'gaze avoidance' which effectively rejects, and thus often manipulates, given adults (Chance, 1962). Infants also exhibit signals of extreme agitation that cause pain, anxiety and even guilt in care providers when caregivers' attention drops off. This is not to mention the embarrassment that a screaming child can cause a parent in a public place. Teenagers remove love from their parents and may even reject all they stand for; they go off on the dope, they get pregnant, they leave home. And in the wider social sphere, those who seek positive recognition through 'doing good' can be devastated by unappreciative clients.

This tendency, it can be argued, applies to, for example, lover relationships (Cf., Dallos and Dallos, 1997), friendship interactions, family dynamics and patron/client politics; or as Hegel (1807) would have said, between 'masters and slaves'. In general, the psychological principle is that the greater the degree of recognition given to individuals, and expected by them, the greater the damage done when it is withdrawn, regardless of socio-economic position. The social/political results can be seen in the more cognitively analysed economic, social or political benefits that the weaker can receive (such as protection, jobs, praise and status) and/or might be able to give the protector (such as labour, votes and adulation). Both sides become emotionally dependent on each other, although the power relationship may be unequal.

The human ability to 'get one's way' through the manipulation (conscious or unconscious) of dependency (helplessness) signals became, we can theorize, a major basis of later human power, status and prestige relationships and of the politics that go along with these. Clients can come to *honour* their patrons, even worship them, because their patron is 'greater' than a neighbouring patron and provides 'fatherly' protection, and show due ritual deference as a result. Patrons can come to take pride in being '*fatherly*' towards their clients, and extend care as a result. I will refer to this process as 'creative subordination'. But this is not a straightforward process; individuals have to become politically skilled to use creative subordination effectively, and it is a competitive business with emotional dangers ever present. Creative subordination is motivated by emotions such as jealousy, envy, rage, love and hate, and by fears of physical dangers (real or imaginary), fears of abandonment and desires for attention. Later on, fears of failure and desires for prestige and status continue to fuel the politics of creative subordination. The subtleties of creative subordination make up a major ingredient of the human 'will to power' and its social manifestation –'human political society'.

But offspring reproductive 'problematics' are not the only ones that moulded human motivators (and politics) during evolution; females and males too had a part to play.

Females: evolutionary 'problematics'

As we have seen, females who evolved means of attracting others to help in the tasks of child rearing would have been favoured by selection. But the reproductive problem starts even before that. Infants were vulnerable at birth but so too were mothers. The move to an upright stance caused problems for females; the ape tailbone impinged into the area of the pelvis in which the birth canal is located, tending to reduce the size of the canal; the rounding and shortening of the ilium in the evolution of bipedal walking also reduced the size of the bony birth canal (Schultz, 1949; Krogman, 1951; Washburn, 1960; Lovejoy, 1988; Rosenberg and Trevathan, 1996, 2001). This was accommodated to some extent by the evolution of a curve in the lumbar region of the human spine and a loss of much of the tailbone, and by human infants being born at an earlier stage of foetal development, and with a relatively small flexible skull not yet completely fused. Nevertheless, most human females suffer physical trauma in giving birth. Before, or without, modern medicine many died (and die) from it; the difficulties are sufficient, for example, that in industrialized countries it is estimated that over 20–30 per cent of births (and increasing) are by caesarean section (OEDC, 2011; Henrickson, 2012).

A human female who had friends around to help and care for her during and immediately after birth probably lived longer and gave birth more times than those who did not. Bonobos demonstrate the existence of such mutual-friendship groups in the primate line. And if breeding was initially taking place among brothers and sisters and other members of friendship groups, there could have been a number of helpers only too eager to participate in the process of child birth and aftercare – maybe even too eager, pulling and shouting and getting in each other's way and screaming at the sight of blood. Nevertheless, gently twisting the baby through the birth canal, cutting the cord, delivering the placenta, cleaning mothers and infants, cooling fevers, closing wounds, feeding mothers and general emotional care may have been of enormous value to human females (Cf., Bunney, 1993; Friedlander and Jordan, 1995; Jordon, 2013).

In areas of the world today where infant mortality is high it is said that: 'Birth complication and neonatal infection are the biggest scourges' (Cowley, 2003); major remedies include 'assisted delivery' and adequate pre- and post-natal care (Titaley *et al.*, 2008; Hays, 2011; unicef, 2013). So, bonobo female mutual-friendship groups may have played some part in the evolution of humans. It may also have been that some of the strength of the bonobo mother/son bond (also manifest in humans) was accompanied by the addition of the development of a strong mother/daughter relationship. This is because it was probably reproductively advantageous for both: for mothers to have a daughter around to help look after siblings, and for daughters to have a mother around when giving birth themselves (not only aiding but also keeping the shouting and pulling and screeching during childbirth under control). This is an aspect of human reproduction from which males have continued to be excluded (except for male members of the modern medical profession, and even there nurses, midwives – mostly females – and family most often take the baby as soon as it is born).

How much offspring are expected to emotionally pay their mothers back for all this pain and bother is hard to ascertain. Certainly, human history is full of the glorifying of motherhood. But is this through appreciation, guilt or fear? Does 'mother love' mean protection or engulfment? On the one hand, there is the notion of fertility goddesses, protective goddesses and Madonnas, but on the other hand there are warrior goddesses and dangerous Amazons. In modern times, there is a whole school of psychology that looks to the extensive elaboration of mothering as being basic to the controlling and undue repressing of human, especially male, personalities (Cf. Bateson, *et al.,* 1956; Laing, 1965, 1971). However, even if there might have been a tendency to 'over-mothering' (because of the evolution of an addictive, emotional, self-aware basis for it), and an emergence of the glorifying of mothering (because of strong emotional dependencies on mothers, feelings of guilt concerning mothers, and a human tendency to spiritualize many of the things powerfully felt), there, nevertheless, seems to also have been resistance to over-mothering. This would have served females' reproductive interests because there can be a reproductive danger for females in over-mothering. A female could have become so extremely dependent on love from her first offspring that her husband left and she, in turn, became even more emotionally dependent on her child, sacrificing her capacity to have several more offspring or numerous grandchildren – especially if her engulfed son or daughter was so dependent on her that they never reproduced.

Even if the male partner did not leave, females could have practised long abstinence to prevent further births. There are a number of cultures, for example, in which sex is not encouraged for some time after the birth of a baby, and there is ample reason to believe that when they could, certain females have chosen abstinence over the dangers of another birth and another child to care for. Choosing abstinence represents a dramatic change from female ape sexuality; it represents a reduction of instincts/desires for sex, giving human females a considerable degree of control over their own and male sexuality (and resultant parenting behaviour). We have already mentioned how the promiscuity of the female bonobo and the female chimp during oestrus was greatly repressed in the hominin line through the evolution of embarrassment, shame, shyness and, indeed, guilt concerning sexual activities. So, although selection seems to have favoured those who glorified birth itself (as a miracle, for example) and children as 'God's gifts', females have not always jumped in with great enthusiasm or abandonment, and males have not always pushed them to do so.

While males may have generally been left out of childbirth (and are not always keen on another pregnancy or a young child getting in the way of their supply of sex) and are even presented with periods of abstinence, they are expected – in the vast majority of societies – to provide economic support and protection for the female to whom they have become attached and to her offspring. This is a pattern almost completely unheard of among primates in general; certainly it is not the case among bonobos and chimpanzees.

So, how did we get here, what psychological forces were at work? In the scenario presented above in which infant/child/youth appearances gave reproductive success through their ability to attract sexual playmates, males may have been attracted to their sisters or female cousin playmates. But as this evolution progressed, and as incest began to have negative effects, those males increasingly attracted to, and turned-on by, additional secondary sexual characteristics among relative stranger females were then favoured by selection. A female who attracted males on the basis of playmate characteristics *and* secondary sexual characteristics, but who could restrain male sexual urges through a clever 'hiding' of her secondary sexual allurements until she could evaluate the attracted males further would have had a good idea who was a 'real' friend or, conversely, who was just after sex. If a female was good at this she might even entice males to bring gifts and not to expect sex for it, at least not straight away. They would be coming to play. Compare this to the female bonobo who generally would have provided sex right away and the male would have been gone before Ms bonobo could say 'hi, what's your name?'

Female attractivity based on child-like characteristics and playfulness would have gone some way towards repressing male sexual promiscuity, opportunism and domination. But a playmate is not necessarily the most reliable or dependable when the going gets difficult; being able to secure more substantial and reliable support would also have been favoured by selection. The problematic would be to attract males that could be evaluated, and then selected ones being tamed and brought into service, so to speak. I suggest that a powerful emotional force that accomplished the above evolution was that females became not just playmates and sexual partners but also – for reliable males – substitute mothers. Attachment to mothers in general is a stronger force than attachment to friends and playmates. It develops strong guilt-inducing dependencies and responsibilities. Powerful instincts for mothering can be found in almost all primate females, especially in chimpanzees and bonobos.

But now the mothering began to be extended to males with whom potential mating was on the cards. The male who responded to mothering could be counted on (well, sometimes) to stay around and be attentive and keep his sexual urges under control, especially when he went out to bring some food home. Certainly, human males (in most cases) have evolved to come courting without sex and to stay around and help with family matters even after sexual frequency has been greatly reduced; or at least they have evolved to provide support and protection to a female for an extended period of time. And clearly human males can be observed to be almost as *emotionally dependent* on their mating partners and on the wellbeing of their children as are females – virtually unique in the primate world.

This evolutionary pattern would have given females a considerable degree of control over male sexuality and parenting behaviour. With the increasing use of 'mothering' as a control mechanism, we would expect the evolution of female jealousy as a motivation for vigilance regarding male loyalty. And indeed, in modern times females demonstrate a considerable capacity for jealousy. There does, however, appear to be a difference – with an overlap – in male and female forms of jealousy (Symons, 1979; Van Sommers, 1988; Wilson and Daly, 1992; Buss, 1994, 2000). Males tend to be extremely jealous of 'their' females having, or even hinting at having, sexual relations with another male. Females, on the other hand, tend to be extremely jealous of the possibility of 'their' male paying attention to and showering gifts or respect or love on another female (or her offspring).

This difference makes sense in evolutionary terms in that a great danger for males in this scenario is being cuckolded and thus spending energy in supporting the offspring of another male, while the danger for a female is that she loses the resources, help and protection of the 'quality' male she has worked so hard at enticing and 'mothering'. She may see it in relatively rational cost-benefit terms, but she more likely will feel shunned, rejected, by her converted 'son'. It is one thing for him to spread his wild oats a little bit, but when it comes to his love, care and devotion and intimate secrets, they belong to her. Female jealousy is a powerful emotional force, not simply a balance sheet from cost-benefit analysis.

In summary, the postulated reduced birth cycle was a move in the direction of r selection, which, however, given the increased degree of infant dependency, was only viable if it was accompanied by considerable post-natal parental investment in offspring – socially a move in the direction of k selection. Thus, selection increasingly favoured female abilities to not only choose mates with good genetic potential but also, probably more importantly, mates that would help in caring for offspring. Selection also favoured the development of female capacities to control times and places in which to become pregnant and to be able to decide whether or not to rear given offspring. We have seen a number of emotional forces which have evolved to aid in this process: post-natal joy or, conversely, depression concerning the birth of an offspring, and shyness, embarrassment, shame, guilt and rational calculation concerning things sexual. Female abilities – unconscious and conscious – to attract dependent males to bring resources and stay around and help care for and protect offspring also worked to achieve the above; providing substitute mothering for an attracted male, and experiencing jealousy of losing his attention and love (as with a son) served her well. All of this, however, created selection pressures for changes in male reproductive behaviours and resultant emotional-cognitive processes.

Males: evolutionary 'problematics'

If the major evolutionary problematics for females involved attracting and being able to choose vigorous, healthy protective males and being able to maximize the male's parental efforts towards their own offspring, the evolutionary challenges for most males were to: (1) impregnate as many females as possible (a product of their cheap sperm and a carry-over from the primate heritage), (2) be chosen by a female, (3) avoid the aggression of dominant males in staying chosen, and (4) avoid being enticed into providing support for females who had been impregnated by other males. Male sexuality, then, evolved as a conflict between the fact that to utilize cheap sperm to best advantage required exercising r selection characteristics (promiscuity with no interest in the welfare of offspring), but to be sexually selected by females required demonstrating k selection characteristics (co-parenting). Put together, evolutionary 'solutions' to these 'problems' seem to have created as many ambiguities, conflicts, confusions and contradictions in male behaviour and consciousness as we have seen above in female consciousness.

It is likely, for example, that many males would have vacillated among being kind, child-oriented, dependent, charming, sexually demanding, aggressive and domineering towards females. They might, for example, promise love, wealth and status but impregnate and run. Too much promiscuity, however, leads to competitions with other males, competitions that most human males lose. So, for some males, being selected by a loyal female while indulging in minor philandering might have been the best reproductive bet, but for many males, in most circumstances, being loyal and dedicated to one female would have produced the best reproductive success. Staying around a female in a caring capacity not only provided additional care for infants, but also for the infants' mothers, who could also be closely guarded, and who may well become pregnant again in a relatively short time after each birth. Also, staying around would have tended to prevent another male coming along and killing his offspring. In both cases, this would have enhanced male (as well as female) reproductive success.

A male's desire for sex could possibly have been reduced by the 'mothering effect' of his female partner, and thus be more easily satisfied by one female, transferring a degree of male energy away from promiscuity and aggressive status encounters with other males into child-care and nest protection. This would depend, of course, on the degree and strength of the Coolidge effect – males desiring new females after a period of sex with a given female (Symons, 1979; Eagleman, 2011). Nevertheless, outside females often employ mothering tactics in order to get a son away from his birth mother (or from a current partner) – with a hint of eventual sex on offer later. This is done by smiling and laughing with him, listening to him, praising him, teasing him, and even stroking him, perhaps with some baby talk thrown in. And if she behaves during foreplay and sex like an appreciative child or play friend being tickled, groomed, rubbed, and if she makes noises of deep pleasure during sex like he does in orgasm, he is in wonderland. Then, to clinch the deal, she might produce from her body an infant, a part of them both blended together. Most usually, at this point, the mother (or current partner) has been beaten.

This works because in primates the mother/son bond is often very strong (as especially manifest in bonobos, for example), and because the emotions that attach sons to mothers are very similar to some of those that attract a male to a sexual partner – at least to a *long-term* sexual partner. But this can cause problems. As the Greeks observed, and Freud pondered upon, at some point adolescent sons can become competitors, usurpers and enemies of fathers. Puberty usually results in a split from family rather than the son killing the father and marrying the mother. Still, in the 'outer world' 'son tactics' derived from this mother/son dynamic can be employed. For example, in courtship an aspiring male might return after initial and/or periodic rebuffs on the part of a desired female – like a child shooed away but continually returning to pester a parent. Frans de Waal (1988, 2003) considered this to be a basic means among primates of 'teaching' potential friends and/or protectors to accept self's offer of subordination.

An approach shows interest, flight from the rebuff shows that self is not dangerous to the approached individual and that self will immediately give in rather than be a threat (the approached target is in charge), but re-approaching ('reverted escape') opens the door for the target receiver to accept the now 'proven' non-dangerous individual as a potential mate or friend (Price, 1988). So, males that recognize a rebuff but keep *humbly* coming back are likely, in the end, to be accepted by a female (up to a point, of course). In terms of modern human mating, for instance, one bit of dating advice suggests that male persistence is important for success, but a male must be subtle and know when to stop so they do not appear to just want sex, to be extremely 'needy' or to potentially become an undesirable stalker (Cf., Nyholm, 2009; INTJ Forum, 2013; Lee, 2014). On a first approach, he should, for example, suggest that she might like to go to a museum or to a movie rather than that she come up to his place for a few drinks.

So, whatever the exact sequence of evolution, for most modern males the tendency to promiscuity has been greatly reduced (as compared with either dominant males in harem primate societies or the more generally promiscuous bonobo or chimpanzee males, for example). To be attractive to females may have been accomplished by becoming a 'friend' (or even a 'son'), demonstrating signals of planning to stay around, of being helpful in the future and in not suddenly becoming threatening. But these mating processes require the evolution of *trust* in order to work; we have seen, however, that both sexes did not exactly have reasons to fully trust each other (or members of their own sex), and that both sexes could be coy and deceptive in their approach to each other. Some very attractive females, for example, would perhaps have been able to pull off a balancing act between dallying with both gift-giving friends and macho studs at the same time, being able to keep the gifts coming in while attracting a wide variety of males whom they could short list for a round of sperm competition. And there would be well-placed males who did their best to take advantage of this – the one-parent family (as in chimps and bonobos) may have had a long history in human evolution (Miller, 2000).

All this would leave a male vulnerable to being cuckolded while dominant males romanced his 'friend/mother' with displays of bravado, wealth, status and power. Overly attentive caring gift-givers, then, could have been caught in an emotional child-like over-dependency trap. So, the female approach described above would have tended to make males (and many females) wary of females and of female sexuality (and of male philandering). Many males would have begun to shun those females most into sperm competition if a more reliable, albeit slightly less attractive, female was offering sex, a bit of 'mothering' and loyalty as a package deal. Even then, males evolved to not completely trust a female and so set out to monopolize her emotional, practical and sexual output.

And, so, jealousy and mate guarding has never disappeared from human emotional-cognitive processes. But jealousy can be a double-edged sword. Strong sexual jealousy can act as a sign of commitment (Ridley, 1993; Buss, 2000), and therefore can work to the advantage of a male. But extreme sexual/love jealousy can be counter-productive; if a jealous person whines, moans and cries about their predicament all the time it 'gives away' the dependent one's 'need' for the other. Such evidence of weakness makes them seem not such a good prospect after all and perhaps causes the already jealous male additional reason to be worried as his female begins to look for better prospects elsewhere. Fits of jealousy also make an individual vulnerable to the use of jealousy-inducing activities on the part of a mate set on getting their own way. At the same time, jealous, aggressive males can threaten other males to the extent that the jealous male is avoided by potential allies and so loses out on the benefits alliances can provide in mate guarding.

One condition for the selection of male trust (and reduced jealousy) would have been for a female to appear to provide a male with a constant supply of seemingly *exclusive*, seemingly mutually satisfying sex (even if she is not personally very interested in sex) and mothering, leaving the male *convinced* that the female is contented and so will not go looking elsewhere, and for a code of honour to develop among males whereby it was a complete taboo to actively lust after or seduce a male friend's female. So, to avoid temptation and stay loyal to a friend, we can speculate that male sexual/love jealousy eventually took the form of males wanting to control the sexuality of not only their female partners but also of *all* females and of *all* potentially competing males through the use of rules, taboos, moral restrictions, codes of honour and laws – formalized (and often enforced) trust – and that this was supported by many females as well as males.

But this level of control and enforced restraint requires considerable energy, resources, anxiety, vigilance and, above all, reliable *cooperation* to maintain, selecting for a socio-political dimension of human existence. So, sexual and political trust could have evolved in parallel with the evolution of creative subordination setting up the dynamics of mutual dependency and constant politicking in human social life. As we have seen, dependency often requires subordination, but it also provides a certain amount of power flow back to the subordinating person. A male dependent on mothering from a wife, for example, may come to equalize if not dominate within the emotional relationship because the wife becomes dependent on mothering him and/or on the co-parenting, status and wealth he provides. Very often, however, a 'political' balance is achieved and in the process the pair can come to trust each other; the relationship can become one of 'mutual admiration' and exchange. The female may love him as she loves her child, or even as a hero/protector in a spiritual sense, similar to the spiritual elevation associated with the notion of *motherhood* in human societies, and he can love her as his supporter, sex partner, co-parent and friend (this is all quite apart from any benefits of a division of labour they develop within a family setting).

The same processes of emotional interdependency can apply to the development of trust among friends and allies. Males, for example, can come to depend so much on the respect of their friends/allies that they begin to *respect* the *'rights'*, above all the 'sexual rights', of each other. Once this is set in motion, a male can tone down his jealousy; his mates will not only leave his woman alone but make sure no outside male sneaks in, and if they do they will help him seek vengeance (as the Trojans discovered when Paris snuck in to steal Helen). Here we have the process of sublimating sexual desire into various *expectations* and *codes* of behaviour, including the ideas of leaving a desired sex object of a sibling or friend alone, extended courtship and of monogamy almost as a spiritual entity.

Dependency-inducing genetic propensities, then, have been passed on in both sexes during the evolution of human mating patterns, and these not only apply to mating behaviour but also to the formation of friendships, cliques, alliances, patron/client relationships, follower/leader patterns and politics. *Creative subordination* derived from these, however, only works if there are those to whom it is worth subordinating. And it is here that we can see male sexuality sublimated into such things as bravado, camaraderie, career ambitions and spiritual aspirations, success in all of which give reproductive success for males. So, males retained capacities for competition and aggression with and against other males, and, indeed, tendencies to be jealous and envious of them in a wide variety of not only sexual but also social/political spheres.

But progressively the competitions have become more controlled, following rules (however vague they sometimes are and often being unwritten), and are for recognition, achievement, status and to fulfil ambitions rather than for dominance through physical struggles.

This process may have led to what we call human civilization, but it has not eliminated ambiguity, anxiety or conflict in human social life. Individuals may equalize, or even reverse, a power relationship, but they may not know it, or they may just about know it, some of the time, in some situations but not in others. They may know it, for example, when they are happy but not when sad, know it sober but not drunk. And, indeed, relationships themselves may alternate, unpredictably, back and forth. Thus, individuals may be greatly insecure about their relationships. In fact, the more equal a relationship often the harder it is to feel secure in it because equality is harder to 'know/feel' than inequality and so individuals can become ever more jealous, envious, guarded and aggressive. The result of all this in most human experiences will be that individuals react in what appears to be very ambivalent, conflicting and often confused manners.

A male may desperately feel that they want the offspring produced by their mate to be theirs, and males have powerful emotional-cognitive potentials which generate rage and deep anger when images spring to mind of their loved one having sex with another male (Cf., Van Sommers, 1988; Buss, 2000); nevertheless they can love a female that already has a child (and the child). They may feel loyalty and love for a mate but also that their own powerful sex drive requires an outlet not being satisfied by their female, and that they may even be doing her a favour to express it, emotionlessly, somewhere else. A male may feel that, although he loves her, and his children, his hard work in providing for them should be rewarded more than it is. Or, he may not feel any of this and be a totally dedicated husband and father while his female mate wonders why he is so boring in bed and why he does not get promoted at work and thinks that he should spend less time on his bird watching and concentrate on his job, or at least on his family, and asks herself if she might be justified in taking golf lessons from the handsome golf pro at the club.

The point is that humans are not motivated by cost-benefit principles of sexual selection as abstract forces, but rather by drives and emotions (which have strong bio-electric, pre-knowledge underpinnings), and by more consciously rationalized desires and fears based on these. And it seems clear that one of the major characteristics of human consciousness is that it is not a synchronized whole, beautifully balanced in some form of perfect harmony with the universe, but rather is a bundle of not always happy or well-integrated drives, images, thoughts, fantasies, desires and fears. Ambiguity and contradictions in human consciousness do not mean that we cannot understand it, just that we have to start from a different set of premises than is usually the case when the 'essence' of humanness is considered to be rational or spiritual.

Evolution and human consciousness: some conclusions

During the evolution of human mating behaviour, besides developing an ability to 'read' various new physical indicators of the reproductive value of others, we developed considerable skills in being able to manipulate others, but also in being able to avoid manipulations on the part of others. This 'condition' selected for abilities to evaluate others and their possible intentions through observation of a number of non-verbal signs, such as facial expressions, postures and gestures (Cf., Izard, 1982; Buck, 1984; Dunbar, 1996; LeDoux, 1998). These indicated the potential social use and/or dangerousness of others even before extensive interaction had taken place. It seems that at some point these abilities came to include an *awareness* of self in relationship to others – and, as a result, the evolution of a capacity for developing independent identities, including sexual identities. Along with this the development and monitoring of trust became a major driving force in the evolution of human consciousness. It is possible to argue that from all this there arose the general *master* ontological conflict of human nature; that is, the conflict between the drives, emotions, desires and fears which motivate individuals towards emotional independence and individuality and those which motivate them towards emotional dependency on others – 'the human condition' one could argue.

Besides evolving abilities for evaluating others, mating competitions selected for abilities to communicate self-value. The best way to communicate characteristics of self-value is to believe that one has them and that they are worth displaying. There was, in other words, an evolution of a continuous elaboration of skills for creating imaginative forms of self-presentation and social display, and for self-delusion. This upped the stakes in the processes of policing trust and evaluating others; claims of worthiness tend to be exaggerated and faked even among apes, and with the elaboration developed by humans became very sophisticated. This whole process likely became a major basis for the selection of both general cognitive intelligence and abilities for verbal communication, and eventually for elaborate daydreaming, fantasy creation and complex thinking.

At the same time, mechanisms for desiring attention undoubtedly led individuals to desire a certain amount of wealth and status to use as evidence of worth. After all, evolving humans increasingly became nest builders and tool users and creators of food stores, and with self-awareness it was often clear that an individual self had accomplished many of these feats. As such, success in achieving wealth and status represented the power of a particular individual in the larger universe, increasing their reproductive prospects (Cf., Hill, 1984; Turke and Betzig, 1985; Betzig, 1986, 1992), and this resulted in a desire for a degree of personal control over self's circumstances ('a will to power' of sorts), thus transforming a common primate desire for attention, status and dominance into a human social product.

A human's 'accepted worth' (prestige, status and power) would represent a general evaluation by others of them as being safe or dangerous, of their individual skills in dealing with people, of their potential support in social situations and of their accumulated wealth for potential gift-giving. At the same time, however, a desire for power, wealth and status can cause conflicts, including conflicts *internal* to self. A desire to gain power over self at all costs can cause a person to have to struggle with their lust; a desire for power over others can generate immediate personal conflicts and the loss of friends and allies; a wish for wealth can conflict with an individual's desire to shower gifts on friends or potential mates, or to provide for offspring or to show off generally; a desire for status may cause an individual to always feel dissatisfied with what they possess.

Paradoxically, often the greater the conflicts generated, the greater the desire to gain control over self and self's circumstances and, consequently, the greater the desire to seek social power, status and wealth. The 'will to power' feeds on itself.

And this has deep-seated unconscious drives, desires and fears to build on. A human infant's ability to draw attention and care from parents and to sustain it for such a relatively long time operates as a largely unconscious will to power. Infants make extensive, unrelenting demands, day and night, wet and dry, soiled and clean, and not only get away with it but profit greatly, not only in terms of physical provision but in terms of emotional comfort and dedicated parental subservience to themselves. Infants pull this off by (when not sleeping or crying or throwing food) smiling and cooing and presenting as 'helpless'; they look straight into the eyes of the carer with their own big, round open eyes. The child's unconscious developing brain cannot help but learn that unending 'pushiness' and constant demanding are effective if they are accompanied by relatively regular smiling and cooing and presentations of helplessness through a full use of 'face power'.

But, just as we can lose friends and allies with too much assertiveness, parents can get fed-up, tired and resentful. Parents have other fish to fry besides a single, or even several, offspring. Smiling and cooing and crying have to be used strategically; there is a time to cry and a time to smile, a time to coo and a time to shut up. The unconscious brain of those successfully reared gives way to somewhat more conscious manipulations and demands on the part of offspring, but parents are usually up to it ('family dynamics'). Offspring come under pressure to become moulded into being showpieces for their parents, are forced to study against their will, bullied into becoming brave warriors, made to take piano lessons, pushed to join the rugby team or cajoled to take an interest in the floundering family business/farm. Offspring tell parents where to get off, sneak out at night, drop out of school and marry the 'wrong' person. In the end (usually) parents give independence and offspring compromise, to the extent at least that they accept their overall emotional dependency on each other. And offspring recognize that a degree of obedience and respect may be required if they are to receive help with their own children, and life events and possibly to later inherit wealth and status, and parents respect the independence of their offspring if they are to be looked after in their old age, for example.

Generally, in terms of the evolution of human consciousness, it seems clear that humans became quite sophisticated (unconsciously and consciously) in abilities for 'creative subordination'; emphasis must be put on the word *creative*. Being born was not an automatic trigger for receiving mother love; it had to be induced (unconscious as this process often was, largely operating through non-verbal signals). Physical prowess was not necessarily a good indicator of a male's long-term success; he had to be humble, even child-like. A female's physical signs of willingness for sexual encounters were not a guarantee for her or a male's 'genes' making it into the next generation; she had to show trustworthiness. To have reproductive success, both males and females had to desire not only physical prowess and genetic compatibility in a potential partner, but also personality characteristics that came closest to facilitating long-term consortships (bonded by relatively exclusive sex and strong feelings of friendship), joint-parenting, food and nest sharing, and mutual trust.

Human mating patterns have been an emerging process (not a rigid form), and this continues to be the case even today. It is a process that has been idealized, prescribed and formalized, but this has not prevented it from being a movable feast within a range of behaviours that can be considered species-typical of humans. It is as much political (with a small p) as it is one based on pure lust or automatic triggers. It would seem from the evidence concerning primate behaviour; the conditions which probably surrounded human evolution; the clear evidence of neoteny in humans; current knowledge about genetic and embryological processes as they affect morphology and behaviour; and from a great deal of the easily observable emotionally charged sex, love, mating and parenting behaviour of current humans that sexual selection may have been considerably more important than an evolution of 'rational' survival behaviour in human evolution. And it is these very real possibilities that can help us understand a great deal of the behaviour that goes beyond sex, mating and parenting behaviour in human societies.

4 Bipedalism, brain growth, language and the development of human sociability

Chapter 3 looked at human evolution in terms of sexual selection. I speculated as to possible (often unconscious) drives and emotions, semi-conscious desires and fears and even some forms of reasoning, which might have evolved during that evolution. In this chapter I will look at how all the above processes could have been related to: (1) the evolution of bipedalism (traditionally considered to be the definition of the human line and to have freed the hands for tool use, thus opening the way for the evolution of advanced intelligence), (2) the enlargement of the human brain and the role of reason in human evolution (supposedly introducing rationality into biology), and (3) the development of human language (often considered to be a major tool for advancing rationality and for passing on information in an accumulative, progressive, civilizing manner).

It will be argued that we will have to reconsider, indeed downgrade, the importance of all three of the above kingpins of traditional scenarios of human evolution in order to gain a clearer understanding of human nature and modern human social life. In the process, our perspective will help us understand why an increase in brain size came relatively late in human evolution, and suggest reasons why it seems to have stopped increasing from about 70,000 or so years ago. It also provides an explanation as to why language is used more for gossip, entertainment, romance, lying, swearing, distortion, deal making and social distancing than for uncovering 'Truth'. Following this, the basis of human sociability will be examined: a sociability that eventually resulted in political states containing several million individuals as citizens.

Bipedalism

However significant it might have been as human evolution proceeded, it is possible that we have made too much of bipedalism as being the key change which unleashed the potential for the human line. Recent studies of apes make it clear that they, especially chimpanzees and bonobos, often operate from a bipedal stance more or less effectively, and with relative ease. They are particularly prone to a bipedal posture when they are displaying for status and dominance or trying to attract members of the opposite sex, but they also do so for undertaking a number of practical tasks. Yet none of these apes have become human-like as an inevitable result. While the postulated crucial factors in the evolution of human nature presented in Chapter 3 do not exactly give us a scenario for bipedalism, a case for its further development from ape capabilities can be made (see this chapter and also Lovejoy, 1980, 1981, 2009). Indeed, it has been argued that because upright walking in the very early hominini *Ardipithecus ramidus* (4–5 myr), for example, apparently 'provided no energy advantage . . . reproductive success must have been central to its evolution in early hominids' (Lovejoy, 2009, p. 74). And, as we shall see, a case for bipedalism based on sexual selection would be consistent with the observed fact that a large brain came long after bipedalism.

Genetic changes (for a reduction in the growth hormone [HGH], or the release of paedomorphic and/or neotenic inducing DNA or SNPs, or the slowing down of growth and maturation through the effects of alterations in HOX genes, as postulated in our scenario) would have meant that final morphological structures were more vulnerable to modification by behaviour during the longer child-like state (Cf., Goodman, 1963; Schultz, 1963, 1968; Daughaday, 1968; Carter, 1980). This is because 'details of form, such as the angle at which a joint surface is placed relative to the shaft of a long bone, are probably responsive to biomechanical use of the skeleton during growth' (Tattersall, Delson and Van Couvering, 1988, p. 96; see also, Zelzer, 2009). Baby apes are extremely flexible and a move towards paedomorphosis (return to child shape) or neoteny (retention of child features into adulthood) would have meant that such flexibility, and thus bipedal ability, could have existed through the stretched out developmental and growth stages when bone and muscle development would have been greatly influenced by a bipedal stance, perhaps especially while playing and displaying, resulting in a final morphology more suited to upright walking. The importance of changes in *existing* muscle structure and function regarding bipedalism can be glimpsed from the fact that one of Jane Goodall's apes became bipedal after suffering from polio (Lawick-Goodall, 1971). The above flexibility also means that despite the relative closing off of the bony birth canal by the primate tailbone and the reshaping of the ilium during the evolution of the upright posture (Schultz, 1949; Krogman, 1951; Washburn, 1960; see also, Aiello, 1996; Rosenberg and Trevathan, 1996, 2001), birth would still have been possible (even later when the human brain grew larger in the pre-natal period).

In so far as bipedalism might have been a behavioural (survival or reproductive) advantage given more immature births, any genetic modifications that supported it morphologically would have been selected for. We have already seen (Chapter 2) a number of reasons why bipedalism could have been an advantage in semi-tropical swamp/fen-like aquatic environments. As a reminder these included orthograde feeding (using the hands for terminal branch and high-level feeding while in a bipedal stance – which may have existed well before human differentiation [Thorpe, 2006]); creating and keeping clear pathways through tropical and semi tropical forests – a task in which branches usually have to be broken and cut for the clearing work to be effective (an unending task given the speed with which growth takes place in semi-tropical, or even very wet, conditions); placing crossing stones, building causeways in marsh, swamp and wet grassland terrain; constructing bridges and small dams; standing up for a better view of both surroundings and the location of others in highly vegetated and/or swampy conditions; building shelters (including on stilts). All these activities would have selected both for bipedalism and for a refinement of manual dexterity including the further development of the opposable thumb so important in careful handwork and in making and using relatively delicate tools.

Some of the advantages of freed hands from bipedalism mentioned in Chapter 2 had to do with reproductive issues, especially regarding the care of offspring. Swatting flies and mosquitoes, building and installing insect nets, picking ticks off mates and offspring (and friends) and making safe beds for the ill were all mentioned. Tool making and tool use would also have been important for gathering, carrying, digging, fishing and, probably much later, hunting. Additionally, the degree of infant dependency postulated in Chapter 3 (and still extant in humans) suggests that bipedalism would have been a behavioural advantage for: (1) carrying dependent infants who were unable to hold on to a moving mother (father, aunt, uncle, friend or older sibling) relatively long distances in search of food, fresh water and shelter – in fact being able to carry more than one offspring at a time (as has been observed among bonobos in a bipedal stance [de Waal and Lanting, 1998]); (2) holding an extremely dependent infant (one that cannot cling on its own) while it breast feeds; (3) carrying food and fresh water back to safe nests – perhaps while also carrying an infant – rather than being forced to eat and defend it where obtained; (4) again, freeing of the hands for constructing tools for use in all the above and for building semi-permanent nests for dependent infants and less dominant consort pairs to hide in; (5) free hands and improved dexterity could also have been important in assisting births given how difficult that became for human females; (6) holding a baby tenderly while smiling in its face and bonding with it by looking into its eyes.

It can be argued that certain advantages of bipedalism relating to sexual selection, status and interpersonal politics would have been equally important. These would have included: (7) seeming larger as an infant and thus being able to dominate and secure more resources and care than less bipedal siblings; (8) being able as an infant to use hands to grasp objects for display and to use them skilfully and imaginatively to draw parental attention (and thus care), and to frighten away less skilled siblings. Modern mothers certainly tell their young children to 'stand up straight' when they want them to be shown off to their best. In adults, especially in male adults, a capacity for display would have been extremely important in achieving sexual, status and political success. Adult apes often use objects for agonistic, status and sexual displays, and when they do it is almost always done from an upright position. It has been argued that this is because it makes them appear larger, more frightening and/or attractive. Whatever the reason, it clearly seems to enhance their status. Given that human females lost visible signs of oestrus and became quite choosy, males who were successful had probably become quite clever and skilful in their methods of displaying. They may not exactly have become song and dance men with hat and stick, but they could have been on the way.

On the other hand, through free hands during stable bipedalism, males would have been more able to hold, carry and otherwise force semi-willing, or even unwilling, females to go with them (perhaps away from potential rivals), dragging them, or even tying them up. And, of course, they could have forced them to have sex. However, bipedalism would have (9) obscured the sexual area of females, reducing long distance signals of their ovulation (and also oestrus generally) thus increasing their ability to choose mates and times of copulation, while (10) encouraging sexual coupling as a face-to-face, 'reaching towards', laying down affair, making female cooperation more essential even with chosen partners. But this exposes both parties to danger much more than is the case in rear mounting (and quicker ejaculating) apes resting on all four limbs (and being able to easily look around) making it likely that the pair have come to some arrangement beforehand to go to a safe place; suggesting that (11) shared child-care may be on the cards. At the same time, an upright position would have (12) exposed a male chimpanzee's or a male bonobo's soliciting erection right away clearly communicating that they were *not* a child/youth-ape coming to play and be mothered but rather a male with only one thing in mind and so to be avoided. In general, free hands are extremely useful in non-verbal communication and, given that humans increasingly used relatively subtle non-verbal signalling in place of more dramatic signs of sexual attractiveness, proceptivity and receptivity, (13) skilful talking 'with hands' may have become a very attractive attribute.

The chimpanzee tendency for males to form alliances and to band together to defend feeding territories, hunt and occasionally seek vengeance was likely an ancestral characteristic maintained in youth-ape males, who may have banded together to defend nests, small islands, walkways, fishing pools and beaches, or to go on the offensive. Given their probable reduced stature and strength compared to non-youth-apes, bipedalism would, thus, have made them (14) seem larger and more threatening, somewhat counter-balancing ape-strength. Additionally, bipedalism would have helped in (15) building defensive barricades, gates, drawbridges and boats, and in using sticks and stones not only as tools but also as weapons. It is important to note that to use such weapons effectively an individual not only has to have free hands, but must also be able to stand and move about in a vertical position, with a high degree of stability, for a relatively long period of time. Throwing a spear or using a bow and arrow (or swinging a baseball bat or golf club) requires stretching a stable vertical body extensively before letting go if any degree of power and accuracy are to be achieved. (Later in human social evolution, the invention of stirrups made horseback warriors considerably more formidable than they had previously been simply because they could stand up on a fast moving horse.)

In the scenario suggested, once bipedalism was set in motion, many of the above activities would have acted as feedback selection for the others, and for bipedalism itself. But, as we have seen, bipedalism was around before anything we might call hominins entered the picture. Even some monkey species can operate as bipeds for short periods of time (Dixson, 1998). Indeed, it has been suggested that, 'bipedalism, manual dexterity and a large brain [relative to body size] are likely to have evolved more than once' (Wood, 2002, p. 135, see also, 2002a). What this suggests is that we may not have a clear line of demarcation between pre-humans and humans in the area of locomotion (or in terms of manual dexterity or brain size). However, the neoteny which aided early bipedalism also suggests the eventual possibility for the evolution of a larger brain in the hominin line.

Brain growth

The adult human skull reaches an allometric shape (size-related shape) equivalent to juvenile chimps with an overall *size* equivalent to an adult chimp (Penin, Berge and

Baylac, 2002; Hansell, *et al.,* 2004). In evolutionary terms, as the adult human skull retained the shape of a young chimpanzee it had considerably less muscular attachment areas on the skull than seen in an adult chimp, suggesting weaker neck and jaw muscles. The skull was more rounded, thinner, with a flatter face. The modern human skull is made up of a number of bones which are progressively fused during foetal, post-natal and developmental stages (humans end up with eight fused cranial bones alone, for example). The finally evolved human skull experiences a much slower ossification of the thinner, non-muscle-attachment bones than that seen in apes, allowing for greatly increased brain growth both embryologically, and very significantly, post-natally during development into adulthood (Biegert, 1963; Ross, and Henneberg, 1995; Brennan and Antonyshyn, 1996; Neubauer, Gunz and Hublin, 2009; Martínez-Abadías, *et al.*, 2011). The human skull does not completely ossify until about 20 years of age (McNamara, 1999). The consequent skull flexibility of this makes birth easier through the narrow birth canal caused by bipedalism.

The evolutionary growth of the human brain was a relatively slow and uneven process, and it is impossible to claim a clear Rubicon demarcation for humans based on skull shape, adult brain size or *type*. Indeed, a relatively large brain (to body size) does not automatically mean 'human', having evolved a number of times before, as in shrews, mice and dolphins (Corballis, 2002; Wood, 2002). Nor, of course, is brain size a completely accurate indicator of intelligence, especially of specific types of intelligence (Roth and Dicke, 2005). The higher apes have relatively large and intelligent brains with a well-developed visual cortex. This seems to have continued in the human line, providing a basis for an emotional, social and cognitive intelligence very useful in being able to 'read' facial expressions and behaviour in order to evaluate individuals and their intentions – to '*see*' if they are safe or dangerous, friend or foe, possible mates or not.

An emotional intelligence for reading and reacting to *social* expressions and intentions also had solid ground from which to grow as the existence of such things as anger, hate, envy, jealousy, lust, friendship and love clearly exist among the higher apes. Also, these apes can be quite curious, including about other individuals, and even exhibit behaviour that can possibly be identified as 'sympathy' and 'empathy' (de Waal and Lanting, 1998, pp. 154, 152–158; see also, Mason, 1979; Patterson and Linden, 1982; de Waal, 1988, 2002, 2005). It is not altogether evident, however, as to what a chimp or bonobo, for example, would have done with a bigger brain or an increase in the above type of intelligence. A human skull on them would have put them at a definite reproductive and survival disadvantage through a reduction in their facial muscular capacity to use their fangs in combat, defence and display activities, or to maintain the powerful large jaw chewing capacity of chimps or bonobos so useful for eating some of the tough fibres in their normal diets. And could a human-like intelligence have displaced – or benefitted – the sexual pull of a huge pink swelling, powerful sexual drive, widespread promiscuity and almost constant sexual capacity of these apes? Probably not.

Nevertheless, the chimp/bonobo patterns of sexual behaviour may represent an example of 'runaway sexual selection' whereby certain sexually attracting characteristics and patterns of sexuality (such as extreme promiscuity) become established and remain long after they serve any survival (or noticeable reproductive) value. Indeed, in runaway sexual selection certain attracting characteristics can go on when they incur a cost with regard to survival (Dixson, 1998; Miller, 2000; Jones, 2002). Among the African apes, the amount of energy put into, and dangers encountered in, striving for dominance in order to have reproductive success, the amount of copulations required for each live birth, the considerable length of time between female pregnancies and the long dependency of just one offspring on a lone female may all have militated against their long-term success as species. Indeed, in the period leading up to and during human differentiation, larger brained 'more intelligent' apes were on the decline while smaller-brained monkeys were on the increase. And the decline of the apes relative to monkeys continues even today.

So, is it possible that the larger than monkey brain, when put together with other chimp behaviours, was counter-productive from an evolutionary point of view – in effect, a very costly socio-political ornament rather than a reproductive asset – allowing possibilities for a new species to take its place? Costly ornaments do evolve and, indeed, some have suggested that in the first instant the human brain is just that, a costly ornament of sexual attraction (Miller, 2000). It certainly is costly. The human brain absorbs approximately 17–20 per cent of the energy requirements of the human body while making up only 2 per cent of its weight (Miller, 2000; Roth and Dicke, 2005). This is much more, as a proportion, than in any other primate (Dawkins, 2004a).

An argument can be made that it can be effectively used to attract mates and to rear offspring in that a larger brain could be employed to creatively display, to charm, cajole and deceive mates, to fool competitors, to 'politic' effectively among potential allies, to cooperate with them in mate guarding and to pair bond and care for and rear a nest full of offspring. But it is not attractive in and of itself; it is only when these behaviours themselves were being selected for that a brain to support them would have been an advantage. And these behaviours could not have been in direct competition with established powerful fangs, aggressive social relations, struggles for dominance, promiscuity, sperm competition and tendencies for males to kill the offspring of a female they might have become interested in. In other words, it is only *after* a significantly new form of reproductive and social behaviour (compared to those of an ape ancestor) had evolved that a larger brain might have been of some use.

Once paedomorphism, neoteny, a much faster birth cycle and extremely dependent infants at birth (and bipedalism) emerged, and were reproductively viable, this, arguably, was the case for humans. Not only was a brake taken off some of the anatomical restraints for the development of a larger brain case with these physiological developments (like the allometric reshaping of the large brow-ridged bony skull/jaw and attached muscle), but also the emotional/cognitive characteristics of a human-type brain could have given reproductive and survival advantages. This, however, does not necessarily mean that an ape brain would have to be much larger, just a little 'rewiring' would probably have done the job. At this point another possibility is worth considering – the idea of 'runaway evolution' (Cf., Wills, 1993). This is an idea that once 'brakes' to the evolution of some characteristic have been removed, even if the characteristic is more or less neutral regarding evolutionary advantage, it might just run away a bit in certain circumstances. In the forest wetlands/rivers/lake environment I have suggested for human differentiation, for example, fish and other aquatic-based products may have been a major source of food. These are full of omega 3 and omega 6 fatty acids and iodine that, it is now known, are essential to brain development and efficient mental functioning (Cunnane, Harbige and Crawford, 1993; Broadhurst, Cunnane and Crawford, 1998; Small, 2002; Cunnane, 2006). The environment may also have been rich in a number of other high-protein brain-enhancing foods, ones that were relatively easy to obtain (Morgan, 1985; McNamara, 1999; Braun, et al., 2010). This means that other areas of the body could still function relatively efficiently despite a runaway brain taking up more and more energy. Additionally, the normal process in modern human infants of the *removal* of neural redundancies early in their lives (Chaugeux, 1980) may not have existed at the earliest stages of hominin evolution.

So, the brain could grow considerably – or at least become more complex. But this raises a problem. If the brain was able to run away, why did it only grow very little and relatively slowly from about 5 to 2 million years ago, but then have a spurt between about 2 million years and 70,000 years and then stop growing about 70,000 years ago (Cf., Washburn, 1960, 1978; Wynn, 1988; Aiello, 1996). I suggest that at the earliest stages of differentiation the previously considered emotional-cognitive processes of sexual selection and child-care were more important than analytical abilities relating to technical survival problems. The former would have involved the limbic-hypothalamic regions of the brain and would not show up in fossils to anywhere near the same extent as forebrain expansion would. These emotional-cognitive processes, however, greatly increased the human capacity for self-awareness and emotionally based evaluative skills, thus establishing, or expanding, brain mechanisms and processes for a later selection for analytical and technical abilities to work on.

And during the early stages of differentiation, while runaway brain growth might have been possible, and to some extent have been taking place – with limbic-hypothalamic brain evolution moving apace – there likely did remain checks on frontal region growth. One may have been that the birth canal problem was not 'solved' for some time. A youth-ape with a big brain case (head) would be selected against because those attracted to them would end up with spontaneous abortions instead of an offspring and/or babies and females dying in childbirth. Another brake on early growth of skull enlargement could have been that, although the powerful sagittal crest of apes would most likely have disappeared with neoteny, the sides of the skull may have remained a strong anchor for relatively powerful jaw muscles – after all, humans were eating food raw, and some may have been relatively hard, tubers, shoots and nuts for example, requiring a certain amount of chew-power. And it does seem that jaw and human dentition evolved separately from neoteny and the allometric shape changes in the human skull (Penin, Berge and Baylac, 2002). So, such bone deployment may also have continued to act as a restraint on skull enlargement.

In support of the above, a mutation has been discovered that greatly reduces the size of human jaw muscles and is dated to approximately 2.5 myr (Hansell, *et al.,* 2004), significantly after differentiation took place. Also, in modern humans the mutation that retards the production of cell-surface sugar N-glycolylineuraninic (a compound that restricts brain growth during the development of an embryo) has been dated to only about 2.5 myr (Schmidt, 2002), suggesting that it was only after that point that genes for a larger brain could spread because any genetic propensities for larger brain growth would not take effect until the post-natal period, thus escaping elimination by the birth canal problem. There may also have been an issue concerning the behaviour of individuals who inherited 'too much' analytical intelligence in the early stages of differentiation. They may not have been as friendly and playful as average youth-apes and have become isolates, been shunned and bullied, with reduced reproductive success as a result.

But environments change, and as a result selection forces change. This is more or less what happened at about the time of *H. ergaster* (2–1.6 myr). Without a large, quite analytical brain, a brain capable of considerable memory and mechanical abilities, and an ability to impose discipline and predictable behaviour, I think that it would have been very unlikely that humans could have survived the changes brought about by the arrival of the last major series of glacial advances (2.5 myr – 10,000 years ago). These brought not only periods of intense cold but also great fluctuations of climatic conditions over periodic cycles, with warmer interglacials and very cold ice ages, and often quite distinct seasonal variations. Sea levels changed, while lakes and rivers came and went, as did land bridges and wet and dry spells (Tattersall, Delson and Van Couvering, 1988; deMenocal, 2004; Adams, 2005; Bobe, 2006; Smithsonian, 2013a). It certainly would not have been possible for early hominins to have been able to migrate into the sometimes extremely inhospitable areas of the globe that *Homo neanderthalensis* and *Homo sapiens* were to eventually occupy (Cf., Trinkaus and Howells, 1979; Trinkaus, 1989).

The difficulties of surviving harsh winter conditions cannot be overemphasized. Only with considerable technology and extensive teaching and experience can modern humans, with our very large brains, survive even the winters half way to the Arctic Circle, let alone an ice age. There is a tendency, however, when considering human evolution to believe that 'put humans in any kind of environment and we would have evolved a brain to deal with it'. This is not, I think, how evolution works. Throw the higher apes or early hominins, even with their much greater cognitive and analytical skills than any other non-human primate (and far more than the vast majority of mammals), into the Arctic and they will not, of necessity, evolve more intelligence to cope with these conditions – they will die. The same point can be made about dealing with greatly increased storms, flash flooding, encroaching desert conditions (especially where escape routes have been cut-off) and pestilence, as well as dealing with relatively rapid changes in the flora and fauna of an area.

But changes in nature are not always this continuously dramatic or evenly distributed, especially not for all groups of a relatively widely dispersed species. By the early Pleistocene (2myr) *Homo erectus* was becoming established, progressively replacing other hominin species and beginning to migrate to many parts of the world. During this period the cooling and heating of the planet was most intense, with glaciers sometimes reaching well into north Africa. Although, as noted in Chapter 2, some populations of *erectus* continued to live in semi-tropical conditions up until at least 200,000 years ago, even as far east as China and Java, many *erectus* groups undoubtedly were led up the garden path of warm interglacial valleys, only to be forced, several generations down the line, into retreating into more congenial environments, or to cope with advancing glaciers, geographical disruptions and varying seasonal climatic conditions, all of which selected for greater and greater analytical intelligence. It is possible that the increased prevalence of meat eating, and the development of cooking at about this time, provided an increase in energy supply, giving any runaway brain evolution that was taking place a boost, or replacing the easy brain-rich foods that set the runaway in motion in the first place (Cf., Wrangham, 2009).

And so it is understandable that the relatively great advance in frontal brain evolution took place at this time, largely among *Homo erectus* populations. Skulls from these humans range from 750 ccs to 1250 ccs, with progression from older dated fossils to the more recent ones. This range runs from slightly more than apes and early hominins (450–500ccs) to modern humans (1350 ccs). If we start with *Homo habilis/rudolfensis* and/or *Homo ergaster*, species which coexisted with *erectus* in the early days (2 myr), and run through to *Homo heidelbergensis*, a species that overlapped with them at the other end (at around 200,000yr), we get a range from 510ccs to 1280ccs (Foley, 2011; Smithsonian, 2013), very close to the entire range from modern apes to modern humans. And if we consider *erectus* to be a flowing, evolving species it may well be the case that when more skeletons are found these others will not be considered separate species from *erectus* but rather minor (localized) varieties of early or late *erectus* (Cf., Lordkipanidze, 2013).

Whatever the exact classifications that finally emerge, the geological and fossil record clearly indicates that evolving humans faced a series of ice ages at the time of the transition from *ergaster* through *erectus/heidelbergensis* to relatively modern humans. And the argument here is that already having a runaway potential for a very much larger brain gave *erectus* through to *heidelbergensis* a survival chance in harsher conditions than those in which hominins first evolved. At this point, we can speculate, genetic propensities that allowed for, or encouraged, runaway brain growth were increasingly selected for, as were those that diminished the jaw muscles that in part restricted skull expansion (original mutation 2.5myr), as well as those propensities that reduced neck and facial musculature. The genetic tendencies for a diminished capacity for cell-surface sugar N-glycolylineuraninic (which retards brain development during the foetal stage) likely also accumulated. And females who found the 'big headed freak' – who seemed to be almost prospering in this cold, to be just a bit more successful at hunting, who had actually built a shelter that stood up to the weather – at least a little bit attractive, were also favoured; and males attracted to the 'big headed female freak' who had made some clothing for herself and her offspring and wrapped her babies up out of the wind, were also favoured by selection.

In this context, selective emphasis would have been on the development of the forebrain that aided in longer-term memory so that it could be remembered where rivers might be crossed, and that winter will come again, and so will the spring floods. And selection would have favoured the kinds of intellectual skills involved in tool making and using, and in surviving longer than from one fishing trip, sexual excitement or baby cuddling event to the next. It was the kind of brain that could see cause and effect, currently and in the future; that could envision battle plans to keep out a band of hunters retreating from the increasingly vicious snow storms of the high mountain valleys, or to plan an invasion of what seemed to be a nice lakeside location, or to find an escape route that did not encounter any hostiles, or to negotiate and trade with those who stayed at the multitude of crossroads that must have existed. It was the kind of brain that could invent and live by rules so that a degree of cooperation and law and order could be developed and maintained. It was the kind of brain that could plan for the future. And in all this the evolving brain most likely enhanced, and systematized, the somewhat older political intelligence of apes and early hominins used in manipulating social and mating relationships, turning these into more formal taboos and boundaries.

But this raises a mystery. If it was so wonderfully useful why did it stop evolving about 100,000–70, 000 years ago? Perhaps the skull reached a level of thinness beyond which it could not go and still protect the brain and/or provide an anchor for jaw and other facial muscles. It is possible that the human brain grew so big before birth that neither the mothers nor the offspring carrying genetic propensities for its increase survived childbirth, thus restricting the spread of these potentials. Perhaps the energy requirements of the growing brain simply became too great for the health of the body generally. An argument can also be made that the type of intelligence served by frontal lobe expansion, and selected for in harsh environments, reached a point at which it was selected against. It is important to note, for example, that even when 'let loose', the evolving more analytical human brain never became the kind of brain that could escape the emotional forces that had evolved along with the processes of playing, friendship formation, child rearing, parenting, parent/offspring conflicts, courtship, sexual relationships, mate bonding, status concerns and politics. Jealousy, hate, envy, joy, elation, depression, shame and embarrassment were not banished; they increased in intensity if anything.

Why did the new – more analytical – intelligence not take over and banish the old? Why cannot rules of friendship, love and trust be 'rationalized' once and for-all, and then be lived by? Why wasn't a brain capable of great analytical skills, of perceiving cause and effect in nature and in human behaviour, with an incredible capacity for memory storage and a strong tendency to abstract thinking – capable of inventing quantum physics – able to become species-typical human intelligence? Why instead did it evolve as a complex, ambiguous, often rather messy, sometimes painful, mixture of analytical abilities together with powerful emotions? A tool such as a sharp stick, for example, became a spear to fight with, and came to be loved, loved like a male loves his penis, and to be waved about with the same pride and glee a male bonobo seems to enjoy when waving and showing off his penis. Friendship, love and trust became moral virtues, spiritual entities, preached enthusiastically and often held to be something to kill and to die for, not simply a set of 'rational' procedures to follow.

It seems that the original emotional intelligence gave/gives the new intelligence motivational force. It appears that 'moral' rules only work if they are adhered to and enforced with the same intensity with which lust, anger, hate and jealousy, for example, compel behaviour. Modern human intelligence can be argued to be a product of powerful emotions 'charging up' a number of cognitive/analytical processes (Vannelli, 2001) – cognitive processes that largely evolved during the spurt of brain growth during the late

Pleistocene. *Reproductively*, this integrated intelligence has, most of the time, worked extremely well for a large percentage of humans. The species has certainly thrived and spread. But this mixture has not of necessity made life a smooth trip. The two intelligences are often in conflict, working against each other, subverting each other, trying to dominate each other. Selection has not edited human intelligence so that it is free from ambiguity, ambivalence and conflict. It is likely that this is because the mix of the two intelligences is a result of a long polygenetic development (Cf., Singer, 2003; Baron-Cohen, 2004; Ulick, 2005; Leslie, 2011) in which the reproductive/social success of the *middle* keeps potentially conflictual elements of both extremes in a population through normal random genetic distribution (I am using two intelligences as a convenience for the discussion, some may want to argue that we have evolved three or four, or more, intelligences).

Cognitive/analytical abilities ('rationality') have a tendency to stereotype, classify and sequence objects and behaviours into relatively clear, often distinct, categories. These categories are frequently treated as causes of other categories, usually in a sequential algorithmic, or in a circular, manner. For many people, discovering these categories and 'causal relationships' equals 'rational' analysis. The brain that is good at this seeks shapes and structures that seem to have clear definitions and appear to be made up of 'functionally interrelated' distinct parts; sometimes different structures (or parts of structures) are perceived as being incompatible – in which case the world is often seen as an opposition between right and wrong, correct and incorrect, rational and irrational, and people with strong views on the above are often very sure that they know which is which (and that their interpretation should be obvious to all intelligent people). There is a tendency with this kind of thinking to teleologically assign a 'greater/higher' abstract purpose to human existence than just surviving, eating and reproducing, and then to treat the seeking of this purpose as the basis of 'morality'. An extreme of this intelligence can lead to the view that the 'higher purpose/essence' is more important than any single human being (including self).

People with a high development of this intelligence usually have a considerable capacity to see and remember details, are prone to repetitive behaviour and obsessive routines and have trouble with imagining out of the ordinary things or events. Near the extreme end, this intelligence approximates the characteristics which have come to be known as Asperger's Syndrome, increasingly seen as a type of intelligence rather than a 'disease' (Baron-Cohen, 1995, 2004, 2011; Grandin, 2005; Ulick, 2005; Lehman, 2010). Individuals with a noticeable degree of this intelligence can be extremely good at being camp memory persons, legend narrators, reciters of the ancient law, platonic philosophers, conspiracy theorists, mathematicians, computer programmers, dedicated musicians, physicists, single-minded career climbers, idealist politicians and generators of bureaucratic roles, rules and procedures.

However, individuals with such an intelligence can be perceived to be 'aggressive, cold, egocentric, impersonal, antisocial, unempathetic, and tough-minded' (Miller, 2000, p. 331; see also, Eysenck, 1992; Baron-Cohen, 2011). Although they sometimes have very good verbal skills, they have a very difficult time understanding other peoples' beliefs or feelings. Such individuals can become extremely sanctimonious and arrogant, seeing the apparent sexual interests and emotional reactions of other individuals as being frivolous, if not 'irrational', and considering ambiguous behaviour as being 'wishy-washy' and cowardly, if not out and out devious; such individuals often love humanity in the abstract but are not good at social interactions with real people; they can be very resistant to change, often having a problem dealing with the idea of randomness, or understanding unpredictable behaviour (Baron-Cohen, 1995, 2004, 2011; Miller, 2000; Grandin, 2005; Ulick, 2005).

Although single-mindedness and step by step approaches to problems can be very important up to a point – those with a sense of purpose are usually more successful at a variety of tasks (including reproductive) than those who have no sense of purpose – nevertheless, when the purpose of solving a mathematical problem, or of fully living up to God's purity goals, or of not indulging in sex until evil has been conquered, for example, becomes stronger than sexual, romantic and parental drives, reproductive success is weakened. At some point analytic and reflexive skills became counter-productive in comparison with an emotional-cognitive capacity to very quickly desire (and fear) specific things, such as being attracted to certain types of individuals, and being aroused by them sexually and/or personally, and wanting to befriend and even strongly bond with them, while also fearing strangers enough to be able to avoid the wrath of potential enemies.

So, a hypothesis can be suggested that frontal brain size stopped developing (70,000–100,000 years ago) because frontal brain activities began to significantly conflict with and/or detract from motivational emotions, desires and fears which were reproductively advantageous. Indeed, for a social species, it is not just reproductive but also survival success that requires a great deal of emotional intelligence. And so its evolution seems to have kept up with cognitive developments. Certainly, human emotionality appears more evolved than even the considerably sophisticated emotional behaviour of the higher apes. Individuals with well-developed emotional intelligence can be extremely sensitive to the feelings of others, giving individuals with such intelligence an enormous capacity for empathy (Cf., de Waal, 1996; Baron-Cohen, 2004, 2011; Hrdy, 2011); it can also lead to sympathy and pity, qualities clearly somewhat absent in the analytical type of intelligence, but which can attract reproductive mates and friends as well as reciprocal appreciation and support.

A capacity for imagining others' reactions to self (empathy) can result in fantasies and imaginary scenarios concerning human relationships, providing a capacity for emotional planning of sorts. All of this provides motivation and skills for making friends, seducing lovers, falling in love, raising and fussing over children, imitating others, playing roles, cooperating out of friendship and loyalty, being admired/respected by others, forming support networks and for politicking self and family (and friends) through social/economic life. Individuals with a strong element of this type of intelligence can make successful story tellers, creative artists, politicians, statesmen, teachers, social workers, clerics, salesmen, loyal allies, friends and visionaries.

But when concentrated at the emotional end of a hypothetical continuum, emotional intelligence too can become reproductively counter-productive. It can, for example, lead to overwhelming and desperate desires to be loved and admired by others, rendering individuals subject to the dependency trap (and inviting disdain). Excessive showing-off for attention may earn the resentment and contempt of others, if not rejection. Or, such individuals may seek to dominate situations and to lord it over others, to manipulate others, and even to humiliate others in order to make themselves noticeable and, they believe, admired. An overabundance of emotional imagination can lead to creating fantasy worlds in which self is the hero, and for which a rather heroic past is invented and great romance, success and recognition imagined for the future (with little basis in objective conditions). Such individuals may come to believe in their own fantasies, created pasts and imagined mythical futures so much that they get a reputation for being not only air-headed romantics but also pathological liars (Cf., Leslie, 2011).

At the same time, this intelligence can sometimes generate emotional combinations that manifest as volatility, outbursts and an apparent lack of 'self-control', with little desire to apply an analytical approach to particular social situations or intellectual issues. When frustrated, such individuals can burst forth with angry displays of envy and jealousy. Some individuals may accept that their behaviour has upset others and apologize profusely, but with a certain lack of sincerity, making people even more suspicious. Overall, individuals with an overwhelming preponderance of emotionally fuelled intelligence may not be trusted. In general, they will *not* be desired as sexual partners, parenting partners, alliance mates, friends, leaders or responsible members of a community. They seem unprincipled and 'self-serving'.

What is being suggested is that individuals with some version of a balance between cognitive/analytical skills (even with occasional obsessive tendencies and lapses of sympathy/empathy) and an emotional and politically astute intelligence (albeit interspersed with fits of 'irrational' love, anger, jealousy and envy, for example), would be the most common in a population. This is because each of the intelligences discussed above in their extreme forms would not have had as much reproductive success as a balance of the two (especially given the evolutionary scenario presented in Chapter 3). However, as noted, the polygenetic nature of these intelligences suggests that a minority of individuals at both extremes will always appear in a breeding population. So, we get some indication as to why an, often uneasy, integration of these 'intelligences' came to be species-typical rather than one taking over the other. And because excessive growth in either one was checked by a lack of reproductive success, why the brain overall stopped growing.

Given the polygenetic input into each of these, and the wide variety of possible experiences any human can have, an individual's overall intelligence will be a complex mixture of the two tendencies (and probably of a number of others – such as creative or spiritual propensities). While conflicts exist, even in the mind of a given individual, in most circumstance the two 'types' of intelligence often more or less support each other, providing a degree of stability and motivational continuation over time for 'average' humans and for human groupings. For example, analytical/cognitive intelligence may discover good cooperative ways of hunting, fishing and gathering, but an emotional intelligence is required to turn potentially competitive individuals into a bonded unit to achieve success with the new techniques. Powerful drives along with emotional intelligence lead to sex and bonding and parenting, but analytical intelligence contributes to understanding the relationship between sex and its reproductive consequences, to formulating marriage rules and obligations and in discovering techniques for successful child-care. An emotional intelligence may generate fairytales, myths and heroes that analytical/cognitive intelligence turns into systematic legends, admired leaders, codes of honour and religions.

The polygenetic nature of these tendencies and the success of somewhat different mixtures in the middle (especially in variable environments) mean that it is highly unlikely that evolution has, or will, 'edit' a completely consistent, thus predictable, balance for humans. Particular balances might serve some individuals most of the time, and some balances might serve most of the people some of the time, but none seem possible that would perfectly serve all individuals all the time (to paraphrase Abraham Lincoln). Evolution is not about achieving a perfect balance (even if there is such a thing), it is about reproductive success; and the emergence of a large human brain has not escaped this fact. Indeed, the recent discovery of the species *Homo floresiensis* on an isolated island (Dawkins, 2004a; Foley, 2011) shows that an evolutionary trend, or even a stable balance, of brain size and function can actually reverse itself. Thought to have evolved from *Homo erectus, Floresiensis* experienced a reduction in both body size and brain size (from about 5 feet 6 inches to 3 feet, and 750cc to 450cc). Although the brain case of *floresiensis* is no bigger than that of a chimpanzee, *floresiensis* was a user of relatively sophisticated tools for the time, suggesting a miniaturizing rather than simple reduction of brain areas took place during this evolution.

Whatever the case, the discovery makes it clear that frontal brain enlargement was not an evolutionary inevitability with bipedalism and a number of other human-like characteristics. Island isolation seemed to have reduced the dangers that might have favoured analytical evolution, even considering evolving *floresiensis'* physical weakness. Selection not only stopped but perhaps reversed the trend for brain enlargement that was characteristic of *erectus* in general. So long as sufficient reproductive emotions, desires and fears were intact, a super-analytical brain was an unnecessary luxury, perhaps in this island isolation even an unwanted burden given that a large brain is extremely expensive in terms of energy consumption.

If human intelligence is a number of analytical skills mixed with an expansion of elements of ape emotional/political intelligence, together with a repression of certain ape tendencies and the addition of some new human characteristics, and if particular mixes of these are rather randomly dispersed throughout human populations (in which both analytical rigidity and emotional basket-cases show up from time to time), what about language? Is language the key to the evolution of a unique, higher level of existence?

Language

To be sure, human language is a complicated phenomenon to explain. On the one hand, it seems to play a key role in the processes of advanced concept formation and, as such, perhaps contains a means for discovering some of the 'ultimate' knowledge and/or 'truths' concerning the universe and the essence of humanness for which philosophers have been searching. At the very least it is a mechanism for passing on cultural information so that humans do not have to continuously learn from scratch every useful bit of knowledge gained, and, as a result, knowledge can be accumulated. On the other hand, language is full of ambiguities, extremely changeable, very useful in deceiving others and is mostly used for self-deception, gossip, image creation, seduction, lying, humour, swearing, making accusations (quite often unwarranted ones), including/ excluding individuals and for politicking. None of these activities are (on the face of it at least) especially knowledge or truth revealing.

As noted at the start of this chapter, I think that language has been greatly overrated in terms of its role in basic human nature. Even advanced language operates very much like the general cognitive processes that (along with specific emotional tendencies and an emotional charging of certain cognitive formulations) make up human nature. As such, language seems to be a special case, or type, of general cognition rather than the basis of something 'rationally' unique. That, however, raises the evolutionary question: why, then, was this particular aspect of general cognition selected for? To start looking for an explanation for the evolution of language that does not rely on its purported rationalistic outcomes, let us first consider the suggestion that the human brain might have at some point become an instrument which drew mates by demonstrating how clever a person was. This only works, however, if some means of communicating this cleverness also evolves.

Being attracted to individuals with big, round heads would have been one means of this communication. But equally likely, or in tandem, selection would have worked on what was already there in terms of modes of communication. And primates certainly already had several forms of communication for selection to work on. These include various movements of the face and body such as 'smiling, touching, holding and kissing' (Dixson, 1998, p. 107; see also, Lawick-Goodall, 1971; Bygott, 1979; Snowdon, 2002; Mithen, 2005). And this has clearly continued, if not expanded, in modern humans; various facial expressions, together with smiling and laughter, and our ability to read them, appear to be universal (Van Hooff and Preuschoft, 2003), being, evidence suggests, deeply embedded in the human brain (Baron-Cohen, 1995, 2004; Wingert and Brant, 2005). These expressions appear to be spontaneous and pleasurable during childhood and remain for life. Their significance and meaning are, in all cultures, elaborated. Smiling and laughing came to mean more than anxiety/submission and spontaneous expressions of joy during play; they came to indicate welcome and friendship and to show a degree of equality, for example, all of which could be communicated not only to the individual being interacted with but also to others round about (Van Hooff and Preuschoft, 2003).

Human eyes are a major means of relatively complex communication (Baron-Cohen, 1995, 2004). We glance down when being submissive but stare straight forward when claiming dominance; from a lowered gaze the eyes move slightly up and sideways to be coy and inviting, and completely away to show a lack of interest. We can wink, stare, peer, squint, glance, open-wide, roll, cry and wipe our eyes. The whites of human eyes – unique among primates – make eye movements very clear to a viewer (Copley, 2001). Eye movements indicate an attitude to the person or object being viewed (or to a sound heard) and indicate feelings about the situation which are conveyed to others watching. Interestingly, children diagnosed with 'autism spectrum disorder' tend to look at the mouth of others or objects behind them rather than, as normal, at their eyes (NHS, 2013; see also, Ehrenfeld, 2005).

And, of course, we vocalize. Among primates, screeching, rhythmic hooting, shrieking and sing/song utterances often accompany non-verbal signals. Bonobos, for example, spend a lot of time in vocal back and forth 'languagelike' exchanges (including the use of grins) that communicate internal states, such as delight and nervousness, and express friendship (de Waal and Lanting, 1998, p. 33; see also, Snowdon, 2002; Van Hooff and Preuschoft, 2003). So it seems natural that the evolving hominin brain was able to relate these vocal sounds not only to various facial and eye expressions, but also to certain activities and individuals, and to remember what went with what, and to copy each other in order to solicit objects, friendship and sex. An ability to assign connections and meaning existed in the primate heritage. We know, for example, that in a rudimentary form apes in general are capable of learning and using various sorts of sign language. Although the degree to which this resembles human language is debatable (Cf., Wallman, 1992; Pinker, 1995, 2007), apes certainly have been observed to demonstrate considerable feats of memory regarding symbolic formations, in some cases of short-term memory retention outperforming human university students (Inoue and Matsuzawa, 2007, but see, Humphrey, 2012).

I have suggested the possibility that human differentiation took place on the basis of some individuals demonstrating an extension of elements of childhood into adulthood, and that what is very similar to playing led to reproductively successful copulations. During playing a lot of squealing, giggling, laughing, pretend anger, fake crying, mock fighting, hiding and chasing takes place. These are all forms of communication in so far as they tell playmates what an individual does not like but also what they are enjoying, what they would like to do more of, who they like to be with and so on. Selection could have begun to favour individuals who were good at such things as: rhythmic/sing song utterances, cooing, mimicking and putting-on-faces because these would have communicated that they were up for play, fun, friendship – that they were not dangerous. Given the runaway brain postulated above, these forms of communication skills may have had a tendency 'to runaway' as well. After all, hooting and shrieking and laughing and making sing/song noises can be fun, and they definitely draw attention (so why not do more of these?).

Certainly modern lovers not only use their eyes in all sorts of enticing ways, they tickle each other and make funny sounds, they smile a lot and laugh at each other's noises and little foibles; they coo to each other; among humans, foreplay and sex share many characteristics with children playing (Kortlandt, 1985), and with play-like verbalizing. Such expressions would have had the reproductive advantage of being play-like sounds that did not bring jealous males running to interfere, as sometimes happens when primate females squeak or make more distinctive noises approximating a climax when copulating. Humans have retained some primate sexual noises, but generally panting, moans and deep sighs have replaced squeaks and cries during actual copulation. These are partially spontaneous but to some degree also controlled for effect and for communication. Where more extensive and dramatic noises are made among humans these are almost always done in relative secret in which giggles, squeals, and cries like: 'Yes!

Yes!! YESSS!!!' are considered to be very private communications (not done when the children are awake or guests are in the house). As private communication these often excite sexual bonding, romance and longer-term pair bonding.

It would not have been far along from this primate background for a group of youth-ape friends to develop a consensus as to what certain facial expressions, movements and sounds meant and to remember them for the next time they played together. And certain signs and sounds could have meant that we can copulate together, and that it is a lot of fun, and 'I like being with you, doing things together'. An elaboration of skills to be able to *control* the above types of communication would have selected for brain mechanisms that would later underpin the development of the more conceptual type of language abilities we now employ. And relevant genetic changes for this were not that complicated, basically depending on just the FOX P2 gene with something like two changes from apes in its 715 amino acids (Enard, *et al*., 2002; Dawkins, 2004), perhaps having taken place sometime around 200,000 years ago (Itzhaki, 2003).

Nevertheless, the fun side of language is still with us, albeit in its more conceptual form. Besides the 'silliness' of romance and sex-play talk remaining strong, the role of elaborate joking, teasing and humour in modern courtship has developed considerably. A glance at personal dating ads, for example, quickly confirms that: 'a GSOH is essential'; indeed, is it possible that we could regard 'Human evolution as romantic comedy' (Miller, 2000, p. 417), considering the euphemisms, jokes, innuendoes and verbal coyness that we associate with the getting together of a mating pair. And we never tire of laughing at the foibles of others in these pursuits, or of gossiping about them and putting them in a multitude of myths, novels, films and television programmes. And even when they get together relatively securely, couples make jokes and engage in teasing about each other's 'sexiness' and 'wicked skills' and 'naughty appetites'. Relatively elaborate communication skills could thus have been selected for because they locked playing/mating individuals together at a time when both parents providing care for immature offspring was also being selected for. And this development could have introduced additional selection forces for even more complex language skills.

As youth-ape consort mating became increasingly important in individual reproductive success, younger males gained a degree of reproductive equality with higher status and older males. But it is unlikely that the issue of dominance went away completely, especially when a degree of infant sterility or sexual inactivity returned and puberty became a dividing line between playing and 'love making' (as an evolution of an anti-incest mechanism). And, as 'love making' began to replace playing, more 'advanced' language skills might have had an advantage over both non-verbal and playful verbal communication because advanced language skills could have provided males and females with an enhanced opportunity to evaluate the value of potential mates. For example, in so far as it became part of increasingly *subtle* means of 'getting together' (and of showing off cleverness) spoken language does not give away intentions like non-verbal signals do. An individual can spin a nice yarn, elaborate an exciting story, tell a good joke, impress ever so deceptively.

So, a young male might ask a female if she would like to take a walk in the woods, share some of his food and chat without necessarily raising her suspicion, or that of jealous/dominant males who might be watching. He then could proceed to tell her stories of his exploits – being careful not to exaggerate too much (indeed, often to be self-effacing) but embellishing just enough in some areas to lift his stature a bit, to make him sound 'interesting' and perhaps to make her laugh). And above all, he might ask her about herself (communicating a willingness to listen to, and likely accept, her verbally created self). Eventually he might begin to indulge in youthful-like suggestive joking and even sweet talk about how nice and attractive she was. Who knows, he might appear worthwhile as a potential mate and she might become curious. So, he might set up another meeting, inching his way, with the aid of language, nearer to a successful mating.

Once established with a relatively permanent partner he could tell her just how dangerous the hunt had been or how hard he had worked that day and so she would appreciate him and be loyal to him. Or, he might use language to tell her that when he came home late at night it was because he had to talk through tomorrow's hunt with the 'warriors', and didn't she realize how many dangers he faced for her. And when she became agitated by his attention to another female he could make up no end of stories about why he was talking to said strange female; and if he wanted to push the point he might tell her that she would have to get her jealousy under control. Females, from the start, would ask questions designed to find out how many of his exploits were in fact exaggerations and what his long-term prospects and intentions might be. If she liked him, in the course of her conversation she would let slip hints of her mothering tendencies and hitherto subdued passion, desperate to be released for the right man. If she was clever with language she could keep him on a string through the ambiguity of language for a considerable amount of time, giving her plenty of space to evaluate him and to keep her eye out for something potentially better; she might even converse with other potential mates. If she were extremely clever (consciously and unconsciously), she might become pregnant by a bigger, more dominant male, who was not interested in child-care, and convince her original suitor, through language, that he was the father and that he should get his jealousy under control.

Some of this scenario may be poetic license gleaned from soap operas (which, nevertheless, do represent many elements of human nature). It would take a certain amount of emotional dependency on the part of a male for this last example to work, for example. But it does show how relatively complex language might operate in the process of sexual selection. Language can be used not only for seduction, but also as a precautionary mechanism. It is significant that when individuals begin courtship they usually do so with the most neutral, non-emotional use of language they can muster. This is for at least four reasons: first, it makes them appear non-dangerous to the person they are attracted to; second, it makes them appear non-dangerous to any jealous individual who might be watching; third, it buffers them from the sting of potential rejection; and fourth, it allows them an escape route before they get too emotionally involved with someone not suited to them or not likely to be loyal to them. All of this is sufficiently important that it has even been argued that 'verbal courtship is the heart of human sexual selection' (Miller, 2000, pp. 351–354).

It is also significant that once mating partnerships and/or alliance groupings are entered into, humans use language to overcome more emotionally felt problems and to cover up what might otherwise be perceived to be very disloyal acts. 'I don't know what came over me, I just lost my temper, I will be alright tomorrow', can paper over a multitude of temporary cracks in a relationship and set the stage for carrying on. Or, 'I do love you *still*, I only stayed out all last night because I was walking the streets, I was so wound up; I was trying to find myself', can hide infidelity and is often enough to placate the seeker of an explanation, any explanation, and thus keep a partnership going. Or, 'I am not interested in your wives, sire, my loyalty to you is unbounded' may be a lie but the chief adviser is making a commitment of alliance and submission to his master, which may serve them both very well down the line (reproductively and otherwise). 'Look. I simply need help with this – one person can't do it on their own', can lead to a saving of face and possibly a change in the division of labour rather than to a fight leading to a permanent rupture.

Both deceit and self-deceit can be positively advantageous when it comes to achieving reproductive success. The little white lie is extremely valuable in cementing social links with significant others. The above deceptions may well have kept reproductively beneficial social relations intact, and, indeed, functioning relatively effectively, relationships that would have been completely destroyed by the bald truth. If our close friends and lovers will not lie to us about how wonderful we look, and that it does not matter that we got drunk last night and made a fool of ourselves, then who will? And we tell our children, 'very well done, that's wonderful' when they show us a bodged-together creation which almost any ape could have accomplished. And we say to the parents of a strange looking baby, 'my, what a cutie'. And, in the wider social sphere, members of sports and business teams tell each other 'well done, don't worry, we will get them next time'; soccer coaches after a 4-nil defeat tell the press 'my lads showed a lot of courage today'; vengeance brothers tell each other how brave they all were in battles with no positive outcomes; political leaders tell party workers, 'they only won because the biased media was on their side', when in all these cases individuals know about a lot of mess-ups – some crucial – on the part of friends/team mates and none of them have any idea how they really compare to competitors, or what the outcome of the next situation will be. These are not 'problems' in some ultimate sense with a fundamental 'truth' attached that language can solve, but rather normal processes of sexual selection and human intercourse in which language is a tool.

Examples such as these come to mind so very easily, clearly making the point that deception and self-deception are fundamental parts of human life, and that language can be a great aid in deception. Non-verbal signals could not have accomplished any of the above. In these examples, the use of language makes individuals feel better but also allows them to reach out for confirmation of their continued inclusion in a team. 'Do you really think so?' from a 'praised' individual would generally solicit a reinforcement of the 'you did alright' messages from the examples above, as there is only one *team* answer to this question: 'Yes', and to really confirm it: 'Yes, I definitely do'. In most human 'understandings' of situations there are no exact criteria to act as a measure of their 'truth', but most often it is the case that whatever happened, maintaining the morale of a team is considered to be a 'higher purpose' than apportioning blame among individual members.

In this sense the 'lies' are benefitting the deceived because they remain included, but also the deceivers in so far as they keep team members/subordinates/clients/workers relatively happy and working. And, in fact, from an evolutionary point of view, it can be argued that human language most often benefits the deceptive speaker much more than the probing or eager-to-learn listener – just the opposite of what 'language to pass on information' theory would predict (Miller, 2000, pp. 350–354). For example, when it comes to information sharing, the wrong answer (or no answer) may provide the most reproductive success to questions like: 'where is the good fishing hole?', for instance. A female's accusation on a first date: 'You only want to sleep with me', answered with the 'truth': 'dammed right, I can't wait', is likely to be counter-productive, whereas the 'lie': 'oh no, I find your interest in Egyptian hieroglyphics fascinating', may give him a chance.

And, in terms of language providing an efficient means for sharing survival information and for cooperation in survival activities, the evidence is almost in the opposite direction. Young boys – originally the vengeance mates, hunters, warriors and herdsmen of their tribe – employ (even with the availability of modern linguistic developments and education) a language of 'minimalist grunting' backed up by 'cryptic, grammatically degenerate mumbling' (Miller, 2000, p. 352). This dialect requires almost no words at all. The tendency for using this 'Early Adolescent Mumbled Dialect' (Miller, 2000, p. 252) remains for life among males. Tom Wolfe (1980), for example, identified a US Army man-on-the-ground 'Creole'. This was 'a language in which there were about ten nouns, five verbs, and one adjective, or participle, or whatever it was called'. This is illustrated by 'a couple of good buddies' in conversation:

'I tol'im iffie tried to fuck me over, I was gonna kick 'is fucking'ass, iddnat right?'
'Fucking A.'
'Soey kep' on fuckin' me over and I kicked 'is fuckin' ass in fo'im, iddnat right?'
'Fuckin' A.' . . . 'Fuckin' A well tol', Bubba' (p. 125).

Miller and Wolfe may be overstating the case somewhat. It is likely that young males use more sophisticated language, at least some of the time, and US Army personnel certainly have more language than this for most situations. But the point is that most survival, cooperative and warlike activities do not require, or often use, much language. There are evolutionary reasons as to why this might be so. A great deal of talking is the last thing one wants to do in a cooperative hunt, vengeance raid or pitched battle. Additionally, learning and acquiring most practical knowledge is often best done through demonstrations, visual illustrations and short commands with the use of a minimum of often-repeated words – such as: LIKE THIS, NOT THAT WAY, STOP, NO, WRONG, BE CAREFUL, YES, GOOD, WELL DONE – with the hands, eyes and tone of voice communicating as much, if not more, information than words. Even with the most sophisticated of languages, telling someone how to operate complex machinery rather than showing them can be dangerous – like teaching someone to drive through verbal teaching alone.

Additionally, although human language clearly can be used to pass on and accumulate information, and it has done so throughout history (especially in its written form), if language had evolved *primarily* because it provided a definite selective advantage through its capacity for passing on crucial technical and survival knowledge, it would not be so full of redundancies, semantic imprecision, ambiguities or be so vulnerable to constant change (Cf., Pinker, 1995; Crystal, 2005; Power, 2005; Coe, 2012); a few very precise rules and a systematic logical system would have served much better (Badcock, 1986; Dunbar, 2004). Language has nothing like the precision of, say, mathematics, musical notation or technical drawings for communicating specific technical information, but very few people learn much mathematics or music, and even fewer are any good at them, and most people's eyes glaze over when they see a complex technical drawing. And the vast majority of us, in our daily lives, do not try to make language more precise, just the opposite. We have an enormous tendency to continually invent new words, change the meanings of existing words and pronounce words differently. Those purists who do try for consistency and precision find that they are fighting an uphill battle.

So, if language has been overestimated in terms of human survival and for facilitating technical efficiency, and if, as it has been estimated, during average adult conversations about 100 words account for 60 per cent of the conversation (and 4,000 for 98 per cent), and that we can get by very well in modern societies with 850, why do we have an ability to learn approximately 60,000 (Miller, 2000, pp. 369–372)? It is possible to argue that most language operates as a means of showing off, courting, seducing, selling, exaggerating, dominating, conning, preaching, joking, bullying, accusing, separating, lying, storytelling, politicking and making life seem exciting and worth living. It might be true that hundreds of words have been invented to name and describe the male penis (in various states), masturbation, female genitals, breasts, human body shapes and to signify and describe the sexual act itself. It might be, as has been argued, that it takes a million spoken words (not all different ones) to get a sexual relationship started, but only about 20 a day to keep it going. It seems to be the case that young adolescent girls interested in romance utilize a lot more words than boys at that age, generally gossiping about relationships, hopes and fantasies, and it is the case that when boys begin to date they suddenly become considerably more fluent (Miller, 2000, p. 382).

And it is not all about sex; thousands of words have been used to describe human personalities, motives, deceptions and wishes, and to talk of love and romance, of heroes and heroines, war and peace, violence and vengeance. And language is very good in aiding in conspiratorial behaviour. From potential lovers getting together without too much risk in the initial stages and/or agreeing how to meet in secret when others do not recognize a particular romance as being legitimate, to making economic, social and political agreements behind the backs of others, language allows for a degree of secrecy that communication through behaviour and non-verbal signals could never achieve. But conspiracies have an antidote within language itself; small talk, joking and gossip are fun; they are light entertainment; they are playing; and they play hell with secrets and conspiracies (most of which leak like sieves).

At the same time, a great deal of evaluating, policing, conflicting and rumour spreading within neighbourhood sexual and political goings-on is done through gossiping about who is doing what to whom (Dunbar, 1996; Harari, 2014). Because:

Gossip is one of the most universal of human habits. No conversation between people who know each other well – fellow employees, fellow family members, old friends – ever lingers for long on any topic other than the behaviours, ambitions, motives, frailties and affairs of other absent – or present – members of the group (Ridley, 1993, p. 321).

Among other things, gossip can be an effective means of mate guarding (an idea attributed to Richard Wrangham, cited in Wilson and Daly, 1992). While a male, for example, is off hunting, his parents (or at least his mother, sisters and neighbours) can keep an eye on his wife. But she may be very clever and secretive, so they can also keep an ear out for the gossip concerning her, and then use vague speculative language to inform the husband of his wife's general deportment during his absence. But more importantly, unless the wife is completely defiant in her philandering, gossip is likely to be the most effective guard against it happening in the first place.

Gossip does not depend on actual proof for its efficacy in regulating social behaviours – almost the *opposite* in fact. Gossip relies on an 'everyone-seems-to-think-that-it-just-might-be-that . . . ' mentality in which actual evidence is a positive hindrance. Gossip spreads like wildfire and there is very little believable come-back against it, and people know this. As a result, gossip can be one of the best antidotes known to humans against extensive lying and secrecy. A first cousin of gossip in the 'no proof needed' arena is the human propensity to make out-of-context, often false, accusations – a favourite linguistic activity of quarrelling siblings, religious enthusiasts, inquisitors, politicians and journalists, for example. It is the 'there's no smoke without fire' mentality that makes it easy for accusers to artificially generate smoke and then let people assume that there must be a fire. Both of these methods, gossip and accusations, tend to keep an eye on work mates, teammates, relatives, neighbours, bosses, celebrities and politicians.

The argument here is that the use of language for: display, seduction, status, reconciliation, inclusion, exclusion, social evaluations, behavioural monitoring, condemnation and having fun (gossiping, joking, teasing) was extremely important in its evolution and remains so in its use today, constantly 'oiling the machinery' of social life. Language did not evolve for listening intently to wise people to find out more about how to prosper and be moral, or to understand more about the working of the universe, or to discourse philosophically on matters of 'Truth'. Rather, thousands of words have been invented to find, keep and fool sexual and/or parenting partners, to enhance spiritual feelings, to keep sports fans happy, warriors fighting and mates loyal. And, as in the case of personal white lies, truth is not usually a major concern. And even when a concern for truth is strong, lying often still plays an important part in human discourse. Churchill is reported to have said, for example: 'in time of war, the truth is so precious it must be attended by a bodyguard of lies' (La Freniere, 1988, p. 239; see also, Leslie, 2011). (And, of course, lying in human relationships selects for an increase in memory and creative capacities in the human brain.)

Once established, the above language skills came to have social implications. To some degree they have become a means of designating an individual's standing in a community. Being a good gossip, for example, suggests that one has insider connections, that one 'knows things', or if one is not too skilled that one is simply a malicious spreader of rumours – even a liar. In some sections of society there is what might be called an 'awe-of-language syndrome' in which the greater the complexity and poetry of language used – regardless of meaning – the greater the value attributed to the speaker, affecting friendship, marriage, network and political opportunities. Individuals who talk alike, use the same jargon, usually get on together. In one study, for example, it was found that people mate with those with similar vocabulary sizes (Miller, 2000, pp. 345, 373).

So far so good. However, the development of a greater complexity of the use of language – as seen, for example, in scientific papers or legal documents such as contracts, constitutions and laws – does not seem to be fully explained by a relatively strong human tendency to multiply the number of words in use, to use complex grammar, just to show off, joke, lie, fire-up, accuse, gossip, exaggerate and/or seduce. Most languages in use today, especially in written form, have added an element of *precision* and *consistency* that is not found in everyday language use. It is tempting to think that language got more 'complex' as human technical knowledge became more extensive and detailed, but we have seen that even relatively sophisticated language is far from the best way to communicate instructive or technical information when compared with mathematics or picture/diagram driven instructions. Here I will argue that a more precise and consistent use of language was given a relatively strong evolutionary boost when humans had to negotiate safety in environments densely packed with other humans (see also, Bickerton, 2002; Snowdon, 2002; Harari, 2014).

As population densities reached a certain level it is hard to imagine that the hominini nature evolving from the primate background could do anything other than lead to constant social fission. There is nothing in the general primate, especially chimpanzee and bonobo (often aggressive) fission/fusion social background, or in our suggestion of youth-ape evolution, that would suggest that humans might evolve some sort of instinctive or emotional force for a high level of universal socialness (such as, an emotionally felt spiritual sense of 'as humans, we are all brothers and sisters' as is often assumed to be what we must 'bring out' in 'brotherhood of man' theories). Rather, as we have seen, early hominin nature most likely evolved in the context of reproducing pairs hiding off from others while living in very small, fluctuating groups of friends, close relatives and/ or alliance mates. And along with this we have had the development of self-awareness with a strong sense of *individual* self-identity, with self perceived as being an active agent in social and mating activities. And, indeed, for much of modern human evolutionary time humans have lived in very small bands where it has been easy – as with chimps and bonobos – to avoid most 'strangers' by bands and families moving away from each other, or at least having to communicate with only a selected few outside one's own immediate circle. Yet it is evident that humans eventually evolved to be able to live in extremely high densities of individuals within huge population sizes, quite beyond anything seen in any of the other primate (or even mammal) species.

So, how did large-scale societies come into existence given the evolutionary scenario postulated? It is feasible to suggest that this was possible because we had already evolved the *capacity* for coping with high population densities (but not necessarily the motivation to do so) by the time humans began to be crowded together. One significant element in this was the evolving tendency to *avoid* emotional and sexual displays, or even interactions, with relative strangers. This was part of an evolution that included a loss of clear signs of ovulation and sexual receptivity; an increased voluntary control of non-verbal signals of sexual arousal and attractiveness; a reduction in sexual promiscuity; the emergence of selective pair-bonding; and the development of shyness/embarrassment/shame concerning sexual and personal matters. Sex and child-care became activities of pair-bonded individuals, often done in secret, 'out of sight' of others. All these elements of sexual selection, it can be argued, provided a possible evolutionary basis for the selection of *emotional separation* among individuals who had not developed long-term emotional attachments.

The mechanisms of this evolution would have included the noticeable human capacity to begin to feel 'separation anxiety' at the age of about seven to eight months when separated from primary carers and to exhibit 'a fear of strangers' syndrome (Cf., Sroufe, 1977; Izard and Dougherty, 1982; Buck, 1984). This would sharply curtail a human baby's earliest tendency to elicit care and attention from all and sundry, allowing their individual experiences to 'teach' them (albeit, largely unconsciously) to be very much more selective in terms of with whom to associate. So, as we grow up, when meeting 'strangers' smiles become restricted, grins and noticeable facial expressions avoided, extended eye contact generally eliminated, suggestive gazes and touching more or less completely repressed; hand movements are reduced to a minimum, as are laughing and giggling. Indeed, in normal circumstances no excited verbal sounds tend to be uttered in the presence of strangers. Modesty-posturing and clothing are employed to hide genitals (and other sexually attractive features, sometimes including heads and faces).

Altogether, then, this evolution could have allowed humans to live in their small families and bands, even when in the midst of thousands of unknown individuals, because the unknown individuals had been rendered emotionally neutral. Nevertheless, to make sure the others remained non-dangerous, and that there was not too much conflict with them over mates, resources, glory and space, it would be useful to make some non-emotional *deals* with at least a few strangers; and why not gain some benefits at the same time (such as securing future mates, cooperation, allies). This seems to have been accomplished through the evolution of relatively *non-emotional* language development and usage, status recognition and the administrative elements of politics. Non-emotional language can not only be used to 'make deals', but also to facilitate status and administrative processes, making these almost a package rather than separate developments, coming to include oaths, contracts and laws. Such language (along with status recognition and administrative politics) allows humans to make deals with others in conditions of high population densities, and can be employed without direct emotional contact or commitment, or without any spiritual sense.

While it is unlikely that humans initially sought to live in high densities, at some point environmental pressures will have driven small bands and even federations of bands together in locations where survival remained possible while it was disappearing in surrounding areas. Definite climatic fluctuations with such potential consequences can clearly be seen in late *erectus* to the beginnings of *Homo sapiens* times, from something like the last 250,000 years or so, and specifically in the Eemian interglacial (from about 130,000 years ago), and certainly continuing through to the end of the last great glacial period (20,000 to about 6,000 years ago). Climatic changes during these times could happen relatively quickly, over decades rather than millennia (Adams, *et al.*, 1998; Smithsonian, 2013a), and they included such things as cycles of encroaching and retreating seas, growing and diminishing glacial fields, monsoon conditions alternating with droughts, and the coming and going of deserts, forests, rivers and lakes. It is not out of the question to assume that hunters, gatherers and fisher peoples were from time to time driven together along particular rivers or onto a specific lake shore, narrowing land bridges or coastal region, for example. It is also possible that some populations, in some locations, were so successful that population growth itself was a contributing factor to an increased density in those areas.

There might have been a lot of fighting among groups (and this has continued into the anthropological present among numerous hunting/gathering and local farming/fishing groups studied by anthropologists). But there were also opportunities to exchange marriage partners, trade goods, form alliances and, indeed, to cooperate over certain activities. Well into the above period, *Homo sapiens* became firmly established, and by the end agriculture had been invented (the Neolithic revolution), and soon large populations were living together and building fortifications around densely populated settlements to keep intruders at bay, or to build water systems to support the enclosed populations and sustain their gardens. History suggests there may have been a lot of bloodshed along the way, but this path has had sufficient historical continuity to have wiped out most of the world's hunting and gathering peoples in the process, and even for many individuals to have come to *want* to live in metropolitan conditions – because that is where parents cannot keep close track of one, where all the nice looking members of the opposite sex are showing off, where booze and sex are on sale, and where there are even some exciting economic opportunities.

So, by remaining relatively expressionless, and by not looking directly into the eyes of the hundreds of people who will pass me in the course of a couple of hours (being fully dressed in roughly the same manner as others are dressed), I can walk down the street of a busy city and avoid all the others. If, however, I run into a colleague from work with whom I do not have a close personal relationship, I will, nevertheless, say something like, 'Hi, nice to see you, crowded isn't it? I'll send you those papers on Monday'. I thus maintain a social link of a specific kind. And my language has implied a contract into the future. I could have said, 'I sent you the papers last week', implying an understanding of a past obligation, or I could have said 'I'll do my very best to get them to you on Monday, but it might be Wednesday because my grandmother could die this weekend'. We have here an unspecified future but a contingency contract to deal with it nevertheless. I can say to the local handyman: 'nice morning. How are you? Can you come round and fix my broken window?', without promising him any emotional part of myself, I am nevertheless hoping to get something from him.

It is extremely difficult to communicate with non-verbal signals, primate hooting, Army Creole or early play-seductive language in a way so that the above types of links can be established and maintained. Our non-verbal, emotionally charged communications tend to elicit play, sex, parental love, aggression, fight, flight, submission or complete physical avoidance. Army Creole may not even take us that far! Early play, seductive, gossiping, teasing, joking language is much more suited to communications in relationships that also have emotional links (or potential emotional links). But with the handyman I want an outsider to help me in my daily activities. I want him to be part of my safe, distant-other, universe of humans. By keeping intonation and non-verbal signals to a minimum, and, further, by using written language (for example, written instructions for the window work and he in turn provides a written estimate and later invoice), communication is turned into a relatively unemotional (not charged) means of agreeing to keep one's distance, to follow generally accepted rules, to take turns, or indeed even to cooperate over specific issues, without entering into potentially dangerous-to-self emotional relationships.

I will not be telling the handyman or my colleague any family secrets (nor probe them for theirs), nor ask them for a loan, but I will enter into an economic or social exchange with them, and may well buy them a drink at the pub, and be in the same Christmas pantomime with them, and my child may get excited by and consequently marry the son or daughter of one of them, opening the possibility of a new type of link with them. I, and my children, can do all of this because we speak enough of the same language to communicate in relatively formal terms, enough to make some deals. I do not have to learn handyman jargon, stereotypes or in-jokes. I do not really care if the handyman sends a reliable competent assistant so long as they do a proper job. As far as I am concerned all handypersons are basically one thing, 'a handyman'. Nevertheless, the language I share with the handyman and my colleague contains certain implied *guarantees* that we will stick by our word; we have given our undertaking, we will live up to our part of the bargain; we have pride, we want to be known as being honest.

In this perspective, non-emotional ('rational') language evolved because to a significant extent it keeps people *apart* emotionally rather than because it brings humans together (opposite to what most theories of human language imply or state its purpose to be). But there are not two languages in opposition to each other, emotional (fighting/binding/loving) and non-emotional (separating/dealing/contracting); from relatively early on the two developments have been roughly integrated into the modern languages of today – the extremes of each only used in specific situations. Remember the example of the use of non-emotional 'small talk' from aspiring males to create an image of being non-dangerous and interesting, and to avoid interference from jealous, higher status males when attempting to entice a particular female. Having gained the attention of a particular female, however, he may revert to emotional language and begin to flatter her and use a bit of 'baby-talk' and start to make silly comments.

He may say he loved her because he was desperate for sex. And if she responded as if she believed it (even if she did not altogether) he may even come to think it to be true because *he did say it* and she *did respond* as if it was true; and she may then come to believe he means it because he really does not deny it and he does seem to be 'genuine'. Deceit sows the seeds of self-deceit, and if a dash of sexual satisfaction, feelings of guilt and growing friendship were to ensue, this exchange may lead to a mating union in which many children are born and a couple come to believe that they are a happy couple – not least because they keep telling everyone they are, and others say they are.

Courtship may be full of giggling, deceit and self-deceit, but at the same time non-emotional language will help individuals 'find out' about each other, and to even broach a post-matrimonial division of labour and the amount of 'mothering/fathering' expected. Both may want a written marriage contract (official wedding) with a very definite set of obligations of care placed on each other, including what happens if it does not work out. (The ultimate use of non-emotional language in this regard, I suppose, is the pre-nuptial agreement.) If a couple seem to be too much in love to worry about these things, their parents may well see to it that some formal words with legal meaning and backup are employed. The same processes apply to non-sexual and/or not overtly reproductive relationships, such as in business or politics. A link to powerful and potentially dangerous individuals can often be achieved through, for example, pledges, oaths and the swearing of allegiances but, to make sure, written contracts, bills of sale, laws and constitutions with their extremely unemotional (and relatively unambiguous) language come to be desired.

And castes of individuals to do the writing and to understand the writing of others soon result (scribes, lawyers). And when some individual or group has the job of elaborating and specifying, elaborating and specifying will be done. Written language has not reduced the human tendency to amplify, elaborate and exaggerate beyond basics, indeed, this was clear early on. Of the 2,500 signs in the ancient Egyptian corpus, for example, 'only a small percentage of these were in common use'. And in 1717 the Kang Hsi dictionary in China had 40,000 characters of which 34,000 have been described as being 'monstrosities and useless doubles, generated by ingenious scholars' (Coe, 2012, pp. 35, 30). Additionally, specialized vocabularies, correct pronunciation and spelling increasingly became important. To be successful one has to play the 'proper-use-of-language' game. If complex language is used at court, so we must master similar language usage to show our loyalty and worthiness of courtly attention. It is not easy to separate emotional and non-emotional language in these examples; certainly, the participants will not usually have done so.

One significant result of modern (integrated) language is the ability for attaching social meaning and value to objects, individuals and groups through naming (Harman, 1988; Dunbar, 1996; Kapuscinski, 2008). We attribute nicknames, titles and role designations to individuals to give them personal and social identities. We also stereotype (name) conglomerations to give a multitude of activities, peoples and things *individual* identities in our own minds. Teachers, for example, are not seen as a vast multitude of personalities, races, ethnicities, with a wide variety of beliefs, emotions, problems, joys and experiences, and with a number of motives, hopes and aspirations, operating in various parts of a society, but as *one participant* in our 'mental map': *Teachers*. This makes it much easier to comprehend our world and to have a feel for it in our own terms, and, of course, to communicate about it. This use of non-emotional language facilitates distinctions between 'us' and 'them', or at least between 'us' and 'outsiders'. So, we, our family members, our friends, some more distant relatives and a few acquaintances make up our band, but our band, in our cognitive map, has linguistically maintained linkages with teachers, bosses, doctors, handymen and pub landlords (for example), while the world of 'named' factories, supermarkets, hospitals and schools (for example) make up the *environment* in which our band operates.

In this use of language *integration* of old/emotional and new/'rational' languages is also clearly evident. Naming designates relationships but also expected closeness/distances among them – mother and son, boss and worker, insider, outsider, for example. Distancing and separating, however, are only effective when words and the way they are expressed have *emotional* force. So, particular types of individuals can be designated as *one entity* by language and then linked in our dialogue and minds with a particular activity or thing that has emotional connotations to indicate the social worth of the designated group. For example, we give disparate individuals doing a particular medical job one name, *nurses*, and equate all nurses with care giving/protecting behaviour, and we address them as *sister*. Or individuals/groups may be associated with body parts or activities we find disgusting and repulsive (assholes/shit), and so we look down on them and stay away from them for fear of contamination. Gangsters can be associated with jackals disembowelling an 'innocent' antelope to emotionally charge the word designating them. On the other hand, we may associate some individuals (our favourite political leaders, for example) with mythological heroic acts or wise decisions (as having 'the wisdom of Solomon'). Classifying humans analytically is one thing, doing so emotionally is quite another, and it is this second element that remains extremely significant in modern language usage and in its relationship to social processes, regardless of how complex language may have become in some instances.

But, given the extreme mutability of human languages, and the role of emotions, desires and fears in human motivations, even this integrated communication capacity has an outer limit beyond which it seems unable to overcome species-typical propensities for small ingroup friendship, pair-bonding, band living, and a fear and distrust of outsiders. And so new 'dialectical groups' constantly emerge. (This has been the case at least until the advent of modern communications and transportation technologies.) In hunting and gathering societies, for example, people live in bands of somewhere between 10 and 50 individuals. These bands are linked within what have been called 'dialectical groups', incorporating somewhere between 500 and 1,500 individuals (Birdsell, 1968, 1973; see also, Binford and Binford, 1966, 1968a; Dunbar, 1996; Dixson, 1998). Bands come together within the confines of their shared dialect, their mobility on foot and their general environmental conditions, for exchanging gifts and other materials, and for celebrating and finding mates. Rarely do any of these activities take place outside a dialectical group. As in many primate species there is considerable fission/fusion with regard to the regular formation and reformation of bands within dialectical groups (with the specific individuals that make up a band also changing over time).

It can be argued that in these societies relations between members of different bands are very similar to my relationships with the handyman and my colleague in the hypothetical examples above. An individual may know that members of the Willow-Grove band speak the same basic language as they do, and that it is permissible to marry one of them, and if you need a really good arrow they are the ones to see, and that their recognized elder is called Bending Tree, but that is about it. All members of the Willow-Grove band are one social unit, 'the Willows'. An individual will also know, however, that the Willows will show up at the spring dance and that they might have some juicy gossip to spread or willow bark to chew and that one or two of them are not bad looking. It may be possible to 'talk them out of something', to make some deals, even to start a flirtation.

Throughout history it has been very difficult for humans to avoid the continuous generation of new dialectical groups and to prevent social splitting. Ancient empires often worked very hard to impose an official language (and gods, religions, uniform rules, laws and cultural values) on conquered people – and their own expanding populations – with mixed success over both the short and long historical terms. When the vast patronage networks and military power of the founders weakened and/or evaporated, a multitude of *ethnicities* (reverting back to their own dialects) sprang up relatively quickly. It has only been with the evolution of the institutions of the modern political state, and with modern communications and transportation technologies, that any degree of larger-than-dialectical-group political unity has been achieved over most of the Earth's surface. Even then the results have been very mixed. Many modern states, for example, have within them a multitude of languages, religions and sub-cultures simmering away, often resulting in competition, conflict, ethnic cleansing and vengeance feuding, especially when a central authority is weak or disintegrates.

And even in more stable states there is a strong tendency to 'organize' into 'tribes', ghettoes, extended families, friendship networks, religious parishes and cults, work places, trade union branches, criminal gangs, clubs, trade associations, community forums and special interest movements. These, of course, operate under the umbrella of local and national constitutions and authorities. These local tribes to some extent develop their own dialects in the form of jargon, in-jokes, special greetings, technical words and specific written and unwritten procedures, codes of behaviour, status designations and authority commands. It is arguable that humans derive most of their emotional satisfaction from their families, friends and close work mates, and a sense of security – and some emotional rewards – within the confines of these tribes, whose dialect they understand and feel safe using, and which give them a sense of belonging to the world in general.

Nevertheless, language (especially non-emotional language) remains one method of crossing the barriers between dialectical groups. Indeed, evidence of a single human nature is strongly suggested by the fact that all spoken languages can be translated into each other regardless of the amount of geographical, historical and political separation that has taken place (Cf., Pinker, 1995; Coe, 2012). As such, language – in its linking capacity – has been extremely important in the evolution of human civilizations. It is unlikely that any type of human social organization for a population much larger than a basic group of between 5 and 40 individuals would be possible without it. Additionally, the ability to live in extremely dense population concentrations is inconceivable unless typical primate non-verbal signals have been repressed and/or rendered more or less under conscious control, and replaced with a relatively easy-to-control non-emotional form of communication, a form of communication which helps keep a psychological distance among strangers but, nevertheless, can be used to 'do deals'.

We should not take this to mean that language is the driving force behind the evolution of civilizations. Language was one of the things that made larger scale social networking possible; it was not the fundamental cause of it. The driving force of human civilization has been (and is) basic human nature with its pre-knowledge, bioelectric (emotional) charging and its standard cognitive tendencies. These lead humans to: compete for mates, try to seduce 'friends', try to monopolize mates, show-off, fear strangers, form friendships, make deals, join gangs, seek vengeance, defend a home base and manipulate each other in pursuit of these ends. Eventually these resulted in tendencies to stereotype the world, and also to elaborate various things and activities into careers, artistic displays, ceremonies, palaces, temples and fatherlands. And even further, there has been a tendency to spiritualize these as representing a higher *purpose* and *progress* and to submit to and worship these ideals as the *reasons* for human existence. Language has been one of the more useful tools in this process, but not the driving force.

And indeed, in terms of personal life, particular tendencies with regard to lust, jealousy, envy, love, hate, shame, embarrassment, guilt, self-image and self-esteem are little affected by the symbolic meanings language deals in. To tell a deeply depressed person, for example, 'that's life, you are fine, you are better off without her' – even when the statement is based on the most solid of empirical evidence – does not *cause* 'fine' or 'better off' in the life of the depressed person. The sublimation of sexuality into love, jealousy, hate, a desire for vengeance and politics predates language. But language makes it possible to tell good lies, and to make up good stories, to gossip and to accuse, to talk about good and evil, to communicate taboos, rules and boundaries. As such, language, like status and politics, makes it possible for humans to live and let live (sometimes) in conglomerations of individuals, families, groups and communities on a scale unimaginable among any other social primate.

Human sociability

Humans clearly are a social species. We do not usually choose to live alone and be without friends or acquaintances. We want to be with, or at least see on a regular basis, some combination of our sexual partner(s), offspring, parents and a selected few friends and acquaintances. The most fundamental basis of human social behaviour is the evolution of human nature-based propensities to lust after and fall in love with certain others, to pair bond and provide offspring care together, to form bonds with siblings and parents (ambiguous as these sometimes are), to play with other children, to form powerful friendships with a selected few playmates (and to maintain these after an individual goes off into a monogamous mating) and to even make a few more friends along the way.

We have a clear capacity to *emotionally interact* with a small number of individuals in a variety of situations, but the individuals involved may not be the same in all contexts. For example, we can have an intense emotional interaction with – and attachment to – our team mates during a game but then switch that to our families when we go home. Moreover, we have an emotional tendency to re-jig even the more long-standing arrangements when things go sour or circumstances change (as they often do). Given human mating competitions and the evolution of self-awareness, these processes entail somewhat more complex patterns of individual dependencies, creative subordination and interdependencies than seen in chimps and bonobos (or any other primate species), and they certainly are more flexible.

When compared with apes we have seen a repression of automatically triggered lust and aggression, in part through the evolution of bonding emotions such as love, shame, guilt and trust. But these are not always reliable, so varying degrees of controlled hate, jealousy, anger and envy are common emotional motivators which help us keep an eye on others. Deceit is not unheard of in friendship, mate and parental relationships; subtle competitions among males, females, parents, offspring, friends, alliance mates and patrons abound; as a result of the escalating nature of the desire and fear processes, enough sex, love, attention and security are, it appears, almost always scarce 'resources'. As a consequence of self-awareness and of these 'social dynamics' we have evolved a significant capacity to dream, daydream, fantasize, think and learn a bundle of *expectations* for ourselves (and for our loved ones), and to be able to reform and alter them as we go along in life. We also have considerable skills and abilities for making, and formalizing – often in a hierarchical manner – social *evaluations* so that activities and relationships are entered into with a degree of forewarning (Cf., Buss, 1994; Whiten, 1997; Vannelli, 2001).

On the basis of these evaluations (often unconsciously made), individuals commonly submit themselves to their friends, siblings, mates, parents and colleagues in order to receive love, care, attention and sex from them. Such *creative dependency* and attempts through evaluations to 'exercise choice' are in a real sense attempts by individuals (consciously or unconsciously) to make safe their own circumstances, and/or 'to get something'. This may or may not involve attempting to directly control the behaviour of others. The social expression of this 'will to power' (Cf., Nietzsche, 1968, 1990) is politics – politics with both a small p and a large P. When this is done in the context of shared living conditions, individuals come to share personal and social evaluations, ways of speaking, certain fantasies and a number of desires and fears, all of which generates the appearance of an 'us group'. It feels safe to be with and to play with individuals from within this group. A significant number of stereotyped social evaluations of outside individuals and groups also come to be shared. When 'us-groups' desire or are forced to join up with others, it requires some form of *negotiated linkages* and/or *contracts* between 'us-groups', very clearly moving politics from the small p to the large P. I will start at the interpersonal end and work up to larger scale politics, considering how all this might work.

Human groups

As we saw in the section on language, there is evidence that in hunting and gathering band societies (at least those studied in modern times) there has been a tendency for bands to remain about the same size. They appear to vacillate between twenty-five and fifty individuals regardless of a number of variations in ecological, technological and economic conditions (Birdsell, 1968, 1953; Dunbar, 1993a, 1996, 2002; Dixson, 1998; Johnson and Earle, 2000; also, Binford and Binford, 1968; Clark, 1968; Hrdy, 2011). Specifically, band size appears to be only vaguely related to the amount of food available in given environments, or to the efficiency of tool use (Washburn, 1968a; see also, Lee and DeVore, 1968; Steward, 1968; Dunbar, 2002). As ecological conditions improve, or become more difficult, groups within this numerical range generally move closer together, or further apart, rather than join together. There is, however, an evolutionary retention of many aspects of the fission/fusion process we saw to be so common in primates (including among chimps and bonobos). In human band societies, when group numbers drop below approximately twenty-five on a long-term basis, there seems to be a tendency for amalgamation (fusion) of groups, or of one taking over another; and when groups increase in membership size above approximately fifty, tendencies for splitting (fission) increase – sometimes, but not always, with considerable aggression and conflict beforehand (Cf., Marwick, 1965; Chagnon, 1979; Boehm, 1997).

Within a relatively stable band it is not uncommon for small family groups to wander off for a time and rejoin the band later. This is especially the case where resources are spread very far apart (Cf., Service, 1963; Lee and DeVore, 1968; Steward, 1968; Hrdy, 2011). It was also noted in the section on language that human foraging bands do enter into 'larger than band' social relations determined by 'competence in speech and mobility on foot' called 'dialectical tribes' (Birdsell, 1968, p. 232, 1973). The numbers which make up dialectical tribes also seem to have a degree of consistency: they, as noted, comprise somewhere between 500 and 1,500 individuals (Birdsell, 1973; see also, Washburn and Lancaster, 1968; Steward, 1968; Dunbar, 1993a, 1996, 2002). It seems that various local bands aggregate into very loose federations and as they do so they come to share a common dialect that is unique to their 'tribe' (federation). They usually give themselves, as a federation, the name *humans* and only exchange marriage partners within 'the tribe of humans'. Dialectical tribes are generally more than linguistic formations and arenas for mate exchanges. Usually they also contain means (patrons, respected elders, rituals, taboos, shared rationalizations, customs and traditions) for arbitrating, and even controlling, conflicts as well as status and power demarcations, among bands. Although in no way can dialectical tribes be considered corporate groups, members do designate the outer edges of their tribe as being a boundary with social significance.

Do the above numbers for bands and for dialectical tribes represent something that can be considered to be 'natural human groupings', or are they something that possibly evolved with humans and persisted through foraging times but then lost relevance for our understanding of human social organization? A consideration of this issue can become rather complicated. To start with, it has to be noted that despite considerable general agreement on these numbers, there remains controversy over their importance. Not everyone accepts the significance of the range 25–50 that a consideration of band societies suggests. Robin Dunbar (1992, 1996, 2002) for example, argues for a 'basic size' of 150 individuals for humans; and the significance of this figure has had a considerable degree of acceptance (in and outside anthropology). This is considered a 'basic size' in so far as it is argued to be the largest group with which a 'set of individuals who have strong affiliative relationships with one another, who know and interact with one another on a more or less daily basis, and who maintain some degree of special and social coherence through time' can exist (Dunbar, 2002, p. 181). This is because it is the cognitive – information processing – limit in which individuals can monitor each other sufficiently to have a stable society, one 'clearly integrated in structural terms' (Dunbar, 1992, p. 490).

This number, he suggests, is the average size of hunting and gathering *clans*, four generations of ancestors, early agricultural villages, military companies, Church of England congregations, typical Christmas card networks and small businesses based on a free flow of information. His conclusion of 150 as a fundamental human group size, however, is largely derived from an attempt to plot mean group size against neocortex/brain/body ratios among primates, including ancient hominins through to modern humans. However, mean group sizes range from one individual to fifty among simians and the 'sizes for monkey and ape groups overlap completely' (Byrne, 2002, p. 171; see also, Dixson, 1998; Dunbar, 2002, p. 180, see also, notes pp. 269–270; Wrangham, 2009). In fact, for any correlation to hold at all, New World monkeys, gibbons, gorillas, orangs and almost all prosimians have to be left out. In using an average of genera (or just one species per genera) a vast amount of variation is ignored. Indeed, when proponents look for examples, the 150 is derived as a rough average from a range of 100 to 250 individuals in various groupings.

In general terms of primate evolution the trend appears, in fact, to have been in the downward direction of group size in relationship to overall brain size and the evolution of advanced neocortical functions – such as information processing, tool use, communication complexity, self-awareness, theory of mind, social manipulating and social politics. This is indicated by large overall brains, neocortex complexity and social intelligence among orangs, chimpanzees, bonobos (and modern foraging *Homo sapiens*) living in smaller groups than seen in a number of small-brained, with relatively weaker cortical capacities, terrestrial monkey and prosimian species (Dixson, 1998; Roth and Dicke, 2005). The contrast between a number of relatively very large monkey groups and more or less isolated large-brained orangs is great.

But the fundamental issue with the above approach, I would argue, is its rationalistic bias. The assumptions are that information processing for the purposes of generating a *stable* and *well integrated* social 'structure' has been the driving force behind human evolution; that for social life to be possible (and congenial) everyone has to interact with everyone else on a relatively regular basis and that somehow there is just the right size for a species when it comes to social group size. To be sure, Dunbar does not explicitly argue this last point, but was searching for a correlation between group size and a neocortex brain ratio, and making an argument for the importance of social intelligence in human evolution. But with the emphasis on cognitive efficiency, social stability and the importance of interacting with others, and claiming that his use of genera averages is justified because it eliminates 'species-by-species ... biases' thus exposing 'the factors that drive evolutionary change' (Dunbar, 1992, p. 477), it seems like it. The 150 has been taken up by businesses, management consultants, social planners and such like as if a magic number had been discovered.

There is, however, no compelling reason to believe that humans find something around 150 people the most congenial (or efficient) group in which to live, other than the idea that many modern (city) people assume that humans *naturally* want to live in 'a community' of about 150 people because everyone would 'know each other' and help and support each other ('a nice English village'). But it is doubtful that there is a village of 150 people anywhere on earth where everyone actually 'knows everyone' – which is different from knowing who some close non-associates are and 'what they are up to'. Can anyone of the 150 ask any other one for a loan? Can, or will, anyone cry and hug anyone else when in a state of personal despair? Will everyone lie for every other one? Will everyone tell every other one about the details of an argument they have just had with their lover? Will everyone argue with and shout at any other one and then cry and make up? Indeed, it is doubtful that there are more than a very few individuals on earth who can do any of the above with anything like 25 individuals let alone 150. Many within a village of 150 will probably hate each other, avoid each other, be afraid of some, disdain some and be indifferent to most, and not 'have strong affiliative relationships with one another ... know and interact with one another on a more or less daily basis'.

Without some combination of lovers, parents, friends, offspring and alliance mates, life for the vast majority of humans is not worth living, and the fate of a larger social grouping is not even a consideration. To be sure, 'villages' (or similar sized groupings) can be market places, locations for alcohol consumption and barn dances, contexts in which to find mates, places for sports contests, religious centres, commuter dormitories and cross-roads where people meet to achieve some of the above, but that does not make a village of 150 or so people a natural human grouping. Whatever the significance of something around 150 individuals for specific human social organizations (in some particular circumstances), the number seems far too large to be the size of a 'natural group' on the basis of being a grouping that naturally forms and consequently keeps reappearing in all types of human societies – regardless of the various functions it might serve in different societies. I suggest that to be considered a 'natural human group' individuals have to be *emotionally engaged* with each other.

Additionally, group size is not necessarily an indicator of social complexity (witness herd animals or baboons, for example, large groups primarily based on instincts). Even with the idea of 'predation risk for most primates' as the main benefit of larger groups (Dunbar, 2002, p. 184), the evidence is far from clear. Group size as a protection against predation seems to hold best when we consider relatively small-brained terrestrial baboons, and still better for even smaller-brained herd mammals, who bunch up on the basis of instincts not burdened by emotional or cognitive considerations. Even if humans moved into more open savannahs early in their evolution (which is highly unlikely – see Chapter 2) we might even have expected a decrease in neocortex size from that of apes so that too much 'thinking', or even emotional conflict, did not confuse herd-like instincts and/or the male aggressive instincts that tend to keep females and offspring herded together in tight bunches, and to keep intruders (especially predators and/or infanticide oriented males) at bay.

Moreover, given the ape/human starting point in terms of diet and bipedal structure, there is considerable reason for groups to have been small (certainly smaller than, say, among deep forest-living bonobos) in order to cover the distances necessary to obtain sustenance, as clearly seen in modern human foraging bands, where it has been noted that: 'The necessity to travel long distances, the high frequency of moves, and the maintenance of populations at low densities are ... features commonly associated with the hunting and gathering way of life' (Lee, 1968, p. 35). In hunting and gathering societies families, or sometimes slightly extended families, often travel together for days gathering and hunting small game before coming back to join other such families to make up what has come to be called band societies. Or small groups of men go off on a hunt while small groups of women forage before returning to night camp to join with members of their band. This is especially the case when foraging and hunting is done in semi-deserts, scrub lands or open prairies (Cf., Service, 1963; Lee and DeVore, 1968; Steward, 1968; Woodburn, 1982; Hrdy, 2011).

As evolving humans began to follow the retreat of the ice north, when the neocortex clearly began to enlarge (at about late *Homo erectus* times), it is still not evident that larger groups would have had an advantage over smaller, more mobile ones given the increased scarcity of vegetation and game (again this is supported by the small group behaviour of modern band societies living in colder, more mountainous conditions). Large-scale human societies emerged well after the onset of *Homo sapiens* and did so in river deltas, river valleys and inter-river flood plains, not on savannahs.

In the absence of convincing evidence of a correlation between the development of a relatively larger human neocortex and group size/complexity, it has been argued that at the level of monkeys and apes *overall* brain growth was more important than specific neocortex enlargement in the evolution of primate intelligence. With the higher primates, the amount of deception practised, and being able to understand it, and successfully react to it, has been argued to be the most reliable predictor of neocortex size (Byrne, 2002; see also, Byrne and Whiten, 1988). Also, it has been shown that in primates the practice of monogamy (as opposed to multi-mating, sperm competition and promiscuity) is better correlated with a relatively large neocortex than a larger neocortex is with general social complexity (Schillaci, 2006), suggesting just how much social interaction, manipulation and personal politics are involved with monogamy.

Additionally, there is evidence that suggests that a 'slow' life history is much more closely correlated to an increased brain/body ratio than the size of a social group. This is because a slowed life history suggests a lot of 'social strategizing' about sex (Schaik and Deaner, 2003), and increased social communication (Tyack, 2003). Certainly, a slow life history and big brains with good memories are found in modern humans, apes, bottlenose dolphins, tooth whales, pilot whales, sperm whales and elephants, all with relatively long lives and complex social interactions among individuals (selected papers de Waal and Tyack, 2003).

If a slow life history is related to a growth in brain size, it fits with the pattern of neoteny, paedomorphism and an extension of childhood and juvenile-adulthood (and with the resultant mating patterns suggested for human evolution in Chapter 3). It seems likely that playing during an extended childhood, leading later on to courting and trying to entice mates to help in childrearing; deceiving and manipulating competitors, possible mates and friends (for example) through emotional and cognitive control of sexual and aggressive drives; making relatively reliable evaluations of social others (mates, friends and potential allies); communicating imaginatively and emotionally; and entering into creative dependency relationships, would all have selected for increased emotional, interpersonal, creative and memory intelligences. There is, thus, a real possibility that (given the starting point of human evolution) the basis of a larger brain, including neocortex, was selected for as a product of *interpersonal* interactions in the context of *small* group living, given that all of the above behaviours are deeply reliant on *emotionality*, which in turn requires personal contact and operates over a very short range.

Even when an argument is made that a larger neocortex could have resulted in a greater and greater conscious/'rational' repression and control of certain instinctive reactions, and that this has contributed to the selection for new types of 'positive' emotional responses (such as romantic love, sympathy, empathy and trust), in all of these *face-to-face* contact remains an essential ingredient. And theoretically, in larger groups, a much larger neocortex with more analytical capabilities could have been a positive disadvantage. Based on too much 'analysis' it can be all too easy to imagine and visualize conspiracies and infidelities, and to remember, think too much about and exaggerate humiliations, antagonisms, envies and hatreds. It might have been too easy to categorize some individuals, or groups of individuals, as being too different, too conspiratorial, too unintelligent, too irrational to play with or mate with. Indeed, a well-developed human cognitive capacity to classify (stereotype) has tended to result in 'strangers' being presented as being dangerous – almost non-human – throughout most of human history, to be avoided, if not destroyed, rather than to join or interact with. Elements of cognitive/analytical intelligence have been employed to organize ethnic cleansing, develop very elaborate fortifications to keep strangers out and/or to conquer them before they do us damage.

Nevertheless, the idea that human evolution resulted in individuals becoming more self-oriented (self-aware) and motivated to selective pair bonding and to favouring small-group living, is certainly in conflict with the general acceptance within social evolutionary/historical thinking that larger and larger groups *must* have been an advantage for humans because it does seem to be a 'rational' development – what with a larger group's potential for a complex division of labour, considerable specializations, 'the sharing of ideas' and increased military power, for example. Besides, modern humans clearly do seem to have 'progressed' from bands to city states to empires to political states (with millions of people in the latter). But this 'rational' explanation puts the cart before the horse. Assumed beneficial outcomes and specific historical outcomes are treated as original causes of both the benefits and the outcomes.

There is no independent basis for the existence of larger groupings in this approach, nor any *evolutionary* reason why motivations for such might have existed. Our explanations of human groupings cannot be based on rationalistic propositions about what humans 'need' in order to achieve larger and larger scale societies because of the presumed benefits of larger societies; or even on a mythological image of a village 'it would be nice to live in'. Large scale groupings must be explained in terms of a 'real' human nature and its natural grouping tendencies as manifest in the contingencies of historical unfolding. I have suggested that anything approximating a natural human group must be based on emotional interactions, interlinked with species-typical cognitive tendencies, and that such groupings are likely to appear and reappear throughout human history, regardless of cultural, technological and political intermixing which result in somewhat different forms. I will now consider how this might work.

In emotionally interactive terms it is very unlikely that any human being gets anywhere near to feeling emotionally close to and fully trusting 150 people – even Birdsell's 25–50 is stretching it because very few individuals can reach 25 in either of these regards. Indeed, Dunbar (1992, 1996, 2003a) recognizes the existence, and even importance for humans, of 'sympathy groups' of about 11–12 individuals, and that larger groups are created by 'hierarchical clustering of smaller cliques', these cliques sometimes being referred to as 'tightly bonded grooming cliques' (1992, pp. 485, 468). Others have suggested the importance of 'sympathy groups' within the range of 10–15 individuals (Buys and Larson, 1979; Buys, 1992), while values as low as 7 (Hays and Oxley, 1986) and as high as 15–20 individuals (McCannell, 1988; Rands, 1988) have also been suggested. And although (as Dunbar notes) military companies are made up of around 150 individuals (although this can vary a lot), divided into platoons of something between 25 and 50 individuals, it must be recognized that these are broken into much smaller combat units, or squads, of 6–10 members in order to generate combat vigour by developing 'a willingness to die for one's comrades' and 'vengeance mentalities' through 'the buddy system'. It is this factor above all that has made militaries effective. This military pattern seems to have been relatively common from at least Roman times up through modern times (Marshall, 1947; Stouffer, 1965; Moran, 2006; Thor, 2011; Wikipedia, 2014).

Sports teams are almost always made up of between five and twelve members. Alfred P. Sloan (1972), creator of General Motors Corporation, argued that to be effective no manager should have more than five people responsible to him. Committees seem to work best with six members (Dunbar, 1996). Irving Janis' (1972) study of 'groupthink' political decision-making suggested that most major political decisions, in modern times at least, have been made by small, very cohesive groups of between five and fifteen individuals who operate largely in secret. There is considerable evidence to support this contention from across a number of modern societies (Cf., Michels, 1962; Halberstam, 1969; Stephan, 1971; Halperin, 1974; Khrushchev, 1977). As noted, in hunting and gathering societies even when bands of twenty-five or fifty members are recognized they do not necessarily have daily significance in the lives of individuals, it not being uncommon in such societies for small family units, or a couple of small family units, to wander off together and only join the band from time to time, making five to ten individuals the effective offspring care and economic/social unit most of the time for most individuals. (The pattern of coming and going, as among bonobos, tends to be in mixed sex groups among humans – Dixson, 1998.) And it is also usual in more settled human hunting and gathering, and early agricultural societies, for families to maintain their own households even when in the context of a band, 'extended family' or tribe (Cf., Evans-Pritchard, 1937, 1965; Service, 1963; Laslett, 1983; Hrdy, 2011). And, of course, in modern industrial societies it is common for a family to be made up of four to eight individuals.

When it comes to emotional bonding and psychological effectiveness, therefore, it seems that a group of something like five to twelve individuals in a 'sympathetic', or 'quasi-sympathetic', relationship makes considerable sense as the basic human social grouping. This would include an individual's five to twelve person close friendship network common in human societies; households and nuclear families (sometimes with odd relatives or friends hanging on); five to ten alliance/hunting/vengeance-seeking mates in hunting and gathering societies; combat squads of six to ten members as seen from the beginnings of organized militaries; the ubiquity of five to twelve members making up sports teams around the world; the core membership of street gangs. It would include Sloan's five subordinates to a manager ('the management team'), Dunbar's six members of a committee within modern corporations and Janis' groupthink political decision-makers of five to ten individuals.

In these groupings self is often on the line and the only protection is one's sympathy network, caught up in the same perceived dangers. So, members tend to be extremely loyal to each other, conspire together to succeed and conspire to hide mistakes; differences of personality are ignored, small quibbles forgotten; the goal is to protect each other, look after each other and make excuses together. When interacting with individuals of our sympathy groupings we usually defer gratification and feel proudly in control of our baser emotions as a result; we accept the sexually exclusive pairings that arise among our numbers and we desperately try to control our feelings of envy and jealousy. We lie to each other in order to make each other feel good; to strangers we lie *for* each other. We sacrifice for each other; we develop strong feelings of guilt for letting members of our sympathy groupings down, and for presentations of self and behaviours that undermine the bases of our binding/bonding.

I will generally refer to this type of grouping as a 'sympathy grouping' (Cf., Buys and Larson, 1979; Wispé, 1991; Buys, 1992; Dunbar, 1998). This is not because everyone necessarily feels sympathy with everyone else (although they often do), but that even when they do not, they feel that they *should*, and they are compelled to express it verbally and act in a sympathetic and supportive manner. Acting sympathetically with real people contrasts with the idea that we slavishly follow abstract beliefs or rationalistic cost–benefit propositions as determinants of our social behaviour. Least of all are we motivated by idealistic visions of how humans should or could live (other than for our abstract discussion in the pub, in the academy or on the campaign trail). Indeed, we find it extremely difficult to love strangers or humankind in the abstract. We feel we should despair for victims of a disaster thousands of miles away, but we do not feel the kind of grief we feel when our child is desperately ill; we do not love 'mankind' as we love our children, our sexual partners, our parents or our very best friends; we want all individuals to be equal but we do not give up our standard of living for the mass of people around the world; we do not defer gratification for the good of poor people; we do not loan money to strangers; we do not sacrifice anything significant for strangers; we do not lie for strangers just because they are human (but rather condemn them for lying for each other); we do not worry about the sexual infidelity of strangers (often dismissing them as probably being immoral); we do not seek vengeance for strangers who have been harmed.

Whatever the words applied, the vast majority of humans have very different – more protective, more caring, more sharing – feelings for their mates, children, friends and comrades than they do for God, humankind or world peace. Yet our powers for sympathy make us profess that we love God, humankind and world peace, even more than we love our own selfish desires. And we do feel a tinge of sympathy and a desire for justice when we hear of great misfortunes and what clearly seem to be injustices; and we give some money to charity to make us feel better, and we condemn violent dictators and drug traffickers and child prostitution and multinational corporations and trade unions and world powers. This might be hypocrisy in a linguistic or philosophical sense, and it may be stereotyping and scapegoating in a rationalizing sense, and attention seeking in a political sense, but that is not the point. The point is that this is just another conflict within human nature that arises from a powerful capacity for sympathy, as evolved in small sympathy groups in which sympathy greatly affects behaviour, but it is a capacity which can also be applied linguistically and abstractly to all sorts of peoples, objects and abstractions, although it is rarely acted on emotionally or practically in these instances. This inner conflict and ambiguity is just one more of the characteristics that make us human.

In terms of the significance of sympathy networks for human social order it can be remembered that an individual can belong to more than one sympathy grouping, even within a given day – success in one (like family or among close friends) can compensate for failure in another (losing a football game/being part of an unsuccessful business decision). Although this size grouping tends to emerge in all types of societies, they are not corporate groups with rigid boundaries, nor are they all bonded with the same strength, nor do they remain bonded the same in all circumstances or for all time. For some family means everything, for others it is desired and its loss would be hurtful but in everyday life it is taken for granted as a house and where a spouse and kids are. Groupthink bonding can be relatively weak if there is no opposition, but very strong if a powerful sense of political danger is felt by members. Bonding of a military squad is very strong under fire, not so strong in peacetime.

The significant point is that sympathy groupings tend to be ubiquitous, and have a size limit as a result of the limited capacity of emotionally bonding, but which can be socially/politically quite effective 'social forces' as a result of this bonding. Also the fluidity of membership and variability of bonding strength gives them great flexibility in reacting to social events. It is to be noted that these sympathy groupings are not being considered as structural units of a larger whole but rather as flexible fluid groupings that, through politics (with both a small p and capital P), and status demarcations, make up human societies. This, arguably, represents human nature at the social level, and is therefore the conceptualization we will have to develop in seeking to understand human social/political behaviour. But this approach has tended to be resisted in modern philosophy and the social sciences. Besides not being as 'rationalistic' sounding as practitioners of these disciplines would like to feel we are as humans, it also goes against an accompanying tendency to conceive of human social behaviour as being considerably more systematic than it really is. Sexual conflicts, marriage dynamics, social competitions, deals, tenuous reciprocity, economic conflicts, patronage, ambiguous relationships and politics have been taken out of the equation in these disciplines, being replaced with *systematic* and *structural* concepts.

Producing, creating, trading, buying, wheeling-and-dealing, fraud, working and skiving, tax avoiding have become 'The Economy' in terms of 'laws' such as 'supply and demand', 'diminishing marginal utility' and 'market corrections' in which rational choice-making rules supreme and mathematical formulations can be produced; creativesubordination, domination, patronage, nepotism, status striving, struggling to maintain prestige and social standing (or to gain some) have become 'The Class System', pictured as geological-type layers of humans in a pyramid-like structure, solid as those in Egypt; sexuality and mating behaviour – the battle of the sexes and oedipal struggles – have become 'Family and Kinship' as bedrock institutions providing socialization, the next generation and happy citizens; politics (of all sorts, including its protecting, providing, rationing, arbitrating, judging, but also its ruthless, hypocritical, unforgiving and hyper-adversarial elements) have become 'Government' or 'The State' with constitutions and law codes and charts of the relationship of government agencies to each other, and diagrams showing the flow of bureaucratic procedures for passing laws; and the whole multitude of social interactions within a political boundary are lumped together as 'Society' self-regulating and *sui generis*, like an automated machine with good feedback mechanisms.

As 'rational' systems, these reifications are often treated as relatively permanent structures. Those who do not like a particular 'system' or structure do not argue that this might be the wrong approach to understanding human social/political behaviour, but rather suggest alternative systems/structures – they call for revolutionary change (in structures, not in social science methodology). These approaches tend to be extremely teleological in that systems are set out on the basis of postulated needs of that system itself, as it fulfils the postulated needs of a society (or the domineering part of the society) in which they exist, and for Progress, as an inevitable destiny of humanity, to take place smoothly. However, a consideration of human nature in terms of emotions, desires and fears suggests that a much more fluid approach is required if we are to match behaviour with human nature and understand the underpinning *motivators* of human behaviour. So, with this in mind, how do we move from sympathy groupings and their fluidity to larger scale social formulations?

Besides sympathy groupings of five to twelve individuals, the figure of twenty-five to fifty also seems to have sufficient widespread existence to have some significance; for example as seen in band societies, functioning extended families in certain rural parts of the world (and for a couple of generations in the cities to which families and individuals from rural areas migrate). And in more modern societies most individuals will have a combined total of somewhere between twenty-five to fifty individuals in their sympathy groupings plus weakly known extended family members and significant acquaintances/colleagues. An individual would have an informal, but definite, relationship with most of these, but an emotional one with only a very few.

I think that the two sets of numbers are compatible and that twenty-five to fifty can be considered to be a *semi-natural* network based on a propensity to link sympathy groupings together. Consider, for example, if during human evolution five to ten females, who might or might not have been related, were raised together and hung out together and shared child rearing and 'how-to-keep-your-husband-honest' information, and provided consolation, sympathy and practical help for each other, and each had a mate and two to four surviving children, we have something close to Birdsell's twenty-five to fifty in a functioning band. Now, all the individuals, especially the adult males, did not have to get along (some may even have hated each other), nor did all the children have to like each other. But males could have got along at least cordially, or avoided each other (while maintaining close relationships with their families of origin and childhood friends outside this band) and the females would have imposed peace and discipline among the children. A clear example of apparent altruism among humans, for example, is the constant process of mothers telling their children to behave and share their toys with other children (up to the point they think their child is being bullied, that is).

The same case can be made for a group of five to ten males, who might or might not have been related. They could have made up a very effective hunting and/or vengeance band, bonded and loyal based on emotional ties of friendship, camaraderie and hero worship – although, again, this is likely to have varied among the individuals with regard to each other. Each may have had a female and two to four surviving offspring, resulting in a viable band of twenty-five to fifty individuals. The females would not have needed to get along so long as they all did not hate each other too much, and anyway, shared child rearing activities and difficult husbands would have given them a sense of togetherness that would have served them well enough. And they could have maintained close emotional relationships with their families of origin and childhood friends. Although emotional bonding may be difficult beyond five to ten individuals, a group of twenty-five to fifty humans could, thus, be relatively well linked as a *functioning* group through the sufficiently strong emotional bonding of a very few individuals. It can thus be considered to be a semi-natural human group size based on this bonding and on its widespread existence as bands, acquaintance networks and as functional groupings in many types of societies.

But there is an evolutionary problem with this. It has been suggested that it would take about 100 male/female pairs to maintain, over a significant length of time, a sex ratio of 50/50 (Washburn and Lancaster, 1968), and we know that humans have had considerable reproductive success through monogamy (and remain very monogamous as a species). But this is not a problem if we consider that some marriage partners and friends will have come from other bands and will still have emotional connections there (remember that young females out-migrate for reproductive purposes among chimps and bonobos, and probably continued to do so among early hominins, and hominin males – in some circumstances – may have done the same). And so a link of sorts among a number of bands probably has existed from the start. And, when bands split in the normal fission process it is likely that certain individuals in a leaving band would still have held parental, sibling, friendship and/or romantic feelings towards individuals in the 'home' band.

These factors would have been especially significant when humans evolved a lack of desire to mate with individuals they had grown up and played with, and so their sexual desires became focused on seeking members from outside their band. We cannot discount the sexual pressures that were involved in driving this process given the likelihood of a frequent scarcity of lovers, wives and husbands for the potentially disruptive adolescents, widows and widowers in a band. Elder males about to hit the 'past-it' mark just when their younger wives were being eyed-up by puberty charged-up young men, or wives seeing an enthusiastic widow eye up their hunter hero's catch and biceps may have been especially concerned to keep in contact with sufficient bands so that a few extra males and females would be available most of the time. Some parents might be similarly motivated – parents who would like a son-in-law or daughter-in-law as a sort of client labourer and/or to help them in their declining years, and to provide grandchildren for the same ends, or just for the emotional reward of seeing their own offspring happy and desiring grandchildren because of the human attraction to infants.

So, linkages with a number of outside bands would have been maintained by various combinations of certain marriage partners having parents in other bands, children having grandparents in different bands, brothers, sisters, cousins and very close friends ending up in different bands. It must be noted that this would in most cases have been a very fluid state of affairs. Throughout most of human history marriage was considerably less stable than we think of it today (Ford and Beach, 1951; Murdock, 1957, 1960; Evans-Pritchard, 1965; Beach, 1977; Stone, 1977; Goody 1983; Hrdy, 2011). Older men died and left widows; females died in childbirth; females were captured in vengeance raids but later escaped, or were stolen by competitors; adultery and prostitution were and remain not uncommon; mates got tired of each other and left or changed. Mountain men living among plains/Rocky Mountain Indians in North America, for example, often married Indian women, marriages which 'might last for weeks, months, years, or a lifetime'. Nevertheless, such men often acted as mediators between the two cultures (Utley, 1998, p. 88). Among ordinary people in Europe it was not until about the 1400s that any form of marriage was regulated or protected in any systematic way, and not fully so until the 1700s (Stone, 1977; Aries and Bejin, 1985; see also, Hajnal, 1965; Laslett, 1983). Nevertheless, throughout, monogamy (although often flexible and serial) remained the most common form of household and the basis for child rearing.

While this degree of pair-bonding flux seems unsettling – and undoubtedly was (and is) for many – it provided a basis for flexible reproductive success in the face of difficult life circumstances, sex/age imbalances and infant vulnerability. At the same time it generated a flexibility of social groupings and linkages maintained by individual sexual, friendship and sympathy desires and fears. At the point that 250 or so adult individuals were involved it would have provided monogamy with a balanced sex ratio. But it is likely that the overall 'field' available for mates and friends would have been somewhat larger. Social relations based on these emotional politics could have extended band linkages quite considerably, reaching something like Birdsell's 1500, with a range of 500 to 1500 being a 'dialectical tribe' based on 'competence in speech and mobility on foot' (1968, p. 232, 1973).

It appears that the natural mutability of language resulted in an inability to communicate when the distance and time of separation reached a certain point. It is likely that geography, economic conditions and past histories of peace or conflict may also have been factors – these elements not being unrelated, of course. Fifteen hundred or so individuals, it does seem, can be considered the outer limits for basic emotional bonding and its derivatives (but is a very long way from personally knowing or being sympathetic with most of the individuals in one's own network). Such groupings would be full of stereotypes and other linguistic designations used in forming non-emotional linkages and for comprehending one's own social environment. As a result, it is probable that many people would have desired a degree of 'political order' (and feared a lack of order). So formalized/regulated exchange of mates and gifts among bands came into existence, rituals and taboos were regularized and longer-term deals made; agreements on the definitions of words were sought; a combination of new language with emotional language to attempt to spiritualize (through ritual, ceremonies and abstract symbols) a larger group as an *entity* would likely have existed. Elders as a stereotyped category, and councils of elders, would have come into existence to do the 'judging' and to rule.

When mates were regularly exchanged among particular bands, or bands combined on a regular basis for some other activities, the linkages would take on the appearance of what anthropologists have called 'clans'. These might form into loose federations of clans making up a dialectical tribe. A means of keeping a federation (dialectical tribe) interlinked was to force individuals to only marry *outside* their own clan, sometimes into specific others, perhaps on a rotating basis. This would keep linkages through marriage, in-laws and grandparents alive. This form of clan exogamy, more or less enforced, certainly has been observed in certain areas among currently studied hunting and gathering peoples (Cf., Radcliffe-Brown and Forde, 1950; Radcliffe-Brown, 1952; Levi-Strauss, 1963, 1969; Fox, 1967; Murdock, 1968).

In conclusion, if one had to argue for natural human groups the sizes would be something like: 5–12 (sympathy grouping), 25–50 (bands) and 500–1500 (dialectical tribes). If one were to maintain that these ranges apply in all types of societies the first two would be easier to defend than the last. Sympathy groupings seem to naturally emerge through parenting, play, friendship, lust, family, love, comradeship, a desire for vengeance, war, team play and 'groupthink' decision-making. The second size is seen in band societies for a very large part of human history and as 'personal networks' thereafter (this usually includes a small number of semi-permanent sympathy groupings plus additional relatives, friendly acquaintances, some in-laws, some local club members and a few neighbours). Linkages of bands may have been developed relatively early in the hominin line as a means of young females and/or males out-migrating for mating purposes. For much of human history this linkage may have been little more than the fact that some individuals in one band had grown up in another, or that some bands came together from time to time for a celebration, gossip and some trading. The last size represents a limit beyond which a linkage of bands could not go for much of human history, but which has clearly been overcome in more recent times, through the development of permanent settlements and what came to be considered civil society (for example, small towns, guilds, local religious congregations, local business organizations, corporations, local trade union organizations, professional associations and political parties), and by the further employment of non-emotional language.

We should not expect the figures in this analysis to be exact. Human brains, environmental contingencies, economic conditions and the life and loves and drug habits of individuals make it impossible for human social life to revolve around an exact number of emotional bonds, especially around sympathy groupings with their inherent flexibility/ changeability; human emotional and social life has not evolved to be a mathematically precise process. One of the advantages humans (and most other primates) have is flexibility of mating and social relationships. In the above analysis of twenty-five to fifty people as a band size being derived from five to ten sympathy bonds, even when groups were relatively stable, individuals would have died, children would have grown up and left, the fertility of each female and male would not be exactly the same, adultery would have caused splits, and conflicts over children, envy and jealousy would have generated tension from time to time, all of which would have altered particular living arrangements. So, it would have been strange if emotional networks or band sizes had remained precisely constant over time, certainly not sufficiently so that they could be pinned down to a mathematical formula.

But what this analysis does tell us is that, emotionally and cognitively, humans naturally operate within identifiable ranges of relatively very small group/network sizes, a fact which is basic to human social behaviour. We spend the vast majority of time with our own families, very close friends and alliance mates, numbers that rarely go above ten individuals in any given context, and we rarely get involved in more than two or three (at most) emotional networks at any one time. For individuals living in large concentrations of people, the rest of the world is largely made up of official linkages and stereotypes.

This fits with our evolutionary perspective as developed in Chapter 3. As a reminder, the likely environmental conditions during human evolution (described in Chapters 2 and 3) included terrestrial living in semi tropical conditions, with, in some places, more transitional forest; the existence of an abundance of small streams, rivers, fens, lakes, swamps and islands; relatively easy-to-obtain food, often in trees with orthograde feeding a common practice; the climbing of trees remaining a potential means of escaping predators (or to feel safe at night). Ready possibilities of social isolation were ever present. In these conditions there is no reason to believe that strict kinship or social rigidity, or large tribal groups, would have been an advantage – almost the opposite. It might have been rare that linkages reached more than 250–500 individuals for a long time, and dialectical tribes of 1000 to 1500 may have had to wait for late *Homo erectus* or early *Homo sapiens* times, and then only in some places.

So, how do we get beyond human federations of 500–1500 individuals? To consider this we have to start with certain additional elements of human nature to those involved in binding as discussed above. As noted, *Homo sapiens'* memory capacity and cognitive abilities, along with new language, usually reduce the world of people outside their own band down to a few stereotypes. There is a recognition that, while one cannot deny that there are seemingly endless individuals moving about in the world, one does not generally remember names above about fifty; but this does not matter – a teacher is a teacher, a thief is a thief, a policeman is a policeman, a lawyer is a lawyer, a government is a government (see also, Dunbar, 1993a). In this example, quite a few (thousands) of individuals and activities have been, for all social/political purposes, reduced to five units. In my cognitive map these five, plus me, my wife, and her mother, and my two drinking buddies, and my three kids, and my probation officer, and whoever it is that writes out my unemployment cheque, make up a comprehensible social world. It does not really matter how many people live in my city/country or what their personalities are individually like, or what their current anxieties are, as long as my social world stays more or less intact I feel relatively happy with it.

But this is only possible because we are largely *not afraid* of all the individuals wandering around, or of the stereotyped groups that surround us. We even like to acknowledge them from time to time. For example, when wandering family/friendship groups in foraging bands come back together there are excited greetings, general excitement and considerable playing (as is often the case among chimps and bonobos). And among modern humans who know each other (even a little) there is a strong tendency at a reunion or 'get-together' to exuberantly let others know 'how glad we are to see them'. Perhaps it is from this tendency that we have evolved desires to hold celebrations and parties on a regular basis, activities in which even relative strangers are sometimes greeted with a degree of enthusiasm. This characteristic reminds us that, however emotionally personal, small scale, and even individualistic, human social behaviour often is, we retain our primate heritage of being a *social* species. We like being in the proximity of other people even if they are strangers – so long as they do not give off threatening signals and as long as we can emotionally avoid them. We get a sense of excitement from living in a city; a football stadium full of shouting fans is exhilarating; we do not really mind shopping in a crowded shopping mall or bazaar (for a short time at least); we would much rather go into a relatively busy pub than a completely empty one.

We have developed ritual forms of greetings and formal procedures to indicate that we will not harm visitors during their visit and that they will be safe during any negotiations that take place. Orations, dramatic presentations, games and communal worshipping have been developed which suggest that in this life 'we are all in it together'. These ritual activities transfer the emotional excitement and joy generated by movement, music, spectacle, competition, exciting stories and superhuman efforts onto symbols and abstract notions of unity. So we have official celebrations, ceremonies, emblems, logos, totems, flags, heroes, national holidays and national myths, for example. All of this is possible, and common, in human societies because the approach to mating behaviour and to the forming of social groupings (as discussed in this and Chapter 3) represents a general trend in human evolution away from an 'agonistic' mode of social behaviour to a 'hedonic' one (Cf., Price, 1988; Chance, 1988a; papers in Whiten and Byrne, 1997; also, Hall, 1963).

Hedonic relationships

In an agonistic mode, social relations, hierarchies and group formations are based on 'reactions to' social others. Aggression, bluff and withdrawal provide social space. Attack, flight and reverted escape are major behaviours that maintain social relations, which are often quite hierarchical in nature. Such behaviours are usually based on relatively specific instincts, or at least on fairly powerful bioelectric drives. Hedonic mode relations, on the other hand, are based on an 'awareness of', rather than on a 'reaction to', social others. In a hedonic mode certain social relations are *sought after* because they are, in and of themselves, pleasurable. Behaviours such as love, affection, friendship, altruism and proximity for the sake of it are common in hedonic-motivated social relationships because they provide visceral and emotional rewards rather than activate danger mechanisms.

For example, it has been argued that love, friendship, mutual dependencies and 'charismatic – dependent relationships' are the bases of social binding and bonding in human foraging societies (Power, 1988). Indeed, it is significant that, although members of immediate families often stay together, there is (as among chimpanzees) very little, if any, relationship between wider kinship and membership of sympathy groups. In fact, there is no absolute consistency of sympathy group membership as they assemble and re-assemble on a relatively regular basis (Service, 1963; Leach, 1966; Chagnon, 1981, 1988; Woodburn, 1982; Power, 1988). Marriage in the vast majority of hunting and gathering societies is exogamous monogamy (outside one's own band or clan) in which there is a considerable amount of individual choice of partner (Murdock, 1968). Overall, within each band and sympathy group there seems to be considerable equality of individuals, considerable flexibility of individual movement and a great deal of individualistic behaviour (Woodburn, 1982).

The selection of hedonic motivators seems natural enough given the relatively high degree of reproductive choice for both males and females based on 'playing' and on 'personality characteristics' as postulated in our scenario of human evolution. Such motivators clearly provide a basis for the evolution of copulation as a *reward* for playing and friendship; such motivators fit in with the pattern in which young youth-ape males were able to win acceptance and become desirable through the use of 'come-mother-me' signals; the evolution of hedonic motivators would have encouraged consortships through friendship on a longer-term basis than found among other primates; and finally, hedonic motivators – such as falling in love, desiring to find friends and to be liked – could have gone a long way in motivating mutual child-care activities. The evolution of hedonic motivators fits in with the fact that human groups have tended to be smaller than found in many other primate species. The smaller a group the better hedonic motivators work.

At the same time, another advantage of hedonic motivators is that if groups become too small and in danger of not constituting viable breeding populations (and possibly of being subject to attack from other groups and predators) it is easier to amalgamate groups through hedonic (and cognitive) motivators than through agonistic ones. Human groups tend to amalgamate relatively peacefully, become absorbed or even take over other groups, and through hedonic motivators members of both groups can perceive (rationalize) these processes as sex, friendship and economic opportunities rather than as political or economic surrender or imperialism. Members of most small monkey and ape groups – more agonistically oriented – seem considerably less able to do the same, and small groups are subject to dying out as a result (Lawick-Goodall, 1971; papers in Tuttle, 1975; papers in de Waal, 2002). Also, it is more likely that when normal processes of fission occur inter-band contacts would be maintained through hedonic rather than agonistic type relationships. The hedonic factor might, in fact, have provided one of the initial reasons why humans have had much greater reproductive success than monkey and ape species.

Additionally, it is much more likely that a hedonically oriented social species would take in strangers than agonistic ones. Among baboons an out-migrating male has to fight very hard to be accepted by another group. Often they do not succeed. And if they get in, the new baboon has to climb a very competitive and hostile social hierarchy before being allowed to breed with any of the resident females. Among chimpanzees and bonobos, lone males have very little chance of being allowed into another group. Even females who normally out-migrate and have sex to offer have to be in swelling and be very cautious in their approach to be accepted. Whereas, among humans an odd stray male may be welcomed as a possible friend when a family needs a labourer, or indeed a mate for one of the young females who has no appropriate mate in sight. The same can be said for the acceptance of odd females, as a potential wife, servant or even as a prostitute.

This pattern also includes, of course, the reverse. Some individuals are avoided, shunned and, indeed, *not included* in a particular group because they are disliked. Patterns of avoidance, discrimination and the formation of 'in' and 'out' groups remain part of the foundations of social behaviour with hedonic motivators, but now, however, on the basis of seeking pleasurable relationships and avoiding or rejecting unpleasant ones, and in being willing to 'make deals' even with those not especially liked, rather than on the aggressive behaviours of threat, bluff, counter-bluff, attack, flight and fight. In terms of group behaviour, hedonic motivators provide flexibility because they are based on dynamic emotional commitments that are often in a state of flux rather than on the more static drives or danger instincts common in agonistic-based social relations.

As such, relatively easy reproductive and survival functional adjustments to ecological, economic and social problems are possible, and common. Among humans a wide variety of individuals are able to seek out their own reproductive and survival interests through personal charm, courtships, creative deceit and anti-dominance mini-conspiracies. Most human hierarchies allow for a degree of social mobility without individuals needing to physically fight their way up and are rarely rigid enough to prevent humans from moving from one group to another, forming new groups or even adhering to no group. Among the western plains Sioux, for example, individuals

felt no loyalty to an individual chief, or a village, or even a tribe . . . the Sioux were loyal to themselves, their families and a set of ideas . . . Families and groups were always separating from the village, drifting off to join another band, or establish a new one themselves. Political economy and social secession, in short, caused no trauma (Ambrose, 1975, p. 49; see also, Larson, 1997; Roberts, 1998).

Hedonic motivators (along with stereotyping) help us to ignore strangers by reducing motivations to agonistic encounters, and, with the help of non-emotional language, even to live among some strangers and do deals with them. But hedonic motivators have included another dimension that greatly facilitates living and dealing with strangers, even beyond the limits of a dialectical tribe. This is the hedonic tendency for humans to display *willing subordination* to other humans and quite often to various *categories* of other humans.

Creative subordination

As we have seen in Chapter 3, modern human behaviour includes a great deal of submission to social others – to friends, siblings, mates and parents – in order to receive sex, care, attention, friendship and practical support from them. It was noted that such submission requires that an individual wishing to submit sends two kinds of signals. The first is that self clearly considers 'itself' willing to subordinate to the other – at least in certain circumstances – and the second is that self will be no threat to the other. For developing social relationships the second is by far the most important. This is because if the second is not present, the first is of little, if any, long-term value to either party; without the second the submissive one still appears too dangerous and possibly too devious to be trusted or accepted. In a number of species, reverted escape – approach followed by flight to signal non-dangerousness, but continual submissive re-approaching of the individual to whom submission is desired – seems to be a major means of getting a dominant, or home-based, individual to learn to accept self's longer-term offer of subordination. This pattern is not uncommon among primates, especially between offspring and their parents. And it certainly is not unusual for human children to continually re-approach an adult, especially a parent, for a favour after being repeatedly rebuffed.

This tendency in primates would have had an evolutionary boost if youth-apes were accepted and treated as playful juveniles by existing adult males and females rather than as competitors for potential mates. As youth-apes remained somewhat juvenile-like and became infertile until after puberty, they created a new category of 'adult-in-waiting' – adolescence – a condition in which males would likely retain a degree of son-like subordination to those females to whom they wanted to get close or be mothered by, but also be accepted as 'hangers-on', if not followers or recognized subordinates, by post-puberty males who did not consider them as threats but rather as juvenile admirers and potential future allies. Later such patrons could provide maturing subordinates with access to females and a degree of social dominance by association. All of this, in feedback, could have extended the period in which young males appeared young, even after they became fertile, firmly establishing adolescence as a subordinate category on the way to adulthood.

Young females looking for alpha (or at least worthy) males would have approached them with submissive and sexually suggestive gestures – almost as a loving juvenile daughter, not threatening to older females. As female choice increased, and females increasingly looked for help in offspring rearing, submissive and sexual gestures would have become more sophisticated and prolonged (again, selecting for the existence of a longer period of adolescence). Progressively, females would have been able to attract a number of males without having to immediately offer sex, the better to evaluate the long-term potential of the males attracted and be able to choose the good ones. Once a male was 'selected', submissive signals would have been directed specifically at him, and even intensified, with the tantalizing 'promise' of definite sex if he showed appreciation and a degree of subordination to her and compliance with her wishes. She would soon have the difficult problem of balancing her 'creative subordination' with his requirements for mothering – a problem for which evolution is yet to provide a completely workable solution.

Among modern adult humans we approach elders/high status people for advice, jobs, loans and aid in life generally. We approach in very submissive ways, showing great respect to those we approach. Deference on the part of young males and females to dominant males and females results in individuals being placed in a hierarchical relationship to each other, and, as we have seen in Chapter 3, this hierarchical relationship places a differential value on individual humans in terms of their 'accepted worth' as potential advisors/helpers/patrons or clients. Creative individuals tend to remember 'their place' if they hope to reap the rewards of subordination or to later improve it. As a result, each social encounter of deference and submission does not have to be repeated every time various individuals meet, and each encounter can take place in a much more relaxed atmosphere.

To a certain extent this ability was already in the primate ancestry. A number of monkey species, for example, live in groups of ranked individuals in which males seek higher ranked males as allies and individuals remember their positions and the ones of others. However, this ability is largely instinctive in terms of individuals responding to specific stimuli and mostly, but not completely, maintained through agonistic mechanisms (Seyfarth and Cheney, 2003). Humans, however, largely *seek-out* relationships of subordination for the *pleasure* of such relationships and/or for the *pleasure* of the hoped-for outcomes of such subordination (subordinating to mothers for care and love, or to adults for sex, marriage and offspring; or following a leader out of hero worship, or from being included on a highly prestigious status ladder, for example).

And specific demarcations can come to be remembered and stereotyped, as in the example above of 'adolescence' becoming a constant *status* demarcation of social deference, or 'elders' becoming a category of dominance because they have been parents and/or patrons.

Status

In general, high status among primates is achieved by the amount of total submissive or admiring attention received by an individual (Chance, 1962, 1988), and we can add that among humans it can be a designated group of individuals that achieves such recognition (an aristocracy, for example). With regard to humans, this process feeds on itself. The more attention and respect given an individual or group the more gifts and help (wealth and power) they are likely to receive, and so the more reason to show them admiration in the hope of receiving favours and protection in return. Because of the reproductive advantages of this, humans have come to both desire attention and status, but also to give attention and status recognition. This allows for flexibility, but it does introduce a balancing act that is not always easy to perform. As a result, humans may or may not work hard to achieve status, they may or may not resent the status of others, but above all they fear losing status because what is at stake is their *inclusion* in a social order: it is their attributed *value* as seen by others, their license to act in certain ways, their acceptance; this is enforced by a powerful human capacity for dreading humiliation, shame, embarrassment and feelings of guilt.

The fear of losing status (often intensified by status ambiguity) can reach a point where individuals and groups feel a 'need' to protect the *status* granted to them at all costs, and so begin to claim a '*right*' to it. In exchange for a 'right' to a specific amount of status, as noted, individuals will usually accept the status claims of others, introducing a degree of social stability and predictability that otherwise might be absent. At higher status levels individuals (and groups) can begin to demand a specific 'higher' amount of attention and services. And when they get these on a regular basis, the designations of aristocrat and *aristocracy* appear. However, dominating individuals and/or groups usually require both allies and the goods and services rendered to them in order to maintain their positions. As a result they must not take their claims to superiority or their demands too far, and usually find it important to make sure they do in fact provide a certain amount of favours and protection to their supporters.

Additionally, high status individuals frequently display 'deference' through a symbolic form of subordination. Leaders/politicians, for example, go to great lengths to claim they are *subordinate* to some higher end or other and are simply the servants of gods, 'the people' and/or the nation; they spend considerable time in attending charity events and major public celebrations, to suggest that the aims of these override their own individual importance. So, even the powerful enter into the hedonic pattern of holding willing submission to be of high value and needing a relatively constant process of reaffirmation in order to make 'reconciliation and reassurance possible' (Chance, 1988, p. 17; de Waal, 1989). At most status levels, most individuals, most of the time, have quite a bit to lose if they challenge (rather than reaffirm) obligations to fulfil allocated duties and/or fail to respect the rights of others.

Indeed, reaffirmation of our obligation to duties and fulfilling them often becomes part of our own sense of personal value. Ignoring them, or not achieving them, generates a sense of failure and feelings of guilt; protecting the rights of others gives us a sense of honour; violating such rights becomes a sin/crime. As a result of all this, status recognition (semi-formally if not formally) affirms, without a personal encounter needing to take place, particular relationships within a society. This is quite unlike agonistically based relationships in which fear, aggression and physical violence all play a major role and which usually have to be reinforced on a personal basis, over and over again. When our social location is given a symbolic designation or is otherwise abstractly represented, it communicates information at a social distance, often well beyond the realm of intimate internal band relations, and even beyond dialectical tribes. One does not have to speak the same language to recognize the king's standard and understand the expected deference and obligations to it, or to be operated on by a 'doctor' who is clearly recognized through accepted symbols as being a doctor.

Status demarcations, even more than new language, make it possible to enter into a live-and-let-live approach with regard to many with whom we have no personal contact or relationship. We officially recognize the status of a multitude of others, and they ours, and then we stereotype and ignore each other. If we can use some of what they have to offer, and our aims do not clash too dramatically, we can get quite a lot out of them, and they us, without entering into any kind of emotional or antagonistic relationship. Because of an individual's status of doctor, for example, we ask them to look at our tonsils; because of my status as plumber, the doctor asks me to fix his or her leaking pipes. Because I value my life more than the doctor does his or her pipes, because many people ask life and death questions of a doctor and almost none of a plumber, we defer to doctors on a regular basis (even on non-medical issues). On the other hand, some hold me in disgust because of the sewers I work in. The doctor has higher status than I. And I accept that.

We have seen the role of language in 'beyond band' interaction. Here we can note that human language is often the language of status attribution. The status value attributed to personalities, charisma, physical appearances, wealth, achievements, luxury objects possessed, members of the opposite sex seduced and fame gained, is communicated and elaborated through the use of language. This is also true of abstract descriptions of status categories and of the power of their symbols, as well as of the duties, obligations, prayers required, homage expected, spiritual attributes and rights attributed to individuals and to categories. In this process, written language comes into its own, not so much as a means of making it all work that much more quickly than simple non-verbal expressions or spoken language would do, but because it does not mutate at the same rate as spoken language, and can be stored unchanged for relatively long periods of time (thus maintaining deals, contracts, instructions, laws and constitutions somewhat more consistently). It can also be translated into another dialect and stored in another society, giving the other society relatively permanent information about the working relationships of the first society, facilitating the making of deals between the two (or intensifying their mutual antagonism).

Status valuations and the relationships based on them, however, clearly require a degree of political negotiation and smoothing-out if status recognition and its 'who does what and to whom' and its 'live-and-let-live' processes are to work. Even then it is not always smooth sailing. Individuals do not always agree on who can do what, or who is owed particular amounts of deference or recognition, or what their own obligations should be, and how much reward they should achieve. And people do not always stick with the agreements they have made, or at least others think they have made. Just how much is a potential sexual mate from another band really worth? What is the value of an alliance with a particular band? Even worse, how dangerous are some individuals, bands or tribes? Given the limited reach of human emotional capacities, 'strange' individuals and other tribes with the appearance of different fantasies, desires, dialects and symbols, are, in many circumstances, perceived as being potentially dangerous. Fear fantasies often make them seem even more dangerous than the anthropologist from Mars would evaluate them to be. Fundamentalists rant and rave about the impurity of 'the other', and even of certain categories of their own societies that they perceive to be tolerant of 'the other'. There are feelings that outsiders must be protected against, if not controlled. And within a 'status society' there are some who see some status locations as exploiting others and stereotype the exploiters as 'evil'. So language and status can cut a number of ways and, as such, can become elements of danger as well as of safety.

Nevertheless, status and the language of status make it possible to recognize linkages and groupings well beyond a dialectical tribe. And status, stereotyping and non-emotional and status language make it possible for people to live-and-let-live in extremely dense populations and to know what services are available from whom, and who to be careful of, who to show special deference to, who can be bullied, who can protect us and so on. But making it possible and providing social information does not necessarily make it peaceful. Nor does it especially provide a basis for social unity. With these problems we are approaching the area of human politics with a large 'P' (as opposed to sexual, friendship or neighbourhood politics). Large P politics are about human attempts to negotiate individual life beyond the confines of band, and even tribal, behaviour. They are about the interplay of the human will to power in the larger social world as humans cope with their particular combination of species-typical desires and fears by seeking safety through vengeance groupings, alliances, military power, ethnicities, religious denominations, trade unions, interest groups, professions and political parties.

5 Desires, fears and the evolution of human politics

Human politics – as attempts to negotiate individual life, from sexual interactions to an individual's place in the larger social world – have had somewhat different (although overlapping) reproductive causes and consequences for males and for females; so I will consider each separately in a quest to understand the desires and fears that underpin them.

Male politics

Starting from the general mating pattern of other higher primates, the change to the human mating pattern suggests a loss of control, privilege and status among dominant and/or older males (and females). This evolutionary change would have come about because the traditional means of maintaining dominance, status and sexual privileges through pure aggression and loud displays, often including the exhibiting of an erection to a status opponent or desired female, did not give reproductive success among youth-apes. Also, new forms of male sexual attractiveness and behaviours based on childlike characteristics – playing, emotional dependencies, friendships and 'personalities' – resulted in an equalizing of status among younger, smaller, weaker or less aggressive males.

However, as differentiation proceeded, the evolution of anti-incest mechanisms would have weakened close in-breeding based on play and opened the door to post-puberty sexual competition, especially if males began to out-migrate and look for females in external bands. So, the mating conflicts among males were not completely eliminated in human evolution. But it does seem clear that conflict has been greatly weakened given that so many human males have reproductive success through monogamy. And despite underlying sexual competition, human males have a tendency to form into gangs and respect the sexual claims of members. Younger, less dominant and/or less aggressive males might have formed playmate/youth partnerships and/or alliances to gang up on aggressive, domineering males. This could have been in order to steal females from them (Hamilton, 1975) or to go on raids for the females of other groups.

Additionally, as a bonded group, males could more easily have defended the interests of each individual against outsiders. Related to this, they would have been more able to exact vengeance against anyone who harmed a fellow member (thus providing a degree of security for a member even when they were on their own because potential harmers would know that – at some point – vengeance, *from a gang*, would follow). Gang behaviour could have aided in the protection of members' families or 'us' groups from 'non-us' groups generally (Cf., Pusey, 2002; de Waal, 2002a, 2003); such togetherness might even have helped in cooperative hunting, building and other socio-economic activities.

As males came to desire exclusive control of at least one female (as suggested in our scenario) they also came to respect their partners' or gang members' 'rights' to specific females. This could also have led to a reduction in the 'Herod Effect' (Ridley, 1993, p. 205) in which males kill the infants of females that are not their own offspring (a practice not unknown among primates – Hrdy, 1979, 1981; Goodall, 1986; Dixson, 1998; Pusey, 2002). Whatever the balance of evolutionary functions or their causes, it seems evident that male alliance behaviours (derived from a common ancestor with chimpanzees – de Waal, 1982, 2003 – and from some earlier primates – Perry, *et al.*, 2004 – and even mammals – Cf., Connor and Krützen, 2003; Drea and Frank, 2003) have played a significant part in the evolution of the human line. It is certainly of great consequence today, considering the near ubiquity of male 'buddyships' and/or gang behaviour, and the relatively solid alliance behaviour (among at least certain human males) in modern life – from army units, sports teams, business 'management teams' to political cliques.

In the beginning, gang behaviour could have been based on the motivators that generated playfulness among youth-apes in general. With the development of juvenile sterility and evolution of puberty, being attracted to some combination of youthful playfulness and friendship with other males likely would have continued and, as with bonobos, would have gone some way to controlling the aggressiveness common among many primate males when they reach sexual maturity. Additionally, in that females were evolving to attract and select helpful partners in child rearing, ones with their lust under control, the potential for the same sexual *restraint* towards a friend's female would have been in place. Male characteristics which females were selecting for mating purposes also served alliance behaviour, such as mutual emotional dependency, ability to defer gratification for a loved one/friend, sharing, caring, loyalty and trustworthiness. Such male friendships and bonding also suggests the evolution of a 'braking system' on male lust/desire when in the presence of a comrade's desired or claimed female in that a male has to weigh up the long-term value of friendship against the short-term pleasure of sex.

But all this was undoubtedly not a smooth, unproblematic process. Males may have resented losing friends to playmates of the opposite sex; or a male could have a tinge of (if not a raging) lust for a comrade's desired female, lust that simply would not go away. But rather than creating male/male conflict this could generate mental confusion, distress/anxiety or guilt, preventing a male from communicating with a friend's female in a seductive manner. Certainly there is evidence that stress dampens sexual capacity in males (Sapolsky, 1994). All of this would act as a selection force for embarrassment, shame, shyness and modesty in general. Modesty certainly dampens the display of sexual signals of both males and females. Modesty makes an individual who has been even accidentally sexually immodest feel afraid of losing face and thus desire to be even more careful in future not to exhibit sexual suggestiveness.

In this process, propensities for establishing more formal sexual avoidances, taboos and rules would have been generated/reinforced. One of the earliest (less formal) human taboos must have been against making status claims and sexual requests by presenting an erection to other males and to females (as is seen in a number of primate species). Close behind we would expect taboos against demonstrating sexual attractiveness or seductiveness in public, so that hiding genitals and the wearing of clothing, for example, became widespread for both sexes. This resulted in the human face becoming the most sexually attractive and informative feature of humans because the face can communicate more than simple sexual receptiveness: the face can say: 'I am really friendly deep down; I respect you and would never violate sexual taboos'. In some cultures, however, fears of both male lust and female sexual attractiveness are sufficient that these cultures try to force females to cover their faces as well as the rest of their bodies.

As evolution proceeded, empathy came to increasingly play a role in social/political life. Empathy can be valuable in motivating the hiding of, if not repressing of, envy, including sexual envy. Self's secret sexual desires for a mate's female can lead one to *empathize* that a mate, or mates, might come to have similar designs on one's own beloved, generating, when emotionally charged, jealousy regarding one's own mate, motivating conscious thoughts which cause most males to leave well enough alone, to guard their own mates and to not tempt fate by attempting to seduce those of others. As a result, we have not only another basis of support for rules and taboos, but also possible motivations for a notion of *honour*. Empathy-fed honour requires that both *specific* but also *types* of individuals and behaviours are off-limits.

And honour has come to be a powerful measure of an individual's standing in a community (and consequently of an individual's own feeling of self-value). Certainly, history is full of conflicts, sometimes to the death, when male honour has been at stake; that is, when males have been cuckolded and/or accused of violating a major taboo, of not having respected or protected female virtue, of having cheated a friend, of having been a coward, of having let a comrade down in battle, or even of having been accused of being dishonourable (Cf., Vellacott, 1956; Chagnon, 1968; Nisbett and Cohen, 1996). Historically, allocation of honour has become a very significant means of providing social and political rewards (status) and resources in human societies. As Max Weber (1958) long ago noted, the distribution of social honour (status) is usually a major ingredient in the formation of human social orders.

Alliances

Whatever the exact form honour takes, it continues to this day to be absolutely essential that males *recognize* the sexual rights of their alliance mates if alliances are to have any duration or effectiveness. 'You don't screw your buddy's wife', said the basketball coach when trying to re-kindle his old team's total loyalty and unity to work for the political election of one of their number (in the film, *The Championship Season*, 1982). Humans remain the only primates able to practice a high degree of monogamy with wives remaining in close proximity, and sometimes contact, with other males (Ridley, 1993; Miller, 2000; but see Strier, 2002). And from this derives a great deal of male politics – and a degree of law and order, cooperation, mutual protection and defence in the larger social world.

But for this to work the human capacity for male bonding has to be almost as strong (if not equally strong) as that for sexual pair bonding or parent/offspring bonds. We are looking for a bonding force more powerful than those based on rational calculation of cost–benefit, or dependent on the guilt generated when friends fall out or parents and offspring come into conflict. We are looking for motivators that engender a powerful sense of camaraderie in its own right (Freud, 1921; Tiger, 1969). We are talking about something approaching obsession regarding intense loyalties, adulation of teammates, hero worship and wanting to be accepted in a network/group. It is what makes football players grab and hug each other and shout wildly when a goal has been scored; it is what makes men get drunk together and talk loudly, and support each other in fights; it is what makes comrades-in-arms risk their lives, and even die, for each other, and it is what makes youth gangs near ubiquitous.

Male alliance bonds may arguably share some of the characteristics seen in the powerful and addictive feelings of lust and love associated with enamoured sexual pairs and devoted parent/child relationships. Is male bonding, therefore, based on many of the same emotional/cognitive processes, fuelled by hormones such as dopamine, oxytocin and vasopressin – hormones involved in the forming of bonds between mates and between parents and offspring (Turner, *et al.*, 1999; Turner and McGuinness, 1999; Fisher, 2000; Fisher, *et al.*, 2002; Szalavitz, 2002; Lim and Young, 2006)? Certainly in evolutionary terms, the fact that sexual attractions bring individuals together is not surprising, and it can be further noted that these processes are extremely vulnerable to the selection process during the evolution of new species; they are in the front line so to speak. If no one desires the 'new look' new genetic directions will not be selected for regardless of how beneficial they may be in survival terms.

At the same time, if changes do take place in what are sexually attracting and activating features that actually result in a new line, it may take some time before a new clear sexual dimorphic recognition is firmly established. In the case of humans, for example, in the context of the evolution of youth-apes as postulated, considering that unisex childlike playfulness, pseudo sex and playful penetration (polymorphous perversion) may have provided the earliest form of 'new look' reproductive success, a clear attractiveness distinction between the sexes would not have yet evolved. Youth-apes would not have developed a complete capacity for being attracted solely to the opposite sex for reproductive behaviour.

It is true that with the selection of juvenile sterility, puberty and increased sexual dimorphism, selection would progressively have made males feel sexually attracted to females, and females to males. But, because of the polygenetic nature of sexual attraction, males (and females) would be a complex mishmash of genetic reproductive propensities for sexual attraction, existing on a continuum ranging from attraction to child-like characteristics at one end and to exaggerated mature adult (almost primate) characteristics at the other end. Certain males, for example, would remain, on balance, strongly attracted to childlike characteristics (to infant and child features), others to slightly more mature characteristics (puberty and just post-puberty characteristics – slight curves, slim, small breasts, emerging pubic hair, child-faces, soft skin, for example). Other males would be attracted to somewhat more mature 'female-type' femininity (definite post-puberty womanhood – noticeable curves, well-rounded behinds, ample hips, firm breasts and noticeable pubic and underarm hair, for instance); some males would find more mature post-puberty womanhood exciting and sexually stimulating (ample breasts, wide hips, 'experienced' looking faces, somewhat more fleshy bodies). At an extreme, a few males would be attracted to larger, muscular, athletic, even hairy characteristics.

The above is an oversimplification of what would be a complex mix of characteristics within a given individual, and with a great deal of overlap in terms of the things that made a particular male attracted to them. It is very possible, for example, to be strongly attracted to both child-like faces and female pubic hair, or soft skin and athletic builds, at the same time. Additionally, individuals attracted to each other do not necessarily become *proceptive* (actively trying to solicit or initiate copulation) or *receptive* (allowing copulation) with each other. For example, being powerfully attracted to child-like characteristics might make one an obsessive parent rather than a paedophile; or adult males attracted to other males can become very close friends rather than sexual partners. This is because, as we have seen in Chapter 2, it is very possible that attractivity, proceptivity, receptivity and consummatory sexual responses are based on somewhat different neural mechanisms – each generated from different genetic materials. Evolution could have moved faster or slower, or at least differently, in each of these processes, and considering the relatively short time since human differentiation it is not surprising that selection has not completely 'edited' male sexual attractions, patterns of proceptivity, triggers of receptivity and consummatory responses to the point where they are working in complete reproductive synchronization.

At the same time, however, a male being powerfully attracted to certain male characteristics as evolution proceeded could have provided reproductive success for males by creating dedicated male partners in life's trials and tribulations; or by motivating attraction to older brothers or friends who acted as protectors, procurers or even left a female to the younger brother or friend when they died; or for subordinating to a gang that protected one from bullies. What this approach suggests is that even with the evolution of a degree of sexual dimorphism at puberty, at least some males (along with females) inherited motivators for being strongly attracted to selected male characteristics and behaviours, and, due to the polymorphic nature of the genes involved (and the variety of individual experiences), many more have some more or less weaker propensities in that direction. And the argument here is that these propensities provide a basis for powerful male/male buddyships and strongly bonded male alliances. It can be further argued that such attractions also result in hero worship and patron/client submissiveness, and when these are put together with partnership and alliance behaviour, longer-term status hierarchies and political networks often emerge.

Such attractions and motivators may also, of course, help explain male homosexuality because in some males the possibility of being strongly sexually attracted and made receptive by other males, and achieving consummatory pleasure with them, becomes theoretically likely. Indeed, evidence concerning male homosexuality can be used to throw light on some of the propositions suggested above. As a cross-cultural phenomenon homosexuality seems to be very widespread, if not universal. In Beach's (1965) sample, it is even socially acceptable (for at least some people) in 64 per cent of societies (see also, Davenport, 1977; Tripp, 1977; Ruse, 1988; Foucault, 1992; Gray and Garcia, 2013). In Kinsey's study in the United States of America (Kinsey, et al., 1948), 37 per cent of males experienced at least one homosexual experience to the point of orgasm between adolescence and old age. Thirteen per cent had more homosexual than heterosexual experience for a continuous three year period between the ages of 16 and 55; 8 per cent had exclusively homosexual experiences for at least three continuous years between 16 and 55 years of age; and approximately 4 per cent were homosexual for the whole period of their lives.

It must be noted, however, that Kinsey's research probably overestimates the frequency of male homosexual behaviour. In the first place, his sample group contained a large number of individuals involved in criminal and delinquent behaviour, often with time in custody. It is estimated that individuals in custody generally have a very high rate of homosexual experiences. Also, many of these experiences only took place at a very young, although post-pubescent, age and were not carried on into later life. Furthermore, numerous more recent surveys (Cf., Weatherburn, et al., 1990; Rogers and Turner, 1991; ACSF Investigators, 1992; Wellings, et al., 1994) report a lesser frequency of male homosexual experience, sometimes quite dramatically so – as low as 1 per cent exclusive homosexuality and only about 6 per cent with some experience in some of the surveys. The lower estimates, however, are also likely to be misleading in that the surveys often relied on self-confession of homosexuality in near face-to-face interviews in contexts where homosexuality was not only stigmatized but also often illegal. Moreover, samples were very small in a number of these studies and it was not always clear whether interviewed individuals who might have had only a few homosexual experiences considered them to count. In one case of a low estimate, for example, the authors themselves point out that their findings concerning homosexual frequency 'should be regarded as minimum estimates' (Wellings, 1994, p. 180; see also, Dixson, 1998; Kauth, 2000; Jones, 2002).

Whatever the exact frequency, however, let us postulate that genetic propensities for sexual attraction, proceptive behaviour, sexual receptivity and sexual consummation are polygenetic (that is, based on a number of genes), and that at least some of these genes are mixed-dominant genes. Given the suggested scenario of human evolution in Chapter 3, these would have been genes that encouraged friendships, youthful play, pseudo sex and real sex among youth-apes, including among members of the same sex. These playfulness and sexual experimenting genes would have become the sexually attracting and activating genes among diverging youth-apes/humans. However, as puberty emerged, attraction to and the activation of sexual receptiveness by more adult-like secondary sexual characteristics would progressively have been selected for. As a result of both these processes, specific bunching of particular genes would produce males only attracted to, and made sexually receptive by (and able to achieve consummation with), other males. These would be dedicated – and very possibly enthusiastic – homosexuals. At the other end of this continuum we would have males only really attracted to child-like characteristics and/or very feminine females, and very easily made proceptive or receptive by even the sight of them. A few of these might be paedophiles, but more would be dedicated 'ladies men', spending more time around and attempting to seduce young women than 'hanging-out with the boys'.

Overall the males in a given society would be on a range running from those who are: (1) dedicated homosexuals (somewhere between Kinsey's 4 per cent to the lower estimates of only 1 per cent perhaps), to (2) males strongly attracted to males and who engage in relatively regular homosexual behaviour, at least during certain periods of time (somewhere between Kinsey's 13 per cent to the lower estimates of about 4 per cent perhaps); then to those who are, (3) strongly to moderately attracted to males but have only a very few homosexual experiences over most of a lifetime (up to something like the high of 37 per cent identified by Kinsey but perhaps as low as 10 to 12 per cent). From there we move to the bulk of the male population, to those (4) strongly to moderately attracted to males but not having any homosexual experience during their lifetime and on to (5) quite dedicated heterosexuals attracted to fully developed post-puberty females. At the end we find another small minority, (6) individuals strongly attracted to and sexually aroused by extremely feminine and even child-like characteristics.

So, most males would be in some degree of a heterozygous condition (a genetic mixture of often indeterminate genes) towards the middle of a continuum for playing, friendship, sexual 'turn-ons', sexual arousal and love propensities. In order for a specific heterozygous balance to remain it has to be shown that the heterozygous is an advantage over both homozygous ends. It seems obvious that dedicated homosexual males would not have much reproductive success. And, although not exclusively homosexual, it is likely that a number of those near that end of our continuum would not be as interested in family life and children as middle-of-the-road heterosexuals – such males might prefer male company to female (as soldiers, adventurers, drinking mates, for example), with masturbation, occasional visits to prostitutes and the use of pornographic material as diversions.

At the opposite end, among those only interested in sex with child-like/very feminine characteristics, we can also see a decided reproductive disadvantage in that children are not fertile (quite apart from current social objections to sex with children and young adults based on an evolved strong desire to protect babies/children at all costs). Even addicted heterosexuals may be so obsessed with female sexual attractiveness that they become overwhelmed with sexual excitement and desperately want to stare, touch, stroke, hold, hug and squeeze the object of desire. As with those addicted generally (for example, to drugs), it seems that a dangerous force has got him under its control – and it is likely that attractive females would give him a wide berth. Such males are stalkers not potential mates, and might even spend a lot of time in prison for sexual violence and molestation. High status, highly sexed, handsome female-obsessed males might be able to be very promiscuous in terms of one-night stands with selected females, but this does not mean that they will have reproductive success, because, as we have seen in Chapter 2, on average human females require regular copulations with the same partner for about three months in order for pregnancy to occur. Males with hyper-sexuality orientations are sexually attracted to and excited by a relatively large number of females rather than being enthused by the possibility of being 'imprisoned' within a long-term relationship.

Some hyper-sexed heterosexual males (like some of the males at the other end of the continuum) might give up successful reproductive approaches altogether. They could indulge in considerable masturbation, perhaps developing a strong tendency to sexual fantasies (Cf., Friday, 1980; Buss, 1994; Dollimore, 1998; Soble, 2002a), or pornography, or nowadays indulging in cybersex on the internet, or be heavy users of prostitutes – all activities as a replacement for longer-term heterosexual relationships, none of which produces much reproductive success. Males at either end of the continuum might find substitutes for their hyper-sexuality in drink or drugs, for example, or even sublimate their sexuality into art or religious callings. Whatever the case, a hyper-heterosexual male is not the handsome, romantic hero of films and novels, but rather an addicted individual largely avoided by the vast majority of females, and at the other end a dedicated adventurer or seaman is not exactly the best potential husband material.

So, men at the extreme ends of our continuum are not likely to have a lot of reproductive success. But both ends will keep appearing in a population if the heterozygous centre is an advantage (Cf., Hutchinson, 1959; McKnight, 1997; Kauth, 2000). The advantage for male reproductive success in the heterozygous condition, it can be argued, is that such heterozygosity most likely would have encouraged many males to split their allegiances, to try to have their cake (sex with a substitute, mother/friend) and eat it too (fun, adventure and mutual bravado with the boys). Arguably, these are the propensities that have provided for the most reproductive success during human evolution because they provide motivations for successful reproductive behaviour, but also for the social behaviour that supports it. And an important element of surrounding social behaviour is male bonding.

Indeed, this can be illustrated by the fact that although weak on reproductive success, sexualities at the ends of our continuum have nevertheless provided a degree of *social* success. For instance, given the common discrimination against them, it might seem surprising that quite a number of high-ranking officials in modern western governments have been homosexual (Tripp, 1977). Bisexuality and/or a capacity to be attracted to other males can lead to survival, and indeed to a degree of privilege in certain circumstances. For those incarcerated in prisons, for example, a relatively high rate of homosexuality is often part of a larger male/male experience. For many it is a forced way of surviving, for others it becomes a way of being granted protection and privileges. For instance, there are gangs in American male prisons that protect individuals but make 'whores' of them for sale to inmates. Those who have sex with the 'whores' consider them to be 'women'. Authorities tolerate this because it provides a degree of order and stability within the prison (Abbott, 1991; see also, Human Rights Watch, 2001, 2001a; Insideprison.com, 2006; Booyens, 2008; Gray and Garcia, 2013); but when released a great deal of this behaviour disappears and individuals can survive to reproduce.

During the Great Depression in America homosexuality was not unknown among travelling hobos as older veterans took along companion boys, often runaways or boys found in bathhouses along the track (Courtwright, 1998). The veteran got sexual release and companionship and the boys got protection, instructions and contacts. In twenty-first century Newcastle it has been estimated that a third of the prostitutes were boys. Many of the clients were married men with children (Doward, 2004). Here the exchange is largely sex for money, but money can lead to personal satisfaction and even heterosexual access. So, a capacity for temporary homosexuality can help an individual survive prison (or a labour camp, concentration camp, hobo camp, male military community or the streets of a large city), and later on reproductive success may well follow. In other cases temporary or pseudo-homosexuality can provide reproductive success by providing a 'leg-up' in later life. In Greece and Rome, for example, male sex (and love) with young, pre-pubescent slave boys was common among largely non-dedicated homosexuals (Veyne, 1985; Kauth, 2000). These were personal, sexual, age and status relationships that provided a sense of power, companionship and release for the older, freeborn men and a degree of privilege, protection and connections for the boys – often penetration did not occur (Ruse, 1988; for more examples see Gray and Garcia, 2013).

It is significant to note that male motivational differences in general life are little affected by the existence of homosexuality. Evidence indicates, for example, that the major (if not only) difference – physical, psychological or social – between homosexuals and heterosexuals is in the objects of their sexual desires (Symons, 1979; Ruse, 1988; Kauth, 2000; but see Dörner, 1976). Homosexuals are attracted to and made sexually receptive by members of their own biological sex, but are not usually very different from heterosexual males in terms of gender characteristics, propensities to male bonding or masculine presentations of self in the larger world (varying individually somewhat, as with heterosexual men). What this means is that homosexuality – long term or temporary – does not turn a male into a 'female'. An exception, clearly, is found among camp homosexuals; these go out of their way to alter, indeed invent, anti-male presentations of self in order to make personal and political statements (Sedgwick, 1985; McKnight, 1997).

It is also important to note that the age at which a great deal of temporary or pseudo–homosexuality takes place is the very age during which partnerships/powerful friendships and alliance formation takes place. This is a time when some individuals (perhaps around 20 per cent) experience a degree of confusion regarding sexuality for some time in their youth (Kinsey, 1948; Johnson, et al., 1994; Weinberg, et al., 1994; McKnight, 1997). Whatever the exact outcomes, our analysis suggests that it is possible (and likely) that there will always be a relatively large proportion of males within a population with a strong attraction to other men but most of these men may not be made sexually receptive by males – at least not most of the time. Evidence for this might include fantasy and gossip among males focused on apparently non-sexual relationships with other men; for example, on the antics, intrigues and heroics engaged in with buddies, brothers-in-arms, team mates, fellow gang members, work mates and political comrades.

And it will be important to ascertain the extent to which male bonding sometimes comes to be seen as 'honourable', almost spiritual. The homosexual/homoerotic relationship of a Greek man of standing and a boy, for example, was meant to be an intellectual relationship in which reason, discussions, learning, love and respect played a major part (Foucault, 1990). It was very often claimed that it was the 'love of the boy' that was important, not so much the sex (Ruse, 1988). The same spirituality can be sensed from the bond of Tennyson with his intellectually oriented 'very special' friends from among the Cambridge Apostles whose death he lamented: 'Two dead men have I loved/With a love that ever will be:/Three dead men have I loved and thou art last/of the three' (*In the Garden at Swainston*). Walt Whitman too saw a strong male bond as representing nobility through dedication, loyalty and comradeship. These views often allowed for marriage with a female as a sort of lesser experience: 'when that is sought as a domestic institution, as it always is among men who want children for helpers in their work and women to keep their households' (Symonds, 1893, 1907 cited in Sedgwick, 1985, p. 211; see also, Brady, 2012).

It seems clear that male/male attractions can be very powerfully emotional and that they share a considerable number of biochemical and behavioural characteristics with sexual attractions, but can, nevertheless, be 'lifted' into the spiritual realm by human emotional/cognitive processes. And it is certainly true that throughout history storytellers, poets and philosophers have noticed and exalted the power of friendship/partnership bonds among heroes and have treated camaraderie among alliance mates as being more spiritual than practical, more akin to romantic or religious love and devotion than to sexual bonding. In the *Epic of Gilgamesh* (1964), for example, 'The great friendship between Gilgamesh and Enkidu ... connects all episodes of the story' (Sandars, 1964a, p. 31), Gilgamesh's attraction to Enkidu was like the attraction usually generated by 'the love of woman' (*Epic*, p. 64). For Gilgamesh, when Enkidu dies it is 'more than he can bear' (Sandars, 1964a, p. 35); 'He began to rage like a lion, like a lioness robbed of her whelps' (*Epic*, p. 93), soon afterward lamenting: 'Oh Enkidu, my brother,/You were the axe at my side,/My hand's strength, the sword in my belt,/The shield before me,/A glorious robe, my fairest ornament' (*Epic*, p. 91).

In the *Iliad* it was only the death of his great friend Patroclus that threw Achilles into a sufficient rage for him to go back into battle, and he did so with vengeance in his heart for the death of the person with whom he himself considered he shared his strongest bond. Virgil, in *The Aeneid*, demonstrates the pain felt by a best friend when his 'beautiful young man' was killed. Certainly, male bonds are considered more important than male/female bonds in the poem. Consequently, as also happens in a number of stories of knights, explorers, crusaders and cowboys, Aeneas abandons the woman Dido when there is god's/male work to be done. The story of the spiritual nature of male partnerships has come down through the ages. Aristotle defined friendship as 'a single soul dwelling in two bodies' and saw it as being essential for the smooth functioning of the polis and for individual happiness (Aristotle, 2002 [350 BC]).

In the middle ages a knight was expected to form a close bond with another knight so that they could protect each other. This same situation is portrayed in the light-hearted modern film *Paint Your Wagon* where, in a completely lawless mining camp, the hero needed a 'partner' because from time to time he got 'melancholy and drunk', at which time he knew he would be extremely vulnerable and would desperately need looking after. In this film, the 'partner bond' was presented as being almost sacred, and for much of the film the partners had little trouble in sharing a wife. In recent times, 'buddy stories' have been a relatively common theme in tales of soldiers, explorers, cowboys and detectives. Often two individuals are seen to complement each other, as, for example, a serious one and a flippant one (*Starsky and Hutch*), a clever, scheming one and a flamboyant, handsome dare devil (*Butch Cassidy and the Sundance Kid; The Sting*); an older, wiser man with a younger, over-confident one was combined with race, geography and culture in *The Heat of the Night* in which a worldly wise, older, small-town white man was matched with a bright, cocky, city black man. Sometimes very different individuals become almost fused into one, more 'perfect' individual, as in the story of a black runaway slave, older, experienced and wise, and a runaway white boy, cunning, and reckless, both of whose social marginality (in fact both bordering on being complete social isolates) fused them into a mutually caring, trusting and even loving pair, an antithesis of the seemingly cruel society in which they had been born and which they were experiencing along the great river (*Adventures of Huckleberry Finn*).

The partner relationship has often been expanded to a small group of comrades. St John attributed to Jesus the words: 'ye love one another; even as I have loved you, that ye also love one another . . . Greater love hath no man than this, that a man lay down his life for his friends' (John: 13: 34; 15: 13 – King James). As a result of such words St Augustine had some difficulty reconciling intense friendship with the love of God. He tried to solve this dilemma by suggesting that humans should love their friends *through God*. Aquinas looked to reconcile human friendship through the *charity* of God and, following Aristotle, had argued that the moral virtues of friendship, loyalty and trustworthiness were essential for the smooth working of social/political life (Bragg, 2006). Indeed, during the evolution of court life following the middle ages sentiment between men had acted as lubricant in many service relationships that are governed by contract today. These were often close to love and never exhibited 'indifference or impersonality' (Aries, 1985, p. 70). During the Restoration period in England homosexual men were able to establish 'molly houses' – taverns and other places – where they met for discussions as well as for sex. In later times, coffee houses became relatively widespread in which intellectual men met in friendship groups to discuss important issues and to 'make deals'; in these, friendship became almost a secular religion (Bragg, 2006).

Throughout the ages, male friendship, loyalty and trustworthiness can clearly be seen in the formation of political and religious *cults*. For instance, as court life was evolving in Europe, humanists of the time developed tight circles of friendship as a means of escaping the 'deceitful' world of politics, increasingly seen by them to exist in almost every royal court in Europe. Actual vows of male friendship were not uncommon among such humanists. This type of cult behaviour was also similar to the life of males in Fascist political cults in 1930s Germany (Sedgwick, 1985; Lively, 2005; Lively and Abrams, 2013; see also, Blueher, 1912, 1917), as well as to the codes of silence, honour and tightness of Sicilian and American Mafia 'families' in the twentieth century (Cf., Reppetto, 2005; Dickie; 2007), and in varying degrees to memberships within a wide variety of revolutionary, intellectual, political, artistic and 'alternative' cults and street gangs throughout history. During the 1500s young men in French towns were more or less expected to form 'heroic friendships' (Olivier, 1985, p. 99), resulting in gangs of young men who sometimes acted as 'moral vigilantes', but sometimes just fought with each other (Roussiaud, 1985). This is a pattern that was very common among young men in pre-empire societies around the world (Cf., Evans-Pritchard, 1940; Chagnon, 1968; Colson, 1974; Ambrose, 1975; Keeley, 1996) and has continued across the world right up to the present time. For example, until relatively recently, Paris gangs, informal masculine networks, were the basis of young male life in certain sections of society. Although not homosexual, there was a lot of sex talk and borderline homosexual proceptive behaviour, including group masturbation. When the young men married there was a clear separation of home life and gang life, which carried on for some time (Lafont, 1985).

It is not just young males, heroes, cultists and political enthusiasts that form strong friendship bonds. Intense friendships were clearly a factor in the Midlands-based development of the industrial revolution in Britain in which a number of major players ('The Lunar Men') found that 'philosophical friendship' increasingly became 'one of the great pleasures, not separate from, but inextricably linked with, the common business of life'. 'Long letters, often highly technical, flew between all the friends, working out problems, commenting and offering advice' (Uglow, 2002, pp. 84, 351). When one of their most sociable and likeable colleagues, (Dr) William Small, died at a relatively early age, they were extremely distraught, spurring budding industrialist Matthew Boulton to write: 'My loss is as inexpressible as it is irreparable, I am ready to burst'; and Erasmus Darwin penned a poem: 'Cold Contemplation leans her aching head,/ ... For Science, Virtue and for Small she mourns' (cited in Uglow, 2002, p. 250).

When given warrior/'defenders of honour' status, such groupings can go down in history and mythology as being something very special. King Arthur's Knights of the Round Table stand out in this regard, as do Spartan warriors who lived and fought in strongly bonded male units (Sedgwick, 1985). The Japanese Film, *The Seven Samurai*, remade as an American western, *The Magnificent Seven*, demonstrates in the form of travelling vigilantes this closeness, along with a strong concept of honour among comrades. And it is not only in times of war, religious fervour or political conflicts that this phenomenon is at work and gang togetherness greatly applauded. Members of sports teams are expected to emulate a degree of togetherness seen only in military squads. Teammates undress in front of each other, shower together, smell each other, their secret attributes on show to each other. They may tease each other but they do not tell on each other in public. The universal popularity of sports and professed importance of 'team spirit' testifies to the extent that, in the right circumstances, we see and applaud virtue in the team ethic of complete loyalty, dedication and effort.

These are just a few examples of the many that could be given. But they illustrate a range of the activities that males get up to through powerful male bonding without those behaviours being specifically homosexual in terms of usual definitions. In this analysis of the underpinnings of male bonding, dedicated and enthusiastic homosexuals remain relatively constant in a population, not because they aid their kin in achieving reproductive success, and thus perpetuate their own genes, but because during human evolution a heterozygous genetic condition regarding sexual attractiveness, proceptivity, receptivity and consummatory propensities gave greater reproductive success than either homozygous condition. As a further result of this, through sexual desires directed at other males being sublimated into male bonding, males have evolved a powerful capacity for partnerships and male alliance behaviour. And the main argument here is that it is the potential political impact of this behaviour that has made it possible for dialectical tribes, status groupings and even larger political units to exist; it is this behaviour that, among others, has made civilization possible.

Alliance behaviour, politics and the rise of civilization

It is impossible to overemphasize just how important alliance behaviour has been in the rise of human civilization. Alliance behaviour among a very small number of testosterone charged-up males with a strong sense of their own invincibility, ignoring possible long-term consequences of their actions (based on very powerful bonds of mutual emotional dependency, loyalty, mutual trust and a willingness to die for each other), has been (and is) the only means, in the final analysis, of enforcing social order (or, on the other hand, of disrupting it) that humans have.

On a day by day basis, during most of human history, the majority of people have been conformists when they are relatively content with their daily lives, or are deferring gratification with the idea of being contented in the future. As long as individuals have some combinations of close friends to gossip and drink with, access to display materials, the possibility of sex and love, a recognized status standing, happy offspring, a sense that their expectations are being met, heroes and gods to worship and be entertained by, a project to become addicted to, a respect for, or a fear of violating, laws, a feeling that there is a patron (government) ready and able to take care of any major problems they may encounter, and their minds are fully occupied in living, there is little reason for the imposition of a youth-force to impose discipline or revolution.

But if individuals are suffering loss (of parents, friends, love, respect, status or wealth), feel that their expectations are being blocked (possibly by a patron or government), and patrons are grossly exploiting clients, and neighbours seem vindictive, and gods and heroes appear to be hypocrites and not producing happiness, and testosterone-charged youth seem to be running wild, it is another story. In these circumstances, no camp leader, clan elder, tribal chief, high priest or patron would have their orders followed or their administrators staying loyal for any length of time unless they had a group of dedicated, fearless retainers willing to follow their orders and to enforce their will. Nor, on the other hand, would the discontented among populations have had their grievances pursued with such vigour as has happened during history when the youth-force has been turned to revolution.

Older men have lost it, wised-up or become cynical (or are in charge); children do not think of it; mothers do not operate that way; young women have easier ways to get their way; older women have become too cagey; church congregations have fleeting impact; community forums and electorates are fickle, and, like church congregations, have no real teeth. None of these can generate 'the force' that male youth power can. Now it is true that as we have lived longer, in better health, and become more organized, with much more technology at our disposal, some older men (and women of all ages) have become involved, not just as leaders and organizers, but also in a few cases as active participants in the front line of both seeking order and in upsetting it. Nevertheless, in the end, modern mayors, prime ministers, dictators, supreme rulers, presidents, revolutionary leaders, cult leaders, religious fanatics and judges rely on testosterone-charged young men to be in the front line (fighting in the fields and in the streets if necessary) for achieving stated aims (often presented as being sacred aims).

The youth force has been variously organized into vengeance brotherhoods, knights of the realm, caste-like police forces and/or armies (armies with squads of six to ten youths just about past adolescence, risking death because they think they are invincible). There is, however, an inherent danger for human social order in all this. When not *defending* order, or when not feeling appreciated, or when seemingly being excluded, young men in gangs or brotherhoods (along with a number of young females) are usually *attacking* it. They like to throw rocks at those who seem to be against them, and to take part in riots, revolutions, wars and bank raids, and to indulge in knife fights and gun battles (with each other, with innocent passers-by and/or with agents of authority).

Human elders have, however, evolved a number of ways to tame the 'youth force', largely through techniques that have also become major ingredients in the development of human civilization. Skilled leaders recruit virile youths (often from low-status backgrounds) into the bottom of what can become extremely complex organizational and institutional structures in which various youth forces are balanced against each other and/or set to compete with each other (both internally and externally to an organization). Such youth forces are often given an honourable status within a larger social order. And when the young warriors begin to lose heart, or to slow down, or to complain too much, they are *individually* expelled or promoted upstairs. Gangs of young men have been given great honour, and often privileges, for defending people in times of danger; this has made members satisfied to sit around and get drunk and seduce women and participate in tournaments when times of peace are at hand. And, of course, there are always new gold, females, horses and foreign lands to conquer when leaders perceive that the troops are beginning to fidget too much.

As human social conglomerations grew larger and more densely packed with people, and where there was relative social tranquillity, and for whatever 'international treaty reasons' external raids and wars were not really an option, young men have discovered, or have been directed to, sports, apprenticeships, crime, youth gangs, coalmines, universities and prisons. But for the acceptable of these to succeed some very skilled work on the part of parents, girlfriends, priests, teachers, policemen, magistrates, social workers, generals and politicians is required. This is political work at a relatively high level. It has to convince young males of the virtues of substituting controlled competitions in place of plundering, riots, coups, revolutions and war; the value of deferring gratification; the value of following rules; and the virtues of respecting elders. Such political work has to be subtle, compromising, clever, convincing, manipulating, democratic and dictatorial – often all at the same time. Achieving success in this work is up against a chimpanzee-like ancestry, bubbling testosterone, oedipal rage and the human will to power (in its know-it-all phase).

Fortunately (from the point of view of elders, mothers and children), political solutions to a particular youth force only have to be maintained for a short time; youth is a condition cured by age. But, of course, a new batch is always coming up. Indeed, while youth forces have to some extent been tamed as societies have become larger and more technically complex, there has also been a tendency for the youth force ethos to seep into later life (into professional sports, businesses and politics, for example). Nevertheless, the marvel might be that we have become so good at dealing with the energies and potential disrupting of, and the controlling of, the youth force considering what we can see among chimpanzees, and what we have seen during human history in the forms of vengeance raids, coup counting, plundering, street gangs, wars, revolutionary guards, suicide bombers and jihadists.

We did have a head start (at least over chimpanzees) in that rudimentary political skills were rooted in the very nature of reproductive success as presented in our scenario of human evolution. In it, humans evolved a considerable capacity (often operating unconsciously, through embarrassment, shame and guilt) to repress lust, especially male lust (compared to male monkeys and apes) and to befriend possible mates and allies. Self-policing, however, requires a certain degree of social memory and conscious manipulation of self's presentation in relation to the behaviour of others. And it can generate its own anxieties which feed the political process. For example, vigorous attempts to control manifestations of lustful desires may generate suspicion of others, but also exacerbate the very frustration and anxiety that led to the politicking in the first place. For instance, a male may realize that his desires are not what they 'should' be, and that his behaviour is sometimes questionable, and feel uncomfortable with such desires. These feelings, fuelled by temptations, can be prime generators of guilt, but also cause displacement rage against 'sexual sinners'.

And at the psychological level there are probably few forces that generate civilization more than guilt (Freud, 1957; Carroll, 1985; Hankiss, 2001) and/or 'raging at sinners'. Both call out for enhanced self-control, taboos, rules and boundaries. But the more such restrictions come to exist the greater the opportunities for anxiety, frustration, rage, shame, embarrassment and depression when self is frustrated by them, or is unable to live up to them, or has inadvertently violated them (and is punished); and the greater the propensity for rage when others seem to violate them and get away with it. The *political* process, then, has fed on itself through continuous feedback, and has been a basis for the evolution of human species-typical intelligence. And the feedback process continues into modern times maintaining us as very political creatures. And the importance of politics and exercising political control remains as strong as ever.

As we have seen, it is likely that a heterozygous condition for many forms of attractions/desires, repressors and bonding gave the most reproductive advantage; but this will always throw up extremes, including of very lustful and/or aggressive individuals, in a population. It also means that a given individual very likely has varying desires and fears activated at different times, in different circumstances, generating potential unpredictability in social/political life. Most males will have more or less learned to present self in different ways in different circumstances – at times as unbeatable studs, at others as caring, respecting protectors of females and children, and at still others as trustworthy comrades. It is not always easy to know what they will do next, or to know if they can be trusted.

We have already seen how partnerships and alliance behaviour can mitigate some of the dangers of sexuality and reproductive failure for males and thus provide a degree of security for them. But alliance behaviour itself can generate a number of tensions, frustrations, jealousies, envies and anxieties that call out for their own policing and politicking. It is not uncommon to see males: (1) being excluded, (2) competing with each other for the attention of other males and for positions in male groups, (3) competing with friends and comrades for the same female, (4) having 'responsibilities' to group members coming into conflict with attractions to potential females and/or conflicts with members' families of origin/mating and/or with offspring responsibilities. In alliances, males have a tendency to hide 'mother/wife-dominated' self, 'over-dependency-on-mates' self, 'cowardly' self, 'selfish' self, 'lazy' self or 'devious' self, while, nevertheless, often continuing to love their mother/wife, be extremely eager to be fully accepted in a group, be afraid from time to time, often be thinking of self more than of comrades, not always be keen to do some hard (or dangerous) work for a comrade, and all the while attempting to manipulate others through presenting somewhat 'false' images of self to cover up all of the above.

The desires (and accompanying fears) that underpin these behaviours are the bases for the generation and emotional charging of more *abstract* status and political notions, including, for example, concepts of honour, duties, rights, responsibility, membership and justice. With the evolution of increased self-awareness, and with greater cognitive abilities, these notions have been greatly elaborated, and spiritualized, resulting in the establishment of relatively clear political/social eligibility criteria, rules, taboos, laws and boundaries. At this stage we are at politics with a large P as individuals argue over specific eligibility criteria, policing methods and the specific dimensions and meanings of taboos, rules, laws and boundaries. In the process, inter-alliance politics can easily become factional political behaviour in which political factions/parties struggle for power and oppose each other in a good versus evil, life and death, manner.

But, at the same time, individuals and factions can also begin to accommodate each other and submit to each other in relatively predictable ways. The next step in this elaborating process involves the establishing of complex political/administrative frameworks within which these politics can take place according to more or less accepted sets of rules, and in which a recognized form of enforcement exists. But this has never reached a state of complete tranquillity, and achieving a 'balance of power' has been the major (if not only) means of maintaining a degree of stability and tranquillity for any period of time. But balances of power themselves can be hard to maintain, and it is not always clear to participants if a particular balance is in fact being maintained. The volatility of youth gangs of self-believing invincibles can remain a problem in the best of political arrangements, as can normal competitions for mates, status, wealth and power; loss for some is always possible, if not likely; fanatic fundamentalists are always at hand to blame a particular arrangement for all problems, or to make accusations of conspiracies set up to do evil; sometimes such 'moral guardians' agitate for riots and bombing as the only way to purify a 'system'.

Agonistic responses among potentially conflicting individuals and groups are always just under the surface of a political veneer in human social life. This is particularly the case with regard to behaviour among groups with long historical antagonisms to each other (based upon ethnicity, religion, race, ideologies, geographical access or economic self-interests, for example). Certainly, history records that considerable aggression and violence has existed between factions within societies, and between authorities and perceived 'obstructionists', 'traitors' and threatening 'outsiders', and between citizens groups and apparently overpowering, corrupt political authorities.

One means of trying to control all this on the part of political authorities has been to take over – indeed monopolize – one of humankinds' most fundamental and universal propensities regarding humiliation and actual harm – the propensity to seek vengeance (Cf., Nietzsche, 1961, 1886; Lopreato, 1984; Kerrigan, 1996; Kapuscinski 2008). This desire, and resultant behaviour, may have evolved at the same time as the evolution of political behaviour, or possibly even before, but its capture by political authorities marks a major transformation in the political evolution of humans.

Vengeance politics and the search for justice

When someone harms us, humiliates us, cuckolds us, steals from us, lies to us, does not live up to their promises to us, or simply disdains us, we want to get even. We want vengeance. But this may not be easy because if they are strong and confident enough to do the above we may have little chance against them. If we have mates who are willing to help us, however, we stand a much better chance. And if the person who harmed us knew ahead of time that we have mates who are willing to help us get retribution, they may not have dared to harm us in the first place. Theoretically, if everyone has mates who are willing to take vengeance for them, nobody harms or humiliates or steals from or lies to an individual for fear of group retribution, and peace reigns over the land.

Unfortunately for peace, human nature and events make this work better in theory than in practice. Male youth gangs, for example, are volatile and often like to go to war just for the fun and kudos of it, and for them 'sticking-together', whatever the details or guilt responsibilities of a situation may be, is fundamental to their individual self-images and senses of honour. In a way this is what makes vengeance seeking effective – knowing that vengeance will be sought for harm to an individual *whatever* the rights and wrongs of the situation. Knowing that there are people both keen and honour bound to seek it, and knowing that it may be done without warning, when not expected, in a gang upon an individual, makes many hesitate before harming anyone.

However, this is not always automatically the case. An individual's mates may not want to support them for fear of starting a war of retaliation over transgressions which they do not especially approve of themselves. Because an individual looks to the support of allies there is always the possibility that regular allies, not having been so emotionally (or physically) affected as the harmed individual, may have cooler heads. They may even think that the individual trying to generate support is a bit of a hothead, or even a bully, or a rabble-rouser. While they may support them verbally, they may not be eager to risk their own lives, limbs and reputations in a war over their mate's claimed suffering. They may also think that, perhaps, asking for compensation might be a better idea, or they may like to ask some of the less warlike elders what they think (or could they, the elders, intervene, they wonder).

As a basic approach to 'justice' in human social/political life, vengeance taking and vengeance feuds have existed for a very long time. First, we have the widespread existence of vengeance and counter-vengeance as a means of providing a degree of mutual security in a number of hunting and gathering and early herding and agricultural societies, the Yanomamo being perhaps an extreme example (Chagnon, 1968; see also, Evans-Pritchard, 1940; Mair, 1962, 1974; Keeley, 1996; Pinker, 2002). Second, throughout at least written history we have had a threat of war and arms races among settled societies which communicate: 'you may attack us, you may be stronger, but the cost to you will be extremely high'; third, we have various political alliance systems that were designed to maintain peace among potential conflicting societies through the threat of combined vengeance against a lone transgressing society (going back at least to ancient Mesopotamia, progressing through such things as the 'Holy Alliance' following the Napoleonic wars and existing today as in NATO).

This balance-of-threat principle can, in some circumstances, evolve into a balance-of-power set of relationships among a number of potentially antagonistic factions. These are relationships in which elders/diplomats tend to replace youth force/warriors. When this happens elders/political leaders have progressively attempted to take vengeance decisions out of the hands of individuals and male youth gangs (and from families and special interests), and to give them instead to elders, powerful patrons, kings, magistrates and courts. And these generally do not like competition in these matters, as it is a direct threat to their authority. Elders (as with parents in general) like to be obeyed, and they desire squabbling children/fighting youths to get along, or at least to leave each other alone. Lords of the manor do not want serfs feuding rather than working and obeying 'his lordship'. One of the major criteria to see if a king is really a king in more than name only is that he has the ability to enforce his 'right' to the exercising of judicial processes and sanctions. And in modern state societies, individual, family, gang or special interest vengeance-taking is not allowed by law at any time whatsoever (although in some places 'crimes of passion' are dealt with more leniently than pre-meditated crimes for profit or fun, and in most places 'extenuating circumstances' are taken into consideration in sentencing).

Anthropologically and historically, in order for a centralizing of vengeance responsibilities to take place there has to be one or more of the following conditions: (1) recognition by youth and elders alike that constant feuding is just too dangerous; (2) one vengeance-seeking faction is able to dominate all others and so demand subservience from other factions; (3) an external power conquers an area full of feuding factions and demands an end to it; (4) a foreign king is invited in to provide law and order among warring factions. In all cases, there has to be a general acceptance by subjected populations that this new state of affairs is better than the lawlessness of constant feuding. But even then any balances of power achieved could well remain extremely precarious. This is evidenced by the observation of the very speedy return to factional, sectarian, ethnic and regional vengeance fighting when central political authority breaks down, including in areas of the modern world previously governed by political states of relative stability. And even within the most developed of stable modern states, central authorities have been unable to completely stop feuding among youth gangs, drug gangs, criminal gangs and sometimes even ethnic communities.

Nevertheless, long-term history has been one of centralizing 'vengeance seeking' (judicial authority) rather than the other way around. In this process, patrons, court officials and general populations have come to a notion of retribution in a more *abstract* sense than one of simply physically striking back at seeming wrong doers. In this we have been able to rely on an apparent innate sense of 'fairness' (see Chapter 7) – fairness not only in terms of an offending action, but also in terms of appropriate retribution. In the process we have come up with theories concerning methods of dealing with unfairness that are more 'political' than simply striking out, attacking or threatening massive retaliation.

Forms of behaviour such as those referred to by games theorists as 'tit-for-tat', 'grudger' or 'even-up' behaviours are examples of patterns of behaviour, found even in pre-human primate nature, which underpin human methods of avoiding out-and-out physical vengeance (Cf., Axelrod, 1984; Dawkins, 1989; Ellickson, 1991; de Waal, 2003). Tit-for-tat, like an eye for an eye, a tooth for a tooth, is a first step to the notion that a 'wrong' has a punishment that is *appropriate* for *that* wrong. It also opens the door to the idea that compensation can be paid for a crime against a person, so punishment becomes a somewhat more abstract process, further removed from the actual behaviour itself. In 'grudger' behaviour (Dawkins, 1989) individuals and groups grudgingly let others get away with a certain amount of misbehaviour before becoming aggressive, then 'forgive' them but consistently punish them for repeated offences. In more modern and elaborated circumstances legal authorities let people get away with certain minor offences with warnings but deal more harshly with repeated offences.

These approaches recognize some of the ambiguities found in human motivations and behaviour. They recognize that 'extenuating circumstances' might be significant in determining punishment, and that individuals can be forced or encouraged to change their ways. The notions of understanding and forgiveness have been introduced. Understanding and forgiveness, along with 'look-the-other-way', are clearly seen in 'even-up' behaviour (Ellickson, 1991). This approach allows minor imbalances in unfair behaviour to exist in anticipation that things will even out in the long run. Certainly, even-up behaviour seems to describe much of our lives in the 'transaction – cost laden, mistake filled world in which we live [today]' (Ellickson, p. 227). In even-up behaviour, when punishments are dished out they are done in a measured sort of way, to restore relationships to what they were in 'better times'. As this becomes elaborated and more formal, an element of 'restorative justice', or 're-education', enters the picture.

Although most studies have a rationalistic 'economic exchange' and/or almost mathematical games-theory bias regarding the evolution of a sense of fairness, it seems more likely that it developed from becoming emotionally aware of the reproductive and sexual 'dangers' of living and competing with other humans in terms of a wide number of personal, sexual, parenting, status and political interactions. This would have been because of an awareness by individuals that if they were perceived to be sneaky, untrustworthy, lying, unhelpful, non-sharing, non-cooperative, bullying, dismissive – in other words, 'unfair' – they might well end up subject to all sorts of rejection, as well as experiencing a lack of sex, friends and partnerships (and as a result would have had a decided lack of reproductive success).

In straightforward cases, recognizing unfairness is relatively easy (for example, in normal circumstances giving a reward to one child and a spanking to another for exactly the same act, or constantly taking from others but never giving back in return seem unfair). In less straightforward cases it is more difficult. For example, who should get what from an acrimonious divorce of two seemingly worthy individuals? But perhaps even more difficult, who should take the blame, which one should lose status, which one should be responsible for the children, who should stay away from whom? In this second case, friends and relatives on both sides will analyse the relationship detail by detail to assess the relative merits of the two adversaries. In the end someone will have to have the authority to decide in those matters that have material or 'rights' consequences if blood feuding is to be prevented. Not everyone will be happy whatever the final result; some will think it fair and some think it unfair. What is important here is that we have a strong sense that some things are fair and other things are unfair in human life, and sometimes it is not absolutely clear who is to blame and who is in the right, but that a decision can be made without insisting upon the idea that someone must suffer physical punishment, complete humiliation or total destruction in order for honour to be served. We have this sense of fairness without needing to be personally involved in particular relationships or incidents in which an apparent unfairness has taken place, and without any recourse to any concept of *morality* whatsoever.

As a result, we are quite capable of analysing human relationships and incidents in the abstract and in passing judgment upon a variety of hypothetical actions humans might undertake. We do this all the time: in gossip, in arguing about the nature of human passions and intentions, in talking about human personalities, in remembering and mythologizing seemingly unfair incidents in the past and in listening to the explanations (rationalizations) of those participants caught 'red-handed'. (Myths, novels and soap operas, for example, thrive on this element of human nature.) Through our capacity to understand and theorize fairness, our desire for vengeance against those who harm and humiliate us has become a more abstract desire for vengeance against unfairness in *principle* (at least among those not directly involved).

But there remains the problem of enforcement. A sense of unfairness is one thing, as is deciding on particular punishments; the ability to enforce it is quite another. However, anthropologically and historically there has been a ready means of enforcement in the form of the tamed youth-force – skilfully co-opted by elders and/or by a community. Thus, the latent threat of physical vengeance never disappears altogether. It remains to prevent transgressors in the first place, but ready to act as vigilante groups and sheriff's posses (under the direction of elders) when transgressions occur, punishments and compensation have been ordered, and the guilty try to resist. It is not a major step from here to the idea that particular crimes are crimes against the elders (general community) rather than against a single individual or a specific group of individuals. At this stage it is common to impose fines (payable to a community) and incarceration so that crime and punishment is even further removed from the individuals actually involved. Police apprehend culprits, courts try and sentence them and prison guards keep them in punishment conditions; and the harmed individual is not allowed to interfere (except perhaps to give a statement and be a witness, *if required*). The seeking of vengeance, thus, has evolved to being the upholding of 'the rule of law' as a general human principle.

The idea of subordination to the rule of law takes us a long way towards the development of civilization. It recognizes that we sometimes have to subordinate our immediate desires for vengeance (as well as our desires for long-term material, status or political gains) to judging/protecting authorities. But we see this as virtue not surrender, and we are admired for it. As Nietzsche observed: 'if you possess a virtue, a real whole virtue (and not a puny drive towards a virtue!) – you are its *victim*! But that is precisely why your neighbour praises your virtue!' (1974, section 21, his emphasis). And this is a 'virtue' that has provided reproductive success because it is a virtue that musters at least some of the power and wealth of a community to taking vengeance on behalf of outnumbered self – the threat of which acts to keep self safe most of the time. Of course, those who administer such justice often have even more reproductive success because to be able to undertake such a role suggests possession of power, wealth and political skills that are themselves sexually attractive (Miller, 2000).

So, whatever form it takes, and however it has been tempered, it is through the human desire for vengeance and the capacity for alliance politics (sometimes acting as vengeance or vigilante groups) that humans became strongly motivated to enforce and follow established codes of behaviour. Without the powerful desires and fears that underpin these behaviours cheats would run rampant and social life would be very different from what we generally know it to be, and obedience would not have become a virtue. The evolution of most of the above behaviours, and their emotional-cognitive underpinnings, does not have to be postulated as a dramatic change, but rather as an evolutionary development from general primate life. For example, on some occasions chimpanzees of both sexes have been observed to engage in 'moralistic aggression' (Trivers, 1985; see also, Clutton-Brock and Parker, 1995) by trying to punish those who have not reciprocated in a social interaction (de Waal, 1982).

However, evolving humans have mitigated some of the ferocity of primate politics in some circumstances, but not in others. We have developed the concept of the rule of law, but at the same time there remains factional and sectarian violence, genocide and a multitude of minor wars. This has been especially noticeable where the youth-force has escaped supervision by elders. Yet, ironically perhaps, it is the continuing attempts at *supervision* and *control* of the youth force that has allowed cities to grow into empires, and for empires and political states to 'defend' themselves and expand. This is because the antics of the youth-force at times has served older, more dominant males in attempts to gain and/or protect privileged positions, just as it has served youth itself through alliance support.

Patronage politics

Among some individuals male–male attractions are sublimated (daydreamed) into hero worship of the male found attractive, and into fantasies of themselves as 'bloodbrothers' or team mates of the hero. And within male alliances there are usually one or more dominant individuals, and the alliance itself often operates in terms of patronage. Here the primate background again has relevance. Among chimpanzees, for example, male alliances are directly related to status positions. A dominant male, or one that aspires to dominance, has no chance unless supported by allies. And there generally are ready clients, usually among younger males, for aiding patrons in seeking dominance. Also, aspiring individuals are likely to give support to a dominant male that may appear to be on the decline, the better to be appreciated by, and be essential to, the dominant male, and to be in a position to take over from them when the inevitable decline happens. Very often the support of females is important, both for making bids for dominance and for fending them off. Males try to entice, if not cajole, females and lower status males into supporting them by being peacemakers among them and by protecting them. Females can be quite choosy as to whom they support, and for how long (de Waal, 1982, 1988, 1989; de Waal and Lanting, 1998; Pusey, 2002; papers in Hamburg and McCown, 1979).

During human evolution, if relatively young and/or weak males were given general protection by aspiring, more established males for help rendered in gaining dominance, and in warding off challenges to that dominance, both could have benefitted. This would especially have been the case for young males if they later inherited access to the dominant male's fishing spot, hunting area, tree shelter or females, for example. The principle of patronage is that relatively powerful individuals attract and elicit aid (tribute) and physical and political support from clients who comply in order to operate in the shadow of the powerful individual for advantages over individuals who would otherwise be their equal, or even superior, and for protection against powerful outside individuals. In patronage, when a client is dealing with his supposed equal (or even slight superior) who is not part of a patron/client network (or is a member of a weaker network), the equal/superior individual knows that in reality it is the client's powerful patron/network that he is being confronted with. In modern times, for example, even the most moronic, illiterate, lazy enforcer for the Mafia can often get his way without using his gun or even raising his voice. In patronage, charismatic/powerful males are provided with a source of wealth, labour and youth-force energies to increase their power base and thus become even more attractive to young, vigorous male retainers, generating a cycle of continuing success for all involved.

And it is not only young males that are attracted to patrons. In the early stages it could have been that very young girls were attracted to the 'big man' and his top retainers; or perhaps the parents or siblings of some very young girls pushed their daughters/sisters towards them, 'to be of service' to them in exchange for protection, resources, privileges and for establishing a valuable social link. Some of these females might have been pre-pubertal, still attractive because of their child-like attractiveness but not yet ready for 'marriage', acting more as household servants than as wives. So the first wife of the 'big man' might not have objected too much. The patron might be saving the young girls for later, or to offer them as future wives to loyal retainers. He may, however, still expect to have the first sexual encounter with them (perhaps for a certain period of time). This may have been the origin of the *droit du seigneur*, which still existed in the Middle Ages in Europe, in which the lord of the manor, or the king at higher levels, expected to have the first sex with the bride of one of his underlings on their wedding night.

This pattern of sexual favours being related to patronage may, in fact, have been relatively widespread during the early stages of human evolution given the degree of status-enhanced sexual access still seen in the chimp (and to a lesser extent bonobo) lines. And during human evolution if sexual success was based on play (as suggested here), being friends with higher status males could have provided more access to female playmates drawn to those friends than being friends with lower status males would have done. But to remain friends with the higher status individual may have required letting him have a degree of sexual access, or even helping him to have access, to one's own 'object of desire'. And, indeed, this pattern still exists in the anthropological literature. Among the Ache of Paraguay, for example, a client often had to let important men have access to his wife if he, and his family, were to receive any of the food shared out after a hunt (Kaplan and Hill, 1985; Hill and Kaplan, 1988; see also, Gray and Garcia, 2013). And as we have seen, it lingered on in the European Middle Ages, and worldwide it is still common that lower ranked men (and women) send their daughters (and sisters and wives) into service (where masters expect a degree of sexual servicing).

But, it can be imagined that this 'system' would generate a degree of male/male conflict. This certainly seems to have been the case in ancient Mesopotamia in the third millennium BC as seen in the epic story of Gilgamesh.

But the men of Uruk muttered in their houses, 'Gilgamesh sounds the tocsin for his amusements, his arrogance has no bounds by day or night. No son is left with his father, for Gilgamesh takes them all; yet the king should be a shepherd to his people. His lust leaves no virgin to her lover, neither the warrior's daughter nor the wife of the noble' (*Epic*, 1964, p. 60).

The wise patron curbs his instincts (as the gods forced Gilgamesh to do) or he has nothing but trouble. The Azande, for example, had a saying that a good prince, or king, does not seduce his follower's wives (Evans-Pritchard, 1937).

Whatever the exact role of sex, and the amount of conflict it could generate, patronage as a process was clearly seen in a number of hunting and gathering societies, such as among Australian aboriginal groups (Service, 1963; Ridley, 1993), among the American plains Indians (Ambrose, 1975; Larson, 1997; Lewis, 1998; Roberts, 1998) and in pastoral Nuer groups with their cowherds (Evans-Pritchard, 1940). But anthropologically and historically patronage was most in evidence where humans were more settled and some males could accumulate noticeable wealth, status and power, such as where fishing, settled animal husbandry or early agricultural life was common (Cf., Benedict, 1946; Kluckhohn and Leighton, 1946; Service, 1963; Harris, 1977, 1978; Fagan, 1986). When we get to the formation of statelets, as among the Zulu, Inca or Aztecs, patronage, along with alliance politics, became a major basis of state power (Service, 1963; Walter, 1969; Harris, 1978).

Patronage suggests selection pressures for a degree of relatively long-term *deferment* of sexual, status and power gratifications among average males (as also would have been the case to a lesser extent with the evolution of alliance behaviour). The client in patronage circumstances has to be very careful indeed as to how he approaches his master's possessions, status, power, wives and daughters (or the master's *desired* possessions, status, power or wives). This can become quite complex because it is not always clear what or whom the master desires, and the patron's offers of benefits can sometimes become relatively ambiguous. To regularize patronage relationships, to eliminate some of the potential dangers in them, a number of traditions, codes of behaviour, taboos, rules and even laws were formulated throughout human history. This was especially the case in the most elaborate developments of patronage politics in the various forms of so-called feudalism that have existed (Cf., Colburn, 1956; Bloch, 1961/1965; Green, 1971; Dumont, 1980).

These have included strict codes of chivalry so that ladies of the court were worshipped rather than lusted after. Codes of honour were established in order to be able to show (and compel) willingness to die for one's patron/or in defence of loyal clients. Grandiose local fortifications and other displays of power to show that a vassal could protect his lord from enemies approaching from the vassal's direction were constructed. Patterns of dress, banqueting and of speaking to show standing and to indicate the respect and deference expected between particular individuals emerged. This included demonstrations of subservience to gods and religious devotion to show humility and reverence for order, authority and hierarchy (even among the high and mighty), a high value was placed on the concepts of duty, responsibility and obedience (at all levels), and a mythical notion of the almost supernatural characteristics and exploits of past warriors, shamans and war-lords emerged.

As with most things human, there was a tendency to exaggerate all of this as patronage networks increased in size, complexity and prosperity. So, the homes/fortifications of dominant males and their courtiers became magical/mystical places (royal courts); they were places where admiring popular attention became intensely focused, emulated; where only the most attractive females were allowed (who became court ladies not servants); where considerable wealth was accumulating and was ostentatiously displayed; where palaces and mighty churches were built; and it definitely was where 'the action was'. Retainers became 'insiders', heroes, swooned over by the ladies and admired and feared by the general populace. They might be showered with gifts and privileges – looked up to, if not worshipped, as being special individuals. Religious leaders (themselves retainers of sorts) often considered that these high placed people and places had spiritual qualities. It became increasingly hard to marry into any level of such insiders unless one was an equal or above.

Certainly mythology and history have highlighted and glorified the admiration and *separation* given to warrior kings and their loyal knights (and to their families and their officials). This, however, raises the question: as patronage developed and patrons and high placed retainers worked to solidify their positions, to make it almost caste-like, how was it possible for ordinary folk to continue to play the game? What prevented only the strong, handsome, beautiful and wealthy having any chance of becoming patrons, retainers or ladies of the court, and of reaping any benefits? The answer is that patrons *required* allies in the form of trusted and loyal clients in order to provide the manpower, resources and threat of vengeance required to maintain their positions and to impose the fear and respect that made their ability to offer protection, judgments and benefits effective. To provide all that was asked of them a client also needed clients, and consequently patronage worked its way down. And, patrons at *all levels* had to win loyalty, not just demand it. Demanding loyalty usually generates more conspiracies and rebels than allies and so is most often a useless tactic. Indeed it is often a very dangerous one. As long as protection and recognized social standing were on offer, however, there were usually many willing, loyal clients, because being a client gave power and status even at lower levels, at least in a local community.

And if those at the bottom were left to get on with their lives, with some guaranteed land or fishing grounds; were able to marry and have children, with enough of their production left over to have occasional festivals; had taverns to drink in and churches to worship in; and if they had communal storages of grains and other foods (managed by the patrons) for times of bad harvests and vicious winters, and protection from wandering youth gangs (robbers), they felt themselves relatively privileged, not prone to rebellion. Lowly serfs, for example, came to expect – and be given – certain *'rights'* to the use of land (for themselves and for their descendants), and protection from their neighbours for good behaviour and labour rendered to their patron. Wealth, high status, privileges and political power might noticeably diminish on its way to the bottom, but protection from neighbours and foreign enemies, and guarantees of a respected place in a community of peers (for self, family and descendants) was sufficient for social stability.

For this to work effectively, however, there had to be a degree of interaction, trust and even sense of comradeship among those on either side of a potential 'fault line' in the hierarchy. A fault line would be a location where a caste line might 'logically' and/or ideologically be formed (for example, between a royal family and a core of knights, or between knights and the king's courtiers, or between a local lord and his tenants/serfs). Trust across potential fault lines was developed on the basis of personal friendships, exchange relationships, sexual liaisons and marriages, and simply by desires to make things work on the part of participants when specific issues arose.

Historically religious institutions often played a part in linking and mediating across such lines; in European feudalism, for example, the local clergy often played such a role, while at the same time often providing some of the ideas for maintaining hierarchy in principle – through teaching the virtues of duty, obedience and deference to authority, for example. The Catholic Church's attempts through the formulation of rules against incest with even distant cousins, and strictures against endogamy and concubinage, and striving to impose monogamy, put pressure on caste-aspiring nobilities to out-breed some distance, preventing a massive concentration of landed wealth in a very few extended families, and so maintained a degree of separation and boundary fluidity in Europe (Cf., Goody, 1983; MacDonald, 1995).

In some circumstances, however, religious notions were the bases for rigid separations, but here other means to cross and to administer to fault lines could be developed. For instance, even in traditional India where a rigid 'caste-system' had been religiously underwritten, even at its highest development there was considerable interaction among sub-castes close to either side of a particular line (often competing for privileges and status even across major caste – *varna* – lines when positioned near it). And anyway many people lived below the caste system as outcasts or untouchables, while at the top, local kings or warlords operating on *patronage* principles were more important politically and socially in maintaining order and protecting people than caste-based demarcations – or their spiritual defenders (Brahmins) – throughout most of Indian history (Hall, 1986; see also, Srinivas, 1952).

Historically, in some locations, top clients were progressively able to have control of great amounts of land and to take on large numbers of clients of their own, and so 'working patronage' came to include a very large number of people over very wide geographical areas. But patronage has its own inherent problems. These include the propensities among top clients to undertake coups, indulge in vengeance feuding with each other and try to establish rival patronages. It is always easy to lose trust in such a setup, to feel that conspiracies against one are brewing, that one has to make a pre-emptive strike or be for the chop. If patrons try to dominate females there is always the testosterone-charged youth-force to contend with; plural wives can become quite disgruntled if they think their offspring are not treated as well as the offspring of other wives; and aspiring clients might like some of these wives, who might give away court secrets and intentions, to attentive aspiring clients who seem to have prospects. As things go to hell and it becomes impossible to trust anyone, top patrons can become quite arbitrary and brutal, losing loyalty all the way to the bottom.

Nevertheless, throughout history patrons and sub-patrons (and their descendants, and/or their replacements) were often secure enough to build great temples, cities, palaces and cathedrals, ones that in some cases took 200 years or more to complete. Whatever the specifics of various patterns, from the time of the development of agriculture to the emergence of the commercial and industrial revolutions and the rise of the modern state, patronage networks provided the basis of political organization right across the world. Indeed, it is possible to argue that, in various combinations, alliance/vengeance and patronage politics have provided the bases of modern politics.

Modern politics

As humans (1) crowded together in large urban conglomerations, and increasingly relied on high levels of technology to manufacture, sell and consume objects and services for survival and for display, and (2) gave allegiance to central governments (at first based on the notion of *nations* of people and then on *citizenship*), a variety of family patterns, gangs, cults, religious congregations and socio/economic organizations (known as civil society) remained as a means of conducting everyday life. And these retained a great deal of the 'old politics'. For example, every sort of grouping within states (religious, cultural, criminal, sporting, trade union, business and government) have been (and are) run by a small clique (largely of males) whose alliance camaraderie is essential to its functioning. The majority of such groups remain hierarchical in which a top boss (patron) and his immediate vice presidents appoint sub-bosses, who appoint sub-sub bosses, right on down to the cleaners of buildings. All down the line individuals expect a degree of protection (job security) and benefits for loyal service. Cleaners, for example, may get a guaranteed wage (possibly health care) and pensions.

Within states factional conflicts and competition have existed as kinship, ethnic and regional feuds, and as competing sports teams, trade unions, businesses, criminal gangs, religious denominations, political factions and political parties. And, as in times past, gangs of young men roam about, only now it is in the streets, pubs and football terraces, and they fight with each other – and with authorities – and elders feel the need to control and tame them. These young men are kept in schools, sent to mines, apprenticeships, assembly lines and prisons, but they are also recruited into police forces, local militias, criminal gangs and warring political factions, or enlisted by government authorities themselves to keep these latter under control. States themselves enter into a process of 'politics among nations' in which each state struggles with others in terms of its leaders pursuing 'national interests'. Alliances and patron/client state arrangements are common in which powerful patron states usually try to keep their client states under some control and to mediate among them (Morgenthau, 1978).

Some new political developments have evolved with the rise of industrial societies and modern political states. The concept of 'the rule of law' (originating from the mitigating of vengeance seeking) has continued, but has been further developed in terms of creating and formalizing more, and more clearly specified, laws to replace customs, traditions and informal taboos; and the formation of laws and their enforcement has been monopolized by state authorities (although local authorities are often left to deal with minor issues). The idea of crimes against 'the people' or against humanity has largely been a development of modern times, although it had its precedent in crimes against elders/the community from vengeance times, and was further developed as religious notions of crimes against God/natural order in patronage times. The rule of law has come to include the notion that every individual citizen should be treated the same as every other in the eyes of the law; every individual theoretically has equal legal rights – a departure from 'high patronage' times when often there were different laws for aristocrats and commoners. What is also new in modern times is that individual identity (based on self-awareness) and individual culpability in terms of 'the rule of law' suggests individual moral worth which has been turned into a secular ideology of *individualism*. From this, concepts of individual (human) rights have been formally stated and written down as 'bills of rights' and as aspects of legal codes, and these cover millions of people rather than twenty-five to fifty members of a band or believers in a particular religious doctrine.

During these developments four major concepts have been employed: *freedom* (*liberty*), *equality, fraternity* and *democracy*. Freedom has not meant that individuals have become free to do just what they want; indeed, in modern societies humans are probably confronted by more rules, regulations and laws than ever before in human history. What freedom has largely come to mean in practice is that humans are free to be able to escape from parental control at a certain age; to be able to choose their own reproductive mate; to live in the neighbourhood they can afford; to wear whatever clothing or jewellery they have the ability to purchase; to buy whatever toys they can afford; to choose their own religion, sports team and political party; to move from one job to another and from one location to another; to set up in business if they can raise the resources to do so; to trade with each other. But these freedoms usually give rise to considerable conflict, envy, jealousy and a constant fear of being 'unfree', of losing out, and dilemmas over 'too much' choice and feelings that one will make the wrong one.

And it often leads to an inequality of status, wealth and power, thus generating problems with the ideal of *equality*, sometimes dramatically so, as when some individuals end up with an enormous amount of status, wealth and power and others with almost nothing. So, there have been those who have argued that *equality* rather than freedom should be the prime objective of a political community. But no one has ever been able to agree as to what equality should mean. Should it be equality of: respect, access to reproductive opportunities, sexual outlets, of child-care responsibilities; or should it be equality of moral value/dignity/social worth (but what do these mean?), or opportunities, treatment under the law, material possessions, or access to political authority? Should these have to be earned or should they be guaranteed as outcomes? Should age, gender, experience and/or effort matter? And given human tendencies to envy, jealousy, fears of failure and potentials for guilt, along with conflicting desires and fears, different individual energy levels, varying degrees of curiosity, different levels of ambition and pure chance, it is unlikely that any of the above 'equalities' could ever be achieved, or that humans could ever agree as to which aspects of equality are most crucial.

But that has not stopped humans from trying to impose 'equality' on others, often through the notion of the '*fraternity*' of humankind. In certain theocratic societies during history, for example, individual display, indulgences of all kind, self-expression and freedom of choice – as seen in the above type freedoms – have been perceived as being displeasing to God and extremely disruptive of God's kingdom (including the human part of it). The only freedom worth having, it has been argued, has been to be able to pray and to obey the 'word of God', and to be able to give oneself completely to God; it is freedom from sin or, in some versions, freedom from desire. Similarly, in socialist parts of the world freedom has been given a notion of 'freedom from' – freedom from starvation, freezing to death, exploitation and 'false consciousnesses'. Once these are achieved, it has been assumed, such things as a tendency to display, competitive feelings, envy, jealousy and 'excessive desire' will simply disappear and in its place will arise a *fraternal* notion of solidarity, that we are all part of a greater moral whole of *humanity*.

Unfortunately, it seems that to achieve the freedom of God's grace or the freedoms from desire or from false consciousness, individuals have to be so stifled, restricted and controlled that any sense of individual *purpose* or *value* disappears; an individual's ability to display to attract mates is greatly curtailed; an individual's ability to be in a disruptive gang against authority is stymied; the whole idea of self-expression and deviation from 'sameness' becomes selfishness. All this restricting usually requires extremely dictatorial authorities in order to be effective. Obsessive, sanctimonious, intolerant dedicated leaders begin to tell everyone how they should live, feel, love and think. Sinners are found everywhere and sent to the stake to be purified, to labour camps or 'the fields' to be re-educated, to psychiatrists to be reprogrammed. But human nature defeats all of these attempts, even when elements of youth power (told that their work is to save humankind and that they are the greatest heroes imaginable) are unleashed against those who are unable (or unwilling) to conform.

Conflicting interests are not easy to suppress. So, certain philosophers have argued that modern states should be run on the basis of enlightenment rationality and *democracy* in which 'real' people use their 'innate reason' to choose leaders and policies that best reflect their legitimate concerns – leaders who can rationally find common ground among conflicting views. However, states that have anything approximating democracy all rely on competing political parties, which are based almost exclusively on patronage, and are designed to compete for power within the confines of political states (that is their reason for existence). These inevitably seem to use every *emotional* argument imaginable; apply stereotype after stereotype; work hard to generate hero worship; and rely very heavily on 'good versus evil' notions of human politics.

In democracies politicking can become intense. Populists promise everything (including the impossible); demigods claim knowledge of the road to salvation and call out to voters with surprising success; moral panics uncover degeneration and wrong doers; mass media turn every event into a question of life and death, 'good versus evil'. In the process of dividing resources questions arise about how to decide who should get what and how much – often with as many answers as participants. In terms of duties and obligations, disputes abound. One person's rationality is another's evidence of irrationality. Defenders of 'true democracy' tend to believe that if everyone had a say in public affairs everyone would believe exactly as they themselves do, and *vote* as they do. But this is not the way it is. Voting can risk a 'tyranny of the majority' (de Tocqueville, 1956), turning every politician into a 'middle-of-the-roader' and cliché master. What's more, voting for leaders means very little if leaders are nominated by a patronage-based party with a monopoly of political control.

To overcome some of the above problems, a number of politicians and philosophers have set out to formalize a concept of (as noted) 'the rule of law' and of 'individual rights', including 'minority rights', and to establish 'separations of powers' so that special interests or power blocks can never monopolize power. This is a sort of approach that recognizes the ubiquitous conflicts, confusions and emotionality of human nature (and resulting socio/political behaviour), and is based on the principle that the role of government is to reduce the harm we do to each other as much as possible and to help people get through life as best as is feasible without infringing rights or causing undue anxiety and fear. However, for a separation of power to work a 'balance of power' has to be achieved. But real balances can be hard to establish and/or maintain. The greatest success in doing so is seen where organizations and groupings are able to *continuously* form in order to advocate particular interests, contend for power and/or to oppose other interests, contenders and authorities, and when there are plenty of 'watchers' and 'watchers of the watchers'. But of course, any of these groupings or watchers can always overdo it and cause more disruption that tranquillity. So, 'balances of power' always remain *political processes* rather than static 'systems'. The result is a constant process of *politicking*. Indeed, an argument can be made that human societies on any scale much larger than wandering bands are hierarchically based networks of politicking individuals and groups.

Rationalists, idealists and spiritualists attack this approach as being the epitome of inefficiency, corruption and a denial of a higher human potential; as a complete giving in to intellectual, social and political laziness (if not disguised self-interest). For many of those not directly involved, politics appears to be a playground for self-aggrandizing politicians who do nothing but lie in order to protect their positions. Politics, they argue, are the natural product of human selfishness and vested interests. Unfortunately for such thinkers, various theological/idealistic/rationalistic attempts to get away from politics have all failed, and indeed have harmed to the point of killing multitudes of heretics, bourgeois exploiters, obstructionists, biologically impure races, immoral sinners and holders of false ideologies. What these approaches refuse to accept is that given human nature and unpredictable events it is not necessarily a good idea to try to do away with muddle-through, but relatively tolerant, human politics.

The focus in this section has been on male politics because during human history alliances, vengeance seeking and patronage have largely been male affairs. But that does not mean that females have not been players – after all alliances, vengeance seeking and patronage emotions evolved as motivators for males to secure and protect access to females. So, it would be surprising if the evolution of female emotions, desires and fears did not contribute to human politics, both as selection pressures for male politics but also in terms of their contribution to political practices themselves.

Female politics

Much more than males, females are able to use sexuality relatively directly for their own reproductive, economic, status and political ends (Symons, 1979; Hrdy, 1981; Fedigan, 1986; Haraway, 1989; Campbell, 2013; among apes see: de Waal and Lanting, 1998; papers in de Waal, 2002). Showing signs of potential receptivity usually gains the attention of most males; it brings a number of them running, and keeps some around competing with each other for a female's attention and favours. As we have seen, females who were able to keep a number of suitors in the frame to see which of them seemed to have good genes, but also might be into deferred gratification and be willing to hang around and provide resources (bring gifts), and even help in future child-care after copulation were favoured by selection. This was because of increased infant dependency and an increased rate of birth during human evolution. This female capacity required political skills which made it possible for females to be able to display their sexual attractiveness and potential receptivity without inviting immediate copulation, but still keep males interested. Once a 'good' male had been identified a female's political work was to monopolize the care and support he was able to provide and lock him into long-term aid in child-care. It was important that she convince the selected male that he would be the father of her future offspring.

Additionally, females could find themselves mothers of three or four very dependent offspring at the same time, unlike female chimps, for example, with just one (less physically dependent at birth) offspring and a gap of five to six years before another. This meant the evolution of a number of care-enticing behaviours on the part of offspring, but also the political skills on the part of mothers to keep the siblings apart, share her attention among them and her mate, wean them, keep dad happy so that he did not take off with some younger eager female, make sure that provisions continued to be provided and be able to establish a network of relatives and friends to help in child-care, material provision and emotional support in difficult times.

Emotional power

Females have evolved relatively complex mechanisms for achieving the above. To remind ourselves: it was suggested that males evolved to become emotionally dependent on females and their offspring, and that females evolved to be able to get the very best out of this. The emotional power of a mother over her offspring, especially her sons, has a long history in the primate experience (clearly seen in the bonobo mother–son relationship). In humans it is a power based on emotional dependency and guilt, not on physical threat, youth force or patronage (Cf., Klein, 1952; Bateson, *et al.*, 1956; Laing, 1965). It takes political skill (often unconsciously applied) to make it work with offspring, and even more to apply it to boyfriends and husbands. In the realm of boyfriends and husbands, a female has to contend with a relatively strong sexual urge in her male, competing females, her mother-in-law, alliance loyalties of her chosen male and his suspicions concerning his paternity.

Thus we get the somewhat stereotyped but not altogether inaccurate notion that females are more in touch with their emotions, feelings, desires and fears than males. For example, 'studies have shown consistently that women are better than men at reading and responding to subtle cues about mood and temperament. They're more trusting, more empathetic and more focused on one-on-one friendships' (Kuchment, 2004); 'on average, women engage in more "consistent" social smiling and "maintained" eye contact than does the average man' (Baron-Cohen, 2004, p. 58). And this starts almost from the very first day of birth, in that female babies stare at human faces much more than boys, while boys stare at mechanical objects more than at faces (Baron-Cohen, 2004; Kuchment, 2004). Females clearly seem (in general) better able to understand the human 'love psyche' than males (Buss, 1989, 1994; Taylor, 2002; Baron-Cohen, 2004).

Like their male counterparts, females have to be able to trust their friends to leave their mate alone. An outside female challenge to a paired-up female threatens the paired female's long-term investment – all the effort she has put in turning a lustful, irresponsible male into an emotionally dependent, reliable 'son'/husband is suddenly jeopardized; she may be left pregnant and therefore out of action for some considerable time with no support while he goes off with his 'new stuff'. In males, respect for sexual partners of friends evolved as a sense of anxiety when around a mate's female and as a sense of honour attached to the sanctity of partnership and/or brotherhood bonds. With females this 'mutual trust' has evolved through reciprocal empathy – gossiping, confiding – regarding the perils of 'keeping a man honest', keeping his sexual demands under control, his mother at bay, the demands upon him by his mates at a minimum and him off the drink. Females find it easy to empathize with each other on these matters, and to respect and 'look after' each other's interests when it comes to 'husband-stealers'. As a result they develop quite strong empathetic emotional bonds with other females, and are able to worry about 'the plight of females' (especially mothers/wives) in general.

Females are certainly very adept at communicating about love and commitment emotions among themselves and even in baring their souls to each other (Taylor, 2002; Baron-Cohen, 2004). For example, studies have suggested that married women are more likely to turn to their girlfriends than to their husbands for emotional support (Kuchment, 2004). A recent survey, for instance, shows that while 45 per cent of men turn to their spouses for emotional support only 21 per cent of women do so, and that 36 per cent of women turned to friends for emotional support but only 17 per cent of men did the same (Singer, Kancebaum and Thomas, 2004). Through evolutionary history these propensities seem to have worked relatively well in getting females to respect the exclusivity of each other's mating and partnership claims, and in providing females with reliable female support networks during post-natal periods – such networks in fact often lasting for a lifetime.

During evolution this would have been very useful if, as many assume, the death rate for males was higher than for females and males married at a somewhat later age than females (as has often been the case historically – Hajnal, 1965; Stone, 1977; Laslett, 1983). Many females would have been left as widows at a relatively early age, with numerous offspring and no new husband in sight. But even before that, useful friendship support for females from females could have included economic contributions, help in giving birth, aid in caring for any offspring that survive and also emotional support during times of stress. Theoretically, it certainly seems that female networks would have increased reproductive success through mutual help in child rearing – considering the dependency of human infants, the frequency of pregnancies, the youth of the mother and the unreliability of males as presented in the evolutionary scenario outlined earlier. There is some primate evidence for this. For example, the female baboons in Kenya's Amboseli basin that had developed the most elaborate social networks also had the best chance of having offspring survive past two years of age (Silk, Alberts and Altmann, 2003; see also, Hrdy, 2011). Among human mothers, of a sample interviewed in 2000, those who got the most support from family and friends delivered higher-weight babies (Feldman, et al., 2000).

In humans, a major factor in fuelling female/female bonding could have been experiences of co-mothering. It might be true that mothers can pick out their own offspring from smell, cries and touch more often than chance would predict (Maestripieri, 2001; Noriuchi, Kikuchi and Senoo, 2008), but it seems equally amazing to note how much effort females will put into looking after a 'stranger' child that seems in distress, and how a group of mothers temporarily accept the role of surrogate mother for their friends' offspring. Although this is not unknown among mammals, and is certainly not uncommon among primates (Maestripieri, 2001; Izar, et al., 2006; Nakamichi, Silldorff, Bingham and Sexton, 2007), for the vast majority of human females it is an almost automatic reaction and seems simply normal.

For females who – as among chimps and bonobos – probably migrated out from their birth group (as opposed to males who stayed in their birth group), there was the acute political problem of *building* or *joining* a support group among relative strangers. A female chimp had to 'worm her way into the favours of the females that already live in her new tribe' (Ridley, 1993, p. 183; see also, Pusey, 1979; De Waal and Lanting, 1998). Female chimpanzees work their way into a 'strange' band dominated by males and resident females by offering sex and being submissive to both males and females. While chimpanzee females do not form groups with any degree of coherence, bonobo females also have to offer sex and to worm their way in, but then become part of distinct, although weak, female alliances. These are clearly more for mutual support, joint childcare and friendship than for securing mates, status or territory. These groups also aid females in preventing males from dominating them, and, indeed, can aid in females personally dominating males, especially sons who seem to hang around their mother for a very long time.

In so far as a process of neoteny was taking place during human evolution (in which a number of evolving human female characteristics were much the same as infant-like characteristics) it is likely that attractions to offspring would have also underpinned attraction to female friends. Genes that caused females to be instinctively and emotionally attracted to cute, helpless, baby-like infants would, with neoteny, have resulted in longer-term attraction and attention to offspring, such as through their teen years, but also to other females (and to males who were less ape-like and who were also attracted to infants). It is probable, however, that the genetic propensities for females being attracted to males, females and offspring existed on a continuum ranging from being attracted to extreme neoteny and caring presentations of self at one end to being attracted almost back to some of the characteristics of an ape ancestor at the other end. So, it must be noted that a given female would not be attracted to another just because she was a female; she would be attracted to particular females who gave off specific signals – especially facial ones – that activated her care giving and sharing emotional propensities.

As with males, female/female attractions existing on a continuum suggest that various mixed genetic propensities near the centre of the continuum – where sexual/reproductive responses would be more flexible – provided the best reproductive success and those at the extremes the least. For example, in some circumstances it would have been reproductively advantageous to be able to be attracted to and made sexually receptive by more rugged, aggressive males (where male protection was at a premium and masculine genes were useful in survival, and where being a second or third wife raising offspring in a harem context was better than being with no husband at all, or even than having a child-like weakling for a husband). At other times, however, it would have been reproductively advantageous to be able to be attracted to more playful, child/friendship oriented males and female friends (when sharing, more caring male and female helpers would have been more advantageous than a warrior, for example).

This possibility would exist around the centre of the above continuum where, for example, females could be equally happy with putting energy into macho males or, conversely, relationship-oriented males, or female friends, depending on the circumstances. And, of course, moving both ways from the centre mark there would be a number of individuals with a high degree of this type of flexibility. In the direction of the attraction to female end of our continuum we might indeed find a degree of bi-sexuality (Cf., Facts, 2013). While a genetic bunching for each extreme would always exist as a product of this heterozygous advantage, both ends would have to suffer reproductively for one end or the other not to become the norm. It is possible to speculate that this was the case. For example, a female strongly genetically bunched at the continuum end of extreme attractions to child-like, feminine characteristics, and also bunched for low proceptivity, receptivity and consummatory propensities, would be almost pre-adult in sexual interests; that is, not sexually aroused by the post-puberty characteristics that the evolution of child sterility and puberty had generated. They would not have emotionally developed beyond the original child-playmate reproductive behaviour that child sterility and puberty had supplanted.

So, at this end, a female's sexual interest in maleness and heterosexual sex would be relatively low, and it could result in such females being attracted to the weakest (non-assertive, still too young) male providers/protectors who were not able to stand up to older, more aggressive males intent on infanticide; these males may not even be sexually mature enough yet to be fertile. Females at this end might be attracted to males who were sexually capable but with whom they were not made sexually receptive, such as seen in platonic relationships with friends, brothers and cousins. In some cases this bunching could result in some combination of 'aunt behaviour', daughters looking after parents, traditional spinster vocations, exclusive female friendship groups and, very occasionally, lesbianism. But this does not mean that there would always be a complete elimination of reproductive outcomes at this end. Females do not have to be sexually aroused (or interested) to have sex and become pregnant. Indeed, lesbians often reproduce *if they have a strong desire* to have offspring (McKnight, 1997). But generally, reproductive output is less at this end than among 'average' heterosexuals.

At the opposite end of this continuum, some females would only be strongly attracted to and made sexually receptive by 'ape-like' studs that were not very reliable in supplying food, protection or aid in child-care. These hyper-receptive females would have a higher dose of testosterone than average for females and be quite sexually enthusiastic, easily aroused (as in the condition of oestrus in ancestral species). Some females near this end might have not have been specifically attracted to 'ape-like' studs, but easily made proceptive and receptive, and easily able to achieve consummatory pleasure with a wide variety of males. However, as with child-oriented, 'feminine' sexuality at the other end of our continuum, enthusiastic and powerfully consummatory female heterosexuality probably did not bode well for reproductive success.

Such females could immediately start to flirt with most males and engage in sex with many of them. They may well be rewarded with powerful orgasms, including multiple orgasms, with each sexual encounter (homozygous consummatory capacity), but if a female goes after almost every male in sight, and presents for sex straightaway with them, she loses any opportunity to evaluate males in terms of overall reproductive value. Males might not mind quick sex but are not going to hang around to be accused of fathering a child that might emerge, or be bullied into providing resources and child-care for a female they know has had sex with everyone in town, and who, they suspect, would continue to do so. Marriages among such females, when they occur, usually do not last very long. And we can be reminded that most reproductive success has been among women who have regular sex with the same male over about a three month period.

Even if a female towards this end of the continuum was more selective in terms of sexual receptivity and more concerned with attractivity and proceptivity – for example, spending a lot of time and energy being attractive through elaborate clothing (that perhaps enhanced and/or revealed key physical attributes), applying make-up (possibly in modern times using cosmetic surgery) and in honing great skills in attempting to be the centre of attention (flirting with a lot of men) – she is highly unlikely to attract long-term male interest or female friends. No female will trust such a female within a mile of her man, and generally will do everything to isolate such females – certainly they will not offer them much support for child-caring or anything else. Moreover, if she does not become receptive after a long proceptive display effort she also becomes contemptible in the eyes of males too (Cf., Buss, 1994).

Anxiety, rejections and social condemnation might lead such females to look for substitutes for sex and/or for consolation for failed relationships. Such substitutes might include alcohol, drugs and, if attractive enough, very flamboyant lifestyles, including blatant affairs with high status males, or, in some cases they may devote a lot of energy to career achievements in place of children. Additionally, all of these possible responses – from hyper-sexuality to flamboyant proceptivity to career orientation – can mean that a female is so interested in sexual endeavours that they neglect child-rearing activities in their quest to be continually attractive to alpha males and to have sex and / or to achieve in a career. None of this attracts male trust or female support. Whatever the exact mixture, throughout history, flirty, sexually forward and promiscuous females have been, and continue to be, held in very low regard, and certainly are not desired by males as wives, as is also the case with high achieving females. Indeed, they have often been considered extremely dangerous, not just to men and women but also to the 'natural order' of the world, general morality and civilization itself, and have been variously labelled as *femmes fatales,* whores, sluts and witches.

As a consequence of the heterozygous advantage, selection has resulted in females showing little inclination to be as promiscuous as males. Even in modern birth control times, with numerous feminists advocating female promiscuity and certain media presentations suggesting that the really modern female can be, should be, deserves to be, as promiscuous as males have traditionally been, there is little indication that the vast majority of females want to be promiscuous. Nor is the condemnation of promiscuous, or even flirtatious, females, by both males and females, any less than it has ever been in most anthropological and historical records; and strong prejudices against career-oriented women continue to exist in many parts of the world.

At the same time, however, in most anthropological or historical contexts, very few females within whole populations have no reproductive success at all. As noted, even lesbians can have, and have had, offspring, and anyway there are very few pure lesbians in society. According to Kinsey and colleagues (1953), for example, about 13 per cent of females had experienced an orgasm from homosexual behaviour by age 45 with 26 per cent experiencing an 'erotic homosexual episode'. Ten per cent or less had more sustained periods of homosexual experience; exclusive homosexuality was very rare, possibly only about 2 per cent. According to some of the more recent surveys cited above for male homosexuality, even this may be somewhat of an overestimation. On the other hand, the low end of these surveys (.25 per cent exclusive female homosexuals and 3–4 per cent fairly regular homosexual behaviour) is probably too low. Whatever the case (that is, whichever data are more accurate), it seems clear that females with low reproductive propensities make up a very small percentage of a population.

Research on homosexuality also makes it clear that female homosexuals as a group are considerably less promiscuous than male homosexuals as a group, who are extremely promiscuous (Kinsey, *et al.*, 1953; Symons, 1979; Wellings, *et al.*, 1994; Ruse, 1988; Dixson, 1998; Kauth, 2000; Dailey, 2013), suggesting a weaker sex drive on average for females than for males, or at least less of a focus on sex alone without consideration of a relationship – or at least some knowledge about, and/or friendship with, potential partners. After all, lesbians do not have to worry about becoming pregnant from a promiscuous encounter, they do not have to restrict promiscuous sexual activities in order to win a husband, they are not under suspicion when as young girls they take their best friend to their room, nor are they usually encumbered by numerous offspring who might distract from sexual activity. Yet their number of sexual partners is just a little more than for heterosexual females, with bisexual females having somewhat more partners, but neither getting even close to reaching male homosexual promiscuity levels.

Female bisexuals and lesbians – especially lesbians – tend to look for at least semi-permanent partners to have sex with, then to engage in sex regularly over a period of time, while male homosexuals very often just look for someone to have sex with period, and try to avoid permanent relationships. This suggests that males are more interested in variety within sexual experiences, and that they have very little interest in courtship, love or romance; they tend to look to find a partner for an evening, have sex and then be gone. A significant difference between female and male bisexuals seems to be that females are equally attracted to males and females while male bisexuals tend to focus on one sex or the other (Van Wyk and Geist, 1995; Lippa, 2006, 2007; Johnson, 2012). This supports the idea that male/male attraction is relatively rigidly focused and quite powerful, and less affected by circumstances than female attractions. It is to be noted that bisexual attraction on the part of both sexes tends to a heterosexual bias in terms of numbers of partners (Weinberg, Williams and Pryor, 1994).

What these data do not tell us is how attracted to each other females are in terms of friendship rather than for sexual reasons, or how strong or precarious female bonds are; and it is this that is significant for our understanding of human socio-political behaviour. We can speculate, however, that in so far as friendship/alliance bonds are based on the same mechanisms as sexual attraction, the above data indicate a weaker addictive, less rigid, type of attracting/bonding force in females than in males. Besides the evidence of sexual behaviour (seen from the above data on homosexuality/bisexuality), mythology, stories and history also suggest a weaker female/female bonding. Throughout history we do not see among female friendship networks the same kinds of codes of honour and duty and intense loyalty as seen among male alliance groups, sports teams and military squads. Nunneries, like male monasteries, are same sex and often show a lack of interest in sex and reproduction, but they do not represent the bonding required for the ubiquitous political or military actions that a significant number of males get involved in. Attempts by feminists in the West during the 1960s–2000s to generate distinct sisterhoods to fight 'male oppression' do not seem to have succeeded except among very small groups of – mostly lesbian – females.

This makes theoretical sense in that given the male sex drive – and desire for mothering – almost all females can attract a male for sex (and reproduction) if they desire to do so, while it is not the case that all males are able to impregnate, or even have much access to sex. A female's problem of reproductive success largely starts after she is pregnant, not before. The opposite is generally true for males, for whom in order to be reproductively successful rampant lust has to be publically controlled, worthiness has to be demonstrated, competitors outdone and cuckoldry avoided. To succeed and/or feel a degree of security in all this, partnerships and/or alliances are made, strong trust and loyalty among friends, partners/alliance mates established, with a powerful concept of honour to 'police' this, and sometimes powerful patrons have to be found and relied upon to make sure honour among clients is maintained. Much of this, of course, represents a sublimation of male sexuality, which in turn goes some way to repressing male lust.

For females, same sex attractions that required the same level of emotional energy and political work that males expend would have been surplus, if not detrimental, to reproductive success. A female knows that the baby is hers, and any help in child rearing they get is good; their genetic heritage is secure whoever does the helping – male partner, family of origin, siblings, friends or servants. For females, relatively weak bonding into loosely woven kinship and friendship networks was much more flexible (and thus reproductively beneficial) than male bonding would have been. For females, friends could change and arrangements be altered as circumstances varied, or if a male left, or a nice new male came along, or if someone died. In terms of male bonding all of this flexibility would have been considered *dishonourable* in the extreme.

Female hedonic politics

Evidence derived from anthropology, history, mythology and psychology supports selection theory in the view that the bond between groups of females seems to be of a different kind than seen in packs of males (Cf., Freud, 1952, 1921; Tiger, 1969; Benshoof and Thornhill, 1979; Kauth, 2000). The male bond as seen in this material is based on a sense of loyalty in the face of danger; trust is everything. The bond is sealed through oaths and a sense of honour; spiritual feelings of responsibility to duty and to comrades are paramount. Heterosexual sexual attraction and energies are sublimated into the above rather than openly indulged in; the ideal of marrying virgins (the inviolability of a mate's desired female and the shame of cuckoldry) is a common theme.

Female bonds of friendship, on the other hand, are based on physical proximity, having children of the same age, generalized caring emotions, gossip, the sharing of practical information, and sometimes on a powerful sense of guilt from feelings of not being a good parent/carer. Historically, female group formation and politics have tended to relate to local (family, encampment, church, neighbourhood and school) life, while much of male politics is involved in politics with a large P and tends to range further afield and to include such things as alliance behaviour, vengeance seeking, patronage, inter-group negotiations and warfare.

Most female networks have been constructed through interpersonal interactions and emotional understandings used to entice mutual support and cooperation in child-care, and in keeping husbands honest and caring. This involves mostly *hedonic* rather than *agonistic* expressions of behaviour, not only in the politics of the home and bed, but also in campfire politics where a certain degree of shared child rearing and a considerable exchange of views concerning child-care is the order of the day. Even in their more aggressive forms such interpersonal politics usually remain hedonic, as in negative gossip about the evils of philandering and non-caring males, barbs about selfish non-reciprocating females, stories and warnings about females who try to steal husbands, and back-chat about insufferable mothers-in-law. Such hedonic politics usually include the rejecting and isolating of uncooperative and untrustworthy females (and males and offspring).

But if females were in the lead in the evolution of a great deal of human interpersonal, hedonic political behaviour, such behaviour would also have served males once unthinking vengeance seeking and brutal patronage were replaced with more subtle politics. During human evolution it was not long before alliance mates and patrons began to 'parent' each other; patrons began to parent clients, to praise them and comfort them and in many cases to feel responsible for them. And clients sometimes acted the child, and demonstrated admiration of the parent (patron) and felt a *right* to a degree of security being provided, as from a parent. Many aspects of male politics were smoothed out and refined by emotional motivators that largely derived from female sexual and campfire politics. Remembering names of clients and clients' children (constituents), charming the wives and daughters of clients (constituents), having a keen ear for political gossip, developing the diplomatic skills of a respected statesman, convincing union and business leaders to sit down together and be friends, giving a speech to open a new library; these all derive not from the exuberance of youth force, or from fired-up alliance mates or boasting 'big men' patrons, but from the female politics of taming males and securing aid in child rearing and developing a bond of sympathy with close and trusted friends.

The skills developed during this evolution are about manipulating others and about maintaining control of self's circumstances. These are political skills that operate through emotional forces in which language usage, negotiation and compromise play a major part. But the greatest contribution of female-derived genetic propensities to politics with a large P has arguably been the introduction of a caring (welfare) sense of responsibility into the political process, albeit often exercised by males. This evolution most likely was built on an ethos of food-sharing, toy-sharing and friendly giving that has existed throughout human history within the confines of a household or camp. Going beyond immediate family and close friends, this has included mothers encouraging their children to share with other children and such things as alms-giving by caring neighbours, or a powerful patron taking welfare responsibilities for serfs, or large organizations such as religions having charity as one of their significant activities. In most of the developed societies of the west, religious charities continue to function on a relatively large scale, but have been joined by a number of secular ones.

For example, a long tradition of philanthropy among the super-rich has developed where vast amounts of money for charitable work have been provided, usually through foundations. These foundations and charities often go beyond simply providing resources for the needy but look to develop education, research, health care and even 'culture' within a society. One of the most famous of the philanthropists was the late nineteenth/early twentieth-century American industrialist Andrew Carnegie, who suggested that any person who made vast sums of money had a moral obligation to give it away, not to their children but for the common good. Accordingly, he gave away the majority of his vast wealth to 'human betterment' projects and foundations before he died (Carnegie, 1901; Nasaw, 2007). All these activities of giving have been praised and, indeed, presented as moral obligations. Welfare emotions have gone to the extent of the development of a sense of failure and guilt among at least some of those who are in positions of responsibility and who feel that they are failing, like a social worker who feels that they have let a client down, or a boss who has let his workers or shareholders down, or a politician who has not brought about a promised reform.

And indeed, throughout history a variety of institutions – such as elders, clan chiefs, 'lords of the manor', churches, monasteries, warlords, kings and 'democratic' governments – have been more or less held responsible for welfare as a *service* to those who submit to their authority. And when powerful patrons forgot their welfare responsibilities religious leaders eventually reminded them of their 'moral responsibilities' and set out to emphasize just how much guilt individuals who do not care for their fellow humans as much as they care for themselves should feel. And when church leaders forgot their welfare obligations, break-away cults and millenarian movements reminded them of the founder's preachings about the sacred duties in this regard, or secular philosophers quickly pointed out the shame leaders should feel for 'looking out for themselves' instead of for 'their people'.

But some argue that giving charity is a form of humiliating those who receive it, a means of claiming status – 'look how much I have, I can just give large amounts away if I want, but you are so useless you have to *take* it'. And certainly giving a lot of charity is a clear demonstration of wealth and power, and a claim for moral superiority, and it imposes on the recipient a moral obligation to be thankful (obedient). So, in the modern scientific, democratic age it is often claimed that a 'rational' society should make sure than none of its members are allowed to live in misery and degradation, and that *public* means are available to ensure that 'the unfortunate' are helped as a *right* not as an object of pity. So, in modern times the institution held most responsible for welfare has been the political community itself. And individuals have been told that they must pay taxes for this purpose; it is their *duty* not a matter of them making moral choices about giving. The moral choice was to pay taxes and to put the community before self.

President John F. Kennedy famously said: 'ask not what your country can do for you; ask what you can do for your country' (Kennedy, 1962, p. 10). It is not clear how much guilt (or action) this generated, but success has been claimed for this approach. Stating that neuroses and psychoses (characterized as bourgeois illnesses) had been eliminated in the Communist China of the late twentieth century, Professor Suh Taung-hwa did, however, note that: 'Many workers, students and peasants feel a kind of guilt towards the Socialist Society. They think that perhaps they have not dedicated enough faith and revolutionary energy to the Socialist construction of China' (cited in Szasz, 1973, p. 252). And, of course, the whole ethos of the socialist state is to put the welfare of the people, all the people, first. But this does not mean that non-socialists do not believe in welfare.

Advocates of free-market economies argue that the welfare of people is best served by everyone being able to freely work, exchange, buy and sell in an open market, but that there should exist a publically funded and organized 'safety net' for those who suffer misfortunes beyond their control. In both cases, leaders of such societies are well aware that, in fact, they are where they are because they have promised a high level of welfare/security as a *right* to all citizens, promises that have been essential for politicians to get into and stay in government in the post-Second World War period right across the world. So we can see that the legacy of female politics, a politics of parenting at the social level, is one that has a powerful hold on modern politics.

Overall, then, modern human politics is made up of various combinations of male and female propensities. However, it is to be noted that while males tend to predominate in male-derived politics (such as vengeance seeking, war, patronage and modern party politics), our analysis of genetics in Chapter 2, and as applied to sexual politics in this chapter, suggests that we should expect a small proportion of females to be quite agonistically orientated and, indeed, involved in aggressive alliance and patronage-type politics. We would also expect that in varying degrees *all* females would have some agonistic characteristics. And this seems to be the case.

The struggle among school girls for 'best friend' and who is to belong and who is to be excluded, for example, can be intense and bitter; competitions for boyfriends and husbands can be vicious; conflicts among village or neighbourhood women (and men) can be inflammatory. Modern female juvenile gangs can be quite violent (Deschenes and Esbensen, 1999; Moore and Hagedorn, 2001) and, among female homosexuals, 'Diesel Dykes' and/or 'Dykes on Bikes' forms of lesbian expression can be noticeably forceful, masculine and even aggressive (Cf., Ilyasova, 2006; Urban Dictionary, 2013). Females historically have taken part in political protests other than over strictly feminist concerns (Cf., Gusfield, 1963; Fry, 1976; Tilly and Tilly, 1981), and have been deeply involved in, if not in charge of, aggressive, competitive market economies in a number of societies (Evans-Pritchard, 1940, 1965; Mair, 1974).

Indeed, females have acted as powerful heads of state and as advisors to heads of state. And even when males were ostensibly the main players in alliances and patronage politics there were often powerful wives, consorts, sisters and/or mothers just behind the throne. Some of these were more politically ambitious, cunning and aggressive than their husbands, lovers, brothers or sons; or at least extremely skilled in politics on behalf of their male protégés. (Perhaps Nero's mother, wife of Claudius, set the gold standard in this area.) Selection would have favoured this approach to female power to some extent because an attractive female, for example, who could provide good sex – or at least fake it – or get a king completely emotionally dependent on her, could become very powerful in relationship to any man (however highly placed) she might encounter. And her offspring, especially her sons, could well end up with not only considerable political power but also have an enormous amount of reproductive success (which, of course, would be her reproductive success too).

But this approach was only open to the most attractive, and perhaps sexually exciting, and certainly cunning, females. In some cases they demonstrated certain 'male' characteristics, such as aggressiveness and a willingness to kill rather than care, at least in their approach to politics. Or they might simply have used female approaches to gaining personal power and privileges more blatantly and systematically (and successfully) than some other females. In either case, they did not bring a distinctive female hedonistic perspective to traditional agonistic politics. Additionally, the courts of even the most powerful female monarchs were often peopled and run by males. And politically powerful females have made up an *extremely* small minority among agonistically 'successful' politicians through human history. And indeed, among males it is only a minority of them who become deeply involved in politics with a large P. Yet motivation for such politics remains sufficiently prevalent that a significant minority have been motivated to participate, and a sizable proportion of a population take some interest in observing, supporting one side or another and cheering when their side wins. Indeed, loyalty to a political faction or party can become as strong as religious loyalties and fan support for sports teams. Heroes and villains exist in politics just as in religions and sports.

Nevertheless, everyday human politics are more underpinned by the hedonic mode of sociability than agonistic tendencies because that is what motivates and gives a degree of security to the vast majority of humans. Most men, like almost all women, spend their time down on the farm tending the crops (although men may do more sexual and war fantasizing and watch more sports – and politics – and try more often to build a career of sorts, than females). Generally both sexes want their economic prosperity, property, status, rights and futures *protected*. Both males and females have an interest in creating taboos, rules, contracts and boundaries that control the sexuality of each other and set out expectations for both. Females do not like aggressive males with only sex on their minds, and males like their alliance mates and over-dominant, greedy/lustful patrons to be under some sexual – as well as economic and political – control.

So both sexes have an interest in traditions, taboos, rules and laws that regulate sexuality and enforce pair-bond obligations and punish sexual transgressions. Both males and females have an interest in their children being protected and cared for, and one day inheriting something of themselves, and so both support traditions, rules and laws that give some guarantees in these areas. At the same time, however, both males and females have an interest in a camp or neighbourhood being safe, and, in so far as both desire vengeance against any who harm them or their friends, they support a youth-force ready to take vengeance and, if outsiders threaten, they support the youth force in any wars that ensue. So, agonistic politics have not completely disappeared, and in some circumstances predominate.

Human politics might have arisen for somewhat different evolutionary reasons from the different reproductive interests of males and females, but the merger of the two politics has been, in reproductive terms, very successful for humans. It certainly seems more accurate to describe them as being blended rather than being two separate antagonistic types of politics. The politics of both, ranging through negotiations, compromises, deals, contracts and physical enforcements, are utilized to secure reproductive ends. But none of this can ever be guaranteed. Circumstances change, disasters happen, different genetic propensities push in different directions, individuals fall in and out of love, demographic patterns alter, new technologies have unforeseen consequences, alliances collide, patrons fret and fight, coups take place. It is the interplay of all these that has selected for a unique form of perpetual human politics – politics which (1) link human nature with personal politics and these with politics with a large P, and (2) which have their own dynamics, and which, as such, (3) continue to have their own feedback on the human evolutionary process. It does so through the interplay of a number of species-typical desires and fears.

Human evolution and species-typical desires and fears

Regardless of later environmental changes that might have selected for the evolution of 'rationality' (analytical skills) as presented in the traditional scenarios of human evolution, the changes in mating patterns discussed in this and in Chapters 3 and 4 *did* take place, and as such, it can be argued, underpin much of the human nature we will require in explaining modern social life. Humans might have taken to seed eating, come to love and cook meat and to make fine and decorative tools, to love to labour and to enjoy food sharing, and to have evolved analytical, 'rational' skills as a result. But if these desires became too important and replaced the sexual, social and political motivators discussed in this and the previous chapters, selection would have held them in check. This would also have been the case with any evolutionary tendency to excessive philosophical reflectiveness that might have been a by-product of the evolution of analytical skills.

The same can be said of the evolution of human language as a 'rationalizing force'. While it can be argued that language has undeniably been extremely important for the evolution of what we think of as human civilizations, and that any type of hedonic-based social organization with a population much larger than a basic group of about 5–15 individuals would be very difficult without it, the sublimation of sexuality into politics predates language. But language, more than any other thing, facilitates doing politics: networking, obfuscating, telling good stories, campaigning, gossiping about trustworthy and untrustworthy individuals, massaging egos, wheeling and dealing, compromising, conspiring, making claims, creating and perpetuating political mythology, debating what boundaries, laws, rules and constitutions might be implemented, Language, status and politics make it possible for humans to live communally on a scale well beyond the capacity of any other primate. As we have seen, this is because these make it possible to emotionally ignore most others while still being able to interact with them; but it does not necessarily make us more rational, or even reasonable.

The lesson for us is that the sublimated reproductive/sexual motivators which generate such things as alliances, male brotherhoods, kinship loyalties, friendship networks, charity, welfare provision, political reciprocity, patronage networks, political administrations and military organizations are not so much based on rational or analytical cognitive processes as on basic human emotional motivators, desires and fears. And once these motivators were in place they took on a life of their own and were/are indulged in for their own sake. For instance, with the *delusion* that becoming a dedicated hunter, an eager competitor, a person with friends and support, a good team player, one trusted by allies and powerful patrons, one who appears to love children and who is looking to improve his rank, *has nothing to do with sex,* many males have become very attractive to females. In being attracted to these males, female motivators too selected for the evolution of politics, but also for a number of characteristics and behaviours we consider to be the bases of human civilizations.

And we tell stories (and write them) about all this; stories about how we are not animals but rather almost gods. We call ourselves Homo sapiens, the wise species, and we 'do' theologies and philosophies and talk about progress, about *inevitable* progress. But there is another kind of story that we tell ourselves. These are the stories about who we are, about a hoped for personal identity, and to create defence mechanisms and formulate expectations as we go out into the world. In these stories a major problem is to overcome fear; fears of being abandoned, of being rejected and shunned, of being humiliated. Within these is often a fear of loss: a fear of a loss of respect, of being a *failure*, of not being able to participate in life. Some of these fears do not show up in an evolutionary analysis as readily as, for example, those sexual fears observable through mating patterns, and so deserve a separate consideration before we set out to systematically list the desires and fears suggested by a consideration of human evolution.

6 Human fears

The story of Oedipus would make little sense as a tragic story if he was not overcome by a fear that he had done completely the wrong thing, and that he would suffer greatly for this, even though he had not purposely or even knowingly done anything wrong. In this story, anxiety and shame turn into guilt, leading to Oedipus's suffering and self-harm. We instinctively understand the fear associated with committing a 'very bad wrong' and the power of shame and guilt to punish; we fear being completely wrong, or inadequate, unlovable, bad, or even unattractive, for example, but we hide these fears lest they become self-fulfilling. Outward manifestations which might indicate deep unconscious fears, such as simmering anger and disgust, are often repressed; the existence of possible indicators such as depression, envy, jealousy, hate and guilt are kept well out of sight whenever possible. We might *feel* shame and try to conceal it, but we are not always *conscious* of our feelings of guilt, and certainly we are often not willing to admit to them.

In Chapter 1 I suggested that to overcome such resistance we look for objects and events that trigger automatic, or at least semi-automatic, flight, repulsion, disgust, terror, weeping, despair, panic, anger and strong avoidance reactions as evidence of fundamental fears. Additionally, conditions which generate a powerful feeling of loss or potential loss, often accompanied by self-blame, were suggested as indications of deep-seated fears, as were historical and cross-cultural definitions of 'evil' and taboos and boundaries which have been established as defences against evil. In this chapter I will focus on these elements of human nature. The chapter will be divided into four main sections: (1) an analysis of the *circumstances* in which flight, repulsion, panic, guilt, desperation, self-blame (guilt) and/or 'striking out' have been common in human cultures – 'primordial fears'; (2) a consideration of human *appearances* and presentations of self which generate feelings of spontaneous repulsion, terror, anxiety, fear-fascination and/or ambiguous sympathy – 'freakishness' and 'freaks'; (3) a consideration of the *behaviours* which have, cross-culturally and historically, tended to be considered the epitome of 'evil' – 'witchcraft'; and (4) a look at human attempts to identify and combat *evil*.

Primordial fears

Humans, especially children, seem to react apprehensively and even sometimes hysterically to some animals, the dark and places from which they might fall (Rachman, 1978;

Ehrenreich, 1997). A fear of falling may be an evolutionary remnant from a likely primate fear of falling out of trees, and thus we often fall in our dreams, a time when more primitive parts of the brain are at work (Sagan, 1977; Cohen, 1979). A child's fear of the dark is as likely to be a response to being left alone as it is to the dark as such (Rachman, 1978). For most people it is relatively easy to sensitize and habituate against at least some of these fears. Indeed, it is common to learn to desire some of them, or some aspect of them, in one form or another. For example, young people get a considerable thrill when experiencing 'scary' rides at a fairground, certain animals become pets, and we come to want to sleep in the dark.

However, the human repulsion, and even panic, concerning reptiles in particular is quite durable and widespread. It might be due to an evolutionary carry-over from certain reptiles (such as crocodiles and water snakes) being a major environmental danger during the period when the mammalian brain was evolving (Sagan, 1977; MacLean, 1978). It might also derive from snakes being a major danger when the earliest humans were gatherers; indeed, most mammals, including primates, are afraid of snakes (Isbell, 2006). The relatively widespread repulsion caused by maggots and some 'creepy-crawlies' (Douglas, 1970) could have been selected for as a result of disease dangers for scavenging humans. A common human avoidance of certain larger animals may derive from the potential dangers they presented as predators or as dangerously wounded prey – a trait also common among mammals generally.

There are a lot of 'might-be's here but the significance of the widespread existence of these fear reactions is not so much that they necessarily motivate a great deal of human behaviour in modern times but rather that they clearly show that fears have a phylogenetic (pre-knowledge) basis which pre-dates human experiences (Cf., LeDoux, 1998; Belinda, *et al.*, 2004). In modern times, such reactions exist even among people who have never experienced a fall from a great height or encountered a snake or lion; they often have to be unlearned, sometimes at great effort, while the much more destructive potential of cars and heavy traffic, for example, is much harder to learn to fear.

There are a number of substances for which repulsion and/or avoidance of them may not have such an ancient pre-knowledge foundation as the above, but it is, nevertheless, very easy to *learn* to fear them on the basis of general feelings of disgust for and repulsion from slimy, sticky, strongly smelling substances: for example, rotten food, mucus, excreta, vomit, sexual secretions, blood and a variety of foreign substances on the body or in the home – 'dirt' (Douglas, 1970). The evolutionary advantage of repulsion to rotting food is obvious, probably having been intensified with the progressive evolution of meat eating. But the advantage of a similar automatic repulsion to excreta, mucus and blood is not so clear. The fear of excreta seems to be a recent development in the primate line. There is an anti-parasite, anti-bacteria survival advantage for most mammals in being as far away from excreta as possible. Most primates, however, are nomads and never use the same sleeping nest twice. Therefore, there were no selection pressures on primates in general to develop mechanisms for strong feelings of repulsion related to excreta. With the development of a home-base, however, it seems that the odour, feel and sight of excreta came to activate disgust/repulsion mechanisms which motivate humans to actively separate themselves from it; this meant removing the excreta from self and nest rather than self from the nest; this process was undoubtedly exacerbated by noxious-smelling faeces associated with meat eating.

This, however, provided an interesting 'challenge' that has had significant consequences for human psychological development. As all parents know, infants do not have the same repulsion to excreta as adults but can, as they are toilet trained, learn it relatively easily. This pattern probably evolved because infants cannot escape their own expulsions and to have a repulsion to them would have generated a considerable amount of distress and trauma and/or have caused them to be in danger in trying to escape it – such as leaving the safety of the nest while parents were temporarily away. So, it seems that genetic information that activates infant repulsion to excreta was selected against in favour of genetic material which provided for a cognitive capacity, with strong emotional involvement, to learn it later. Also, from an evolutionary point of view, in so far as parents have to clean an infant, those who realized the dangers but still did it were selected for on the basis of the degree of *commitment* to both creating hygiene and child-care that doing so represented.

In other words, selection pressures were for adults, not infants, to be repulsed by excreta, but for adults to be able to come to terms with, and control, the repulsion, especially with regard to their own offspring. It certainly seems to be the case that we are considerably more repulsed by the excreta of a stranger, even that of a friend, than by our own or that of our own offspring, and parents are certainly helped in this by the fact that baby excreta does not have the same repulsive effects as excreta from individuals eating more solid foods. Whatever the exact pattern, these reactions, as Freud observed, have become involved in the process of mind becoming conscious of itself in terms of trying to understand both self's relationship to its body and to the controls imposed upon self by the external world.

A baby is not averse to playing with its own excreta, and it is often pleasant enough for them to produce. This, however, can confuse processes of self-discovery. In 'normal' development, self (awareness/consciousness, self-definition) increasingly perceives a separation of mind from the physical body, and mind and body are perceived as a sort of 'team' distinct from the external world. With 'toilet awareness' it becomes clear that a part of the body is leaving, a part of self is becoming external, separated away; it is being expelled. As such, self's excreta can be emerging self's gift to the world as 'pay back' for the loving care previously afforded one; after all, a dirty behind (nappy in modern times) had in the past received a great deal of reward for self in the form of parental smiles, hugging, gentle cleaning and emotionally invested attention; so should not this offering be gratefully accepted? However, at the very time children become aware that their own excretions come from inside themselves parents begin to encourage children to be *disgusted* by these very excretions, especially by faeces. The child is told that the gift is bad, it is filth and must be avoided and banished at once – every trace of it must be eliminated. So, to the child's dismay, and possibly anger, the gift is *rejected*, buried, flushed down the toilet. And the child is taught to control its expulsion and do it in out of the way places. Additionally, toilet training mistakes by the young learner come to represent not only a lack of concentration and failure on the part of the child but almost, to parents, a defiance of adult authority.

Well, so much for the kindness and appreciation of the loving, care-giving world, and so much for the sweetness of the inner-self; does it all mean that self is inherently *polluted*, that self is rotting away from the inside? The lesson seems clear; self contains filth which is dangerous. However, there is another lesson: it is that self can expunge itself of this danger; it can flush the impurity down the toilet; it can wash it away. It can cleanse itself and seek to regain purity. And this can be done by self's own efforts, by controlling those body parts that rid the body of filth/dirt and by thoroughly cleaning them. And great praise is given to an individual's efforts to cleanse itself and barely concealed disgust for failure. And so shame and embarrassment for failure quickly 'teach' consciousness to feel guilty about failure, to hide mishaps, and to feel apprehensive about any bit of dirt left behind; after all, remember the reactions of parents to it. So, we very much want to excrete and wipe away all traces in private, because then there is no evidence of our rotting inner self for the world to see – only the cleansed self is presented to the world. So it can be argued that the fear of impurity (excreta) is a basic fear which helps motivate us to learn to want to be private, to want to hide certain, especially unpleasant, things from others; and, probably as a *quid pro quo,* we quickly learn to respect the privacy of others.

However, the contamination keeps on coming and our insides seem to keep on generating pollution. So, we are motivated to continually wash ourselves, to perfume ourselves and to want to present self in a purified and perfumed way; as a result, it also gives us a means of judging others in terms of their commitment to purity. Reactions to the results of male masturbation are very similar in their effects. For many males it is easy to feel that the result of masturbation has dirtied self; the evidence is hard to get rid of: it is slimy, it sticks to everything, to skin and hands; it changes the colour of clothing; it seems almost a regression to childlike playing with faeces or nose mucus. And if the evidence is discovered it tells others that self has 'played' with, and gotten great pleasure from, the very private parts that mothers and fathers have said to 'leave alone'. Additionally, after ejaculation, a great bioelectric let-down, and possibly feelings of shame, can result in feelings of *guilt* – especially when males continue to masturbate at a later age. In a number of societies there is a certain amount of joking and teasing about excessive indulgence and stories which suggest that masturbation causes blindness, stuttering, feeble-mindedness, if not complete insanity, and certainly personal sexual inferiority; it can be seen as a loss of a vital life force; overall, the lesson is clear: 'control yourself'.

Circumcision (especially during adolescence) as practised in a number of societies reinforces this lesson (Badcock, 1986; Jones, 2002); we cut you once, we can do it again – next time we may cut off more; in fact castration has been used as a punishment in a number of societies in the past, and a minority of those who want to reject sex completely have themselves castrated (Jones, 2002). But male urges to masturbate can be very strong, and excitement and pleasure intense. And so we do it in private and carefully clean up the results, because if the result can be eliminated, if the slime can be removed, perhaps the effects can be mitigated; cleanliness eliminates (or at least reduces) shame and guilt (or at least chances of being found out). However, a certain amount of shame and guilt is usually inevitable – for indulging, for being secretive and for lying about 'what you were doing'. So, as a result of feelings of being polluted or of risking dangers to health, and of suffering guilt and social condemnation, sexual pleasure *itself* increasingly becomes a tabooed arena.

For females, the physical evidence of masturbation is not so dramatic, although a considerable amount of wetness can be generated causing a degree of spoiling of clothing. But for females the extent to which they worry that they will be considered 'deviant' and 'unnatural' if they are found out may be considerably greater than for males. Female masturbation is generally a taboo subject in terms of personal and social discourse; historically, female masturbation was said by some to cause physical disabilities – as in males – but also in females to be a cause of 'hysteria' and nymphomania. All of this makes for strong motivations for females to masturbate in private. So, as for males, those females who have a powerful desire to masturbate, and who get very excited by it, have strong possibilities for feeling guilty; and they really do not usually have anything much to wash away to help rid themselves of the guilt. Again, the whole process lends itself to making sexual pleasure a tabooed area.

Where practised, female circumcision is even more emphatic as a statement to females than is male circumcision that sexual pleasure is not for them and that they had better stay well clear of it; often female circumcision is specifically designed to reduce if not eliminate female physical pleasure altogether. This is quite apart from the pain and trauma that often follows circumcision, so that in the mind of the female victim sexual areas of the body hurt and cause pain; and, of course, so does childbirth; both conditions relating sex to a consequence of great, often prolonged, pain. So sex can come to be seen as dangerous (for both sexes – Cf., Foucault, 1990). Even the intensity of pleasure can be seen as dangerous; orgasms, for example, can be equated with seizures and spasms and even heart attacks, and certainly as evidence that self has lost control of body and mind.

Similar variability – and ambiguity – can clearly be seen in our fear of blood. It was probably an evolutionary advantage to learn relatively unemotionally a fear of blood rather than it being a strong automatic attraction (such as to a wounded animal or enemy, both of whom could still be dangerous), or, conversely, something to run from automatically (such as a personal wound or a wound to an offspring or a friend, or a freshly killed supply of food). A relatively unemotional fear of blood and/or a loss of blood suggests the possibility of both dangers and opportunities, to be treated relatively immediately and with some intensity but with caution and some thought rather than with an over-excited panic reaction. The same sort of process seems to apply to disease in general. It is amazing how non-fearful many individuals can be of the threat of diseases given that objectively they are probably the greatest cause of death among humans; we usually make fun of individuals with a hygiene fetish, and many people simply do not go for regular check-ups, and others refuse or forget to take medicine or to stop smoking or drinking, for example.

And while during major epidemics the terror of diseases can be quite powerful, afterwards it is not something that then sticks in our minds for long – as evidenced by the refusal to undertake 'safe-sex' in many parts of the world and the progressive move away from 'safe-sex' after the initial phases of the AIDS scare. From an evolutionary point of view it is likely that if we were over-concerned with disease or accidents we would be immobilized, and our immune systems would not have developed as they have. So, those who were active and not too worried about getting ill not only got more done but also ended up with more robust immune systems. Diseases do not give the same danger signals as rotten food, maggots or a loss of blood; they are extremely varied in their causes and effects; they seem to creep up on us, and so there was little in the way of indicators for selection forces to work on. Selection concerning illness did work on the avoidance of certain things (for example, rotten food/faeces), but more commonly selection favoured means of dealing with symptoms – such as keeping an individual warm and providing care, attention and comfortable conditions during their time of illness.

To motivate us in this regard we evolved a considerable capacity to fear *suffering* (Lerner, 1997; Hankiss, 2001), which through sympathy and empathy, and a feeling of 'I might be next', has enhanced a cognitive capacity to recognize that things can be done to try to prevent diseases and to help when an individual gets sick. And, in the long run, this cognitive capacity has led to theorizing about diseases and causes of diseases; about those invisible, unpredictable, devastating evil forces which invade our bodies and devastate them. So, diseases have been attributed to cold winds, evil spirits, sin, God's wrath, evil humours and – eventually with modern medicine and scientific knowledge – to germs and genetic malfunctions.

Once self-awareness includes a notion of a clean self (able to be achieved and maintained through thorough cleaning of bodily impurities) and a healthy self (one well cared for), it is very easy to learn to fear such things as menstrual blood, mucus, vomit, puss, body odours, sweat and grime. And a fear of these makes it possible for certain body areas, activities, physical conditions and presentations of self to be emotionally charged with negative feelings. For example, a connection between female sexuality and pollution can easily be made through the link of female sexual organs with menstruation; vomit can be equated with over indulgence in food and drink as evidence of an individual out of control; and a connection between sweat and hard labour can be equated with stupidity of mind. This means that a push for 'cleanliness' and purity can be a major tool in teaching the value of being able to control passions in a number of areas besides sex, and to suggest 'good manners and social habits', and to intone that there are 'proper' ways of presenting self generally.

This fundamental process can take a number of seemingly contrary cultural patterns. For example, purity has not always meant scrubbing the body vigorously with soap and water (Ashenburg, 2008; see also, Curtis, 2007); at times in history it was thought that too much washing opened the pores too much and thus let in diseases or other impurities, such as evil air; at other times, it was thought that individuals needed to bathe to open pores to let out bad vapours. During the time of Queen Elizabeth I it was felt among the aristocracy that the way to a clean body was to always have clean linen rather than to wash the body itself. And in washing should the water be hot or cold? Some considered that hot water excited unhealthy sexual desires, especially among virgins, and that cold baths could be a good antidote to such desires, while also strengthening individual fortitude. And what about before and after sex – is it imperative to wash at these times?

And should we soak in a bath (in our own filth) or take a shower (running water) to get properly clean? And how much perfume should we apply? To what extent should bathing be a private act, and to what extent should sexual organs be deodorized as well as cleaned? In modern times it is felt by some that every surface of a bathroom or kitchen has to be sterilized in order to keep them clean and free of disease.

However, there has also been the view that too much washing or cleansing generally is a result of a vain concern with the physical body at the expense of the *spiritual* health of the soul. Early Christians, for example, were against bathing as being too self-centred and self-indulgent; saints and monks often went about in dirty rags and lived in conditions that others would consider to be filthy, the better to attend to spiritual activities; in modern times some hippies took a similar view. And what about various means of cleansing the inside of ourselves; besides prayer and exorcism, is bloodletting a good method? There has always been an issue of what is really more important, physical washing or ritual cleansing. So, as in most things human, there has been a tendency to characterize all of this in abstract 'higher' terms; so pollution becomes theorized into concepts of sinfulness and evil (for example) and saintliness comes to be equated with purifying (with whatever methods are culturally prescribed); and pride comes from success in keeping clean and pure (however culturally defined); and people and acts come to be categorized accordingly. This process is enhanced when taboos concerned with patrolling cleanliness and purity are used to regulate sex, social interactions and social hierarchies, for example.

This all makes evolutionary sense because a 'clean' self does, other things being equal, have a greater propensity to survival and reproductive success than a non-clean one. This is because, besides preventing diseases, it take more dedication and determination to keep self (and self's baby) clean and healthy than not to; but above all, an ability to strive for purity indicates that an individual is able and willing to control their more physical attributes, including their sexual drives and desires, and to live with other humans, characteristics which have had considerable reproductive value in the human mating pattern as discussed in Chapter 3. Also, the more dedication a person has to their presentation of self the more likely they will be to alter their physical appearance, their personal behaviour and their social comportment in the face of changing circumstances, a characteristic which also has both survival and reproductive value.

But, as we have seen in Chapter 3, reproduction has its own dangers which generate a number of human fears (although not exactly the same for offspring, females and males). To remind ourselves, in order to survive human offspring had to draw care and attention to themselves in a context of increased sibling rivalry, and for a longer period of time, than was the case for other primates. Then, to have reproductive success, they had to break away sufficiently from parental control and parental interests to become breeders in their own right, but not so much that parental support for them and their offspring would completely dry up.

Females – more often pregnant than is common among primates – who were able to entice help in providing care for themselves and for the extremely dependent offspring they produced, had in many cases more reproductive success than those who did not. At the same time, however, females benefitted by selecting genetically worthy males, males who quite often had status and power in the wider world. These 'requirements' did not always match up, however; genetically worthy, high status males may have been interested in promiscuity instead of provisioning and fathering, for example, and so a female had a balancing act to perform in her choice making. Males had reproductive success if their genes were spread widely and resultant offspring reached reproductive age. But male promiscuity and the survival of resultant offspring did not always coincide, and anyway many males were not able to be promiscuous because they had little chance of sexual access when up against alpha males. Most males were better served by breeding in pair-bonded relationships while having the protection of a number of alliance mates who respected and helped protect their monopolistic claims to a particular female.

In Chapter 3 I looked at a number of the emotional, cognitive, social and political formulations that seemed to have evolved as an evolutionary balance among these conflicting reproductive interests. Here I wish to look at evidence from modern life to see what support might exist for some of the fears suggested there. Additionally, modern sources might identify fears (and desires) not thrown up by the speculations based on the principles of sexual selection as applied in the analysis of evolution. To start, although the polymorphous perversion of offspring still draws parents and infants together, almost as 'child-play lovers', in some cases it works so well that incest and attractions to children for sexual purposes still exist in human societies, despite strong religious, moral and scientific arguments against doing so. But at the same time we have also evolved very strong feelings of protection and love towards the care-enticing characteristics of babies and very young children – care and protection which includes cleaning them, preventing hurt to them and keeping them *pure* and *innocent*; sexual relations with them seem to be a contradiction to all of these things; so we fear incest and paedophilia because of the elaboration and spiritualizing of our child-care desires and fears; additionally, our drives to purity of sexuality cause us to put this fear into taboos and laws and to greatly condemn sex with children.

However, a tendency to 'strong' parenting often leads to conflict with offspring, and to negative feelings of failure on the part of parents, even in 'modern' societies with all their psychology and science telling us that 'normal' children will rebel (especially as teenagers). And offspring still feel guilty and suffer a degree of unconscious anxiety/fear of potential abandonment and of being ignored. From this dilemma we have not only developed elaborate cognitively based desires for independence and fears of engulfment, but also fears of not being able to live up to parental/social expectations, of being considered worthless, of not being 'a good person'. So, it is little wonder that considerable feelings of ambivalence, guilt and depression are generated throughout the processes of parenting, growing up and going out into the world despite the existence of priests listening to confessions and advising, child psychologists teaching, counsellors counselling and 'how to parent' manuals flying off the book shelves.

For adults there is often an anxiety among mating pairs that they cannot trust each other – an anxiety witnessed by the abundance of gossip, stories and jokes we have generated concerning this potentiality, and the envy, jealousy and resentment (conscious and unconscious) involved in human experiences of 'love and romance'. While human sexuality has been greatly repressed compared to chimp or bonobo sexuality, mitigating the above lack of trust to some extent, repression itself can cause anxieties and fears. With enhanced self-awareness and cognitive capacities, and the spiritualizing of sex and marriage, it is easy to end up feeling impure, immoral, frustrated and resentful, and worried about which of these is worse or what to do about them.

Overall, then, sexual and reproductive feelings have become quite ambiguous and ambivalent during human evolution. For example, in our minds, love and hate often get somewhat mixed up. We may desperately love our partners and our children, and they us, and we like and trust our friends and allies, but sometimes we may resent, if not hate, all of these. But in thinking about our feelings we feel guilt about our resentment and tinges of hate, and we cannot really imagine ourselves without these 'significant others' being part of our lives, and so our anxieties about driving them away with our ambivalence, occasional resentment, and even 'hate' can become intense.

Storytellers have taken these conflicts to extremes in order to provide us with exciting stories. In the *Oresteia*, for example, 'love/hate' is a powerful force throughout the trilogy, Orestes kills his mother; Clytemnestra kills the father of her children; Electra drives her brother Orestes to kill their mother; Agamemnon sacrifices his daughter. Arguably, hate can be made more intense by the power of original love when the love has been stopped or rejected; Agamemnon rejected Clytemnestra by sacrificing their daughter and by coming back with Cassandra as his mistress; Clytemnestra rejected Orestes by sending him away early in his life and then alienated both her children for a new man (Greene and Lattimore, 1953; Sommerstein, 2008). One can see parallels in the story of Hamlet who kills his uncle and causes the death of his mother, whom he thinks colluded in the killing of his father – indeed, soap operas thrive on these conflicts. We find these stories compelling because we know – unconsciously if not consciously – that there is some truth in them; we understand the conflict because we share the same fears as the protagonists.

What is common in these stories are the powerful psychological consequences of loss and/or rejection. These can come from a loss of mother love, of significant other love, a loss of being cared for, of respect, of status, of power. Given the vulnerability of human infants at birth it was possible to argue on the principles of sexual selection that the basis of such consequences was the existence and enhancement of separation anxiety. Support for this can be seen in the psychologically debilitating results of deprivation as seen in studies of child development and depression. Such work informs us that infants suffer not just from separation but also from the loss of care and the loss of attention and affection from care-givers (Cf., Klein, 1952; Winnicott, 1965; Bowlby, 1969, 1973, 1980; Baron-Cohen, 2011). Long-term effects of deprivation as seen in many of these studies strongly indicate that adults do not necessarily grow out of these anxieties, and that they continue to be greatly affected by more developed fears concerning loss and rejection (Cf., Mitchell, 1975; Brown and Harris, 1978; Belle, 1982; Gilbert, 1984, 1989; Kahneman, 2011).

It seems that this type of anxiety/fear often leads individuals to want to monopolize the attention of relatively few 'significant others', sometimes even just one other. For example, a child may want constant undivided attention from a parent, or a parent may seek unending devotion from a child; this can extend to pathological jealousy on the part of a member of a pair bond. Attempts to monopolize affection and attention, however, often generate a sense of powerlessness and/or total dependency in the recipient. For example, young Alexander Portnoy, remembering his mother in his early years, recalled that: 'She was so deeply imbedded in my consciousness that for the first year of school I seemed to have believed that each of my teachers was my mother in disguise'. Try as he might he could not run home fast enough to catch her in the process of transforming; she was always there waiting with milk and cookies. What fantastic powers she must have had, he concluded (from Philip Roth's 1969 novel, *Portnoy's Complaint*; see also, Horney, 1932; Winnicott, 1965). But, as he also learned, mothers are not beyond using this fantastic power for the purposes of 'emotional blackmail' and guilt as weapons in their endeavours.

This theme has run throughout history, from mythology to modern novels, films and treatises. In the 1942 book *A Generation of Vipers* by Philip Wylie – which sold in hundreds of thousands – 'momism' was blamed for most of the ills faced by humans and a typical mom was pictured as: 'the homely, bonbon-eating middle-age monster who enslaved father and son and administered at least symbolic castration on a daily basis' (Buhle and Wagner, 2002, p. 349). But as mythology and novels like *The Brothers Karamazov* (Dostoevsky) testify, fathers too are important in the emasculating process, especially when offspring think that they are not favoured or that they have been failures in the eyes of their fathers. Fathers can also be engulfing, as in the case of Emma Woodhouse's father in Jane Austin's *Emma*. It can be difficult for offspring to directly attack parents, so, as noted, a fear of engulfment is most often expressed as a desire for *independence*; this is the basis, one can argue, for a human 'will to power', but with a conflict between attachment on the one hand and independence on the other hand extending into adulthood, and into human social and political experiences. This, I have argued, represents a basic, fundamental, universal human dilemma concerning self and others: to be free and independent, but to be loved and trusted at the same time. This dilemma throws up a potential for feelings of loss, at a number of points, and with enhanced self-awareness and contemplation can generate powerful feelings of guilt, which in turn can have profound motivating effects on human social behaviour.

John Carroll (1985), for example, has argued that there are two significant types of guilt in human personal and social processes: moral and dispositional. Moral guilt is simply the breaking of a moral code and feeling bad about it; moral guilt is generally policed by shame and external punishment (or a fear of external punishment). Dispositional guilt, on the other hand, is buried deep in our unconscious minds, its causes are not usually known by individuals who experience it, but its power to cause deep anxiety, restless agitation, depression, projection of feelings and sublimation make it extremely significant for our understanding of the evolution of human behaviour. Although its causes are not exactly known by an individual, it arguably arises from the dilemma of acceptance/inclusion being in conflict with freedom/will to power.

As offspring strive for power against parents, the child tends to project their aggression outward, blaming parents for their own lack of immediate gratification, but then they come to fear their own power: 'this fear is that power will annihilate love' (Carroll,

1985, p. 34). As a response they might turn their aggression back onto self and so develop, or enhance, a degree of 'moral guilt' – an attack from inside brings them to alter their behaviour. But they may project aggression outward onto social others, or they may sublimate it into a wide variety of social activities generating what we have come to call civilization. Some of the mental aggression generated by the oedipal struggle, for example, may be projected outward against parents, but throughout history it also has been directed 'onto an outside person, group [or] object' (Carroll, 1985, p. 35). This can mitigate generational conflict but in the process external groups are often stereotyped as being dangerous if not evil. Another projection/sublimation has been to undertake 'hard disciplined work' (Carroll, 1985, p. 39), a constant striving for achievement and/or for order and control of environments (physical and social) in order to claim 'success', or to get attention by making the world a 'better place' and/or to enlighten (and save) humankind. But when we dedicate ourselves to hard work and achievement we suffer the problems of never knowing what success is, and of ever-escalating goals running always just out of reach, and so the potential, thus fear, of failure at every step of the way looms large.

Indeed, in some cases the anxiety about the possibility of complete separation and a loss of love or failure may be so great that self fears that they are not *worthy* of love and attention, that self is unlovable and that any loss is one's own fault – depressive guilt manifestation – which tends to be immobilizing (see also, Laing, 1965, 1971). According to Carroll (1985) the two basic universal stories concerning depressive guilt manifestation are *Hamlet* and *The Ugly Duckling*. The ugly duckling, because of an early loss of love, feels that he is unworthy of love, destined to be 'unloved'. The result is depression and low self-esteem. Hamlet is so distraught at the loss of both his father and his mother's love that he becomes deeply depressed and immobilized. Not only has his mother rejected him but she has played a large part in betraying his father/ her husband and gives her love and body to his killer! For Hamlet there is the agony of feeling rejected by his mother and feelings of self-contempt for his own inaction. Any possible hatred he might normally feel towards his father's killer is compromised by his distress and anger towards his mother for abandoning him. Yet his intense love for his mother turns the anger inward, to a feeling that he has not been worthy of her love.

It is worth remembering that Orestes had been abandoned and then rejected by his mother, as had Oedipus by his father; the Karamazov brothers did not seem to have a mother to remember, and could never feel that their charismatic father had any time for them. Not all cases are as clear-cut as these, and the reality of human existence means that the types and manifestations of guilt are experienced in a variety of mixed up ways. But there are plenty of opportunities in human existence for the dilemma between inclusion and exclusion, being embraced in love and being scorned/humiliated, praised and ridiculed, between obedience and disobedience, seeking adventure and security, success and failure, for unconscious guilt to be generated, however it is manifest in different circumstances.

A common theme in the above stories is the loss, or the potential loss, of love. Quite apart from conflicts between inclusion and rejection as seen in stories, and in everyday life by priests and psychiatrists, clinical studies in general strongly suggest that depression is often set in motion by 'a threatened or actual loss' of a significant other or of social respect or standing (Brown and Harris, 1978; see also, Seligman, 1975; Mitchell, 1975; Belle, 1982; Gilbert, 1984, 1989). When the loss is self-perceived to be the fault of self these studies show just how debilitating resulting depression can be in its generation of feelings of failure and worthlessness, and in perpetuating depression itself. So, fears of loss may be developmentally based on separation anxiety and anxiety concerning rejection, but these can lead to a considerable elaboration and exaggeration of *fears* of loss in human social life generally, so that actual loss becomes quite devastating.

And actual loss will happen, for everyone. For adults, there is the long-term inevitability of losing the oedipal struggle. Parents try desperately to make their offspring into achievers of the potential they had fantasized for themselves, very often to be disappointed. At the same time the young seem to have access to everything fun and pleasurable that the parents did not have, or at least no longer have, and the young have a lifetime of further fun and potential achievements to look forward to. Once offspring leave home any semblance of control previously exercised by parents is usually gone. The old are confronted by a loss of physical, personal and social power on a regular basis with decreasing chances to recoup them, and are increasingly treated as children as their physical and mental capacities deteriorate.

It is easy for adults to progressively learn to fear both old age and 'the young'. Older people begin to see an 'end of civilization' because the young in general no longer seem to respect their elders. These fears often motivate older people to hold onto their status, wealth and power using all available means; it is their only method of control left. Elder males, at least some of them, try to monopolize the judicial and political processes. They progressively become more conservative and cynical – less convinced that utopias are possible. Their utopias become not youth fantasies of a golden age to come but rather of a past age in which the elderly were respected and considered to be wise.

We have seen some of the 'personal political' conditions that likely surrounded human mating patterns during human evolution, generating a number of the fears (and desires) which resulted in the repression of human sexuality. Among other things, when compared to primates in general, females became considerably more choosy in terms of the male they would mate with; this, however, was not always easy to achieve. Thus we can postulate the evolution of a female fear of non-courtship-induced sex; in modern (and historical) times the most dramatic expression of this is the powerful, visceral female fear of rape. Anti-rape campaigners usually make it very clear that it is not sex that is objectionable, but rather that a male has taken physical control, denying a female any control of her own sexuality (or reproductive future) – thus the claim that rape is 'about power not sex'.

It was speculated that many females recognized early on that to maximize female choice in mating partners, they would have to avoid exciting seemingly powerful, automatic male sexual urges, including by not exposing their own sexual characteristics (the evolution of modesty). From this perspective it was predicted that females evolved a capacity to easily learn to fear, or at least to be apprehensive about, both male and female sexual promiscuity, if not sexuality itself, and to fear males who seemed to have only 'one thing on their minds'. Such fears have resulted in females developing considerable desires to control their own sexuality, even in marriage, such as by restraining their sexual attractiveness and proceptive behaviours, by over-mothering, by feigning illness, by accusing husbands of being animal-like without 'caring' feelings and simply by withholding sex (Cf., Symons, 1979). Literature on hunting and gathering societies and mythology certainly suggests that females have long fought to be able to have a say in whether they wanted to be a third wife, accept an arranged marriage or be exchanged into a particular lineage.

During this process, as civilization was progressively 'imposed' on the physical nature of sexuality and reproduction – in the form of extended courtship, love, romance, marriage, and family – additional possibilities for feelings of shame, moral failure, rejection and loss came to exist. Female sexual urges, when put together with romantic desires, can be strong, and so the repression of female sexuality, through modesty and fearing uncontrolled sexuality, can be an anxiety-generating, potentially painful and guilt-inducing process. In an intense form it can lead to persecutory aggression projected onto parents, boyfriends, husbands, 'male lust' and/or 'patriarchy'. If accompanied by something like separation anxiety and strong repressive teachings, extreme sexual anxiety might lead to depressive aggression directed inward onto a female's own sexuality.

Carroll (1985) argues that *Jane Eyre* is the classic example of sexually motivated persecutory aggression/guilt being projected onto a father figure, but also turned inward onto self's own sexuality. Rochester is 'powerful, authoritative, worldly, wealthy, aristocratic, respected by all, he is ruggedly handsome' (Carroll, 1985, p. 46). Jane has strong sexual passions, she desires him, but when they fall in love she becomes frightened and runs away. This is in part because of her confusion concerning his exposed existing marriage, but also because of her fear of her own passion. When Rochester's insane, passionate wife sets fire to the house and blinds Rochester, Jane takes this as 'proof' of the dangers of passion but is now able to return and marry him. There is no chance at this point that he will arouse her own dangerous passions; she has control.

In many cultures and historical periods females have been taught sexual fears at a very early stage, and they enter reproductive age with the potential for considerable sexual guilt in situ. In certain other cultures a degree of sexual experimentation among infants and adolescents is allowed, if not encouraged. In these societies, however, there is no less desire among females to control their own and male sexuality at the point of reproductive age, a time when romantic arousal and marriage itself become much more important than sex. At that point, and/or at the birth of a child, female sexual favours are largely restricted to their chosen partner (Symons, 1979; Gray and Garcia, 2013). Ironically, perhaps, permissive societies can be more difficult and anxiety-creating than non-permissive societies because of an increased fear of being sexually inadequate or being left out (for both males and females). Also, in such societies females can especially fear that a promiscuous female will steal away a male she has worked hard to 'capture' and tame, especially at the point she most needs him as a helper after the birth of a child (or two or three) when she herself is least attractive or interested in sex.

This may be a major reason that prostitutes are universally condemned by females, or at least held in very low esteem, rather than being seen as role models. Prostitutes can make more money than the average married or working woman; some do it because they enjoy the excitement and non-routine nature of the work, and being a prostitute can give them considerable independence; it certainly can give them control of their own sexuality (Winn, 1974; Roberts, 1992; McKeganey and Barnard, 1996; Gray and Garcia, 2013). This is especially the case historically with some temple prostitutes, courtesans and ladies in waiting, and in more modern times with expensive call girls, female employees in massage parlours, escorts, lap-dancers, female frequenters of middle class swingers clubs, and aspiring starlets. Even God had to woefully lament of Jezebel: 'And I gave her space to repent of her fornication; and she repented not' (*Revelations* 2: 21).

To be sure, some prostitutes are under the exploitive control of pimps and/or are drug addicts, and might have come from very deprived backgrounds and/or have been trafficked. But there are also a lot of females under the arbitrary control of unsympathetic bosses or husbands, and there are many females from deprived backgrounds working at very low level, unpleasant jobs – especially girls in service in past and modern times – and there are female drug addicts at all social levels, but none of these are considered with the same contempt and condemnation as prostitutes. Many may even be praised for working hard and living under very difficult and/or harsh conditions and be seen almost as the opposite of female prostitutes, as heroic examples of womanhood and/or as victims.

Considering seemingly uncontrollable male sexual urges female prostitution under safe, healthy and otherwise decent working conditions could almost be an ideal, indeed rational, means to drain off these urges while giving females an income they are in control of and an escape from some of the circumstances described above. The widespread existence of female prostitution throughout different cultures and historical periods shows that, in fact, this is a solution often adopted, or at least tolerated, and it is a clear indication that in most places throughout history there has been a ready market for the sale of sex by females. But it is usually a reluctant solution on the part of wives and policy makers, and even where prostitution is most tolerated prostitutes themselves are still held in low esteem, and the 'occupation' of even high class call girls is considered disgusting, if not repulsive, to most females (and to many males).

So, why are prostitutes held in such low esteem? The reason prostitutes are despised is that the activity of a prostitute symbolizes a number of female fears concerning their own and male sexuality. Female fears of lust-driven sexuality motivate them to desire (and demand) men to control their sexual urges, not to let them get away with enjoying sex without responsibilities or needing to show any emotions. Females desire men who will bring home the meat, not run off with it to the first whore to be found just because her sex is up-front, no emotional or parental responsibilities asked. Females desire that females 'play fair', that they ration sex in exchange for love, care and support over a long period of time. The contempt that female prostitutes suffer, then, is a result of a female fear of not controlling both male sexuality and their own, and also of female competitors who might steal affection and resources away from them (call girls, escorts, swinging club frequenters and aspiring film starlets are often physically very attractive); these fears are often accompanied by resentment from seeing their own restraint being almost ridiculed as a waste of time.

As noted, these fears also lead to the glorifying of love and romance (and to the importance of children and family), a glorifying that is almost the opposite of giving any value to physical sex itself. Concepts of love, romance and family life represent ideals about the control of sexuality that make females safe from their own, and from male, sexuality. Nevertheless, romantic love can itself be a danger for females and as such can stimulate fears of another kind. There are few men who can completely live up to romantic expectations, and the fear of having been wrong in mate selection and/or the fear that a chosen male wants to escape can be strong for females. These fears can make it difficult for a female to project blame onto her partner, because she chose him after all and because the partner remains in part at least an element of her own earlier fantasy. So, aggression for the loss of the romantic ideal is turned inward into self-blame; the result can be depression and an increased vulnerability to loss. If the man leaves or has an affair, for example, it will be even greater 'evidence' that she made a bad choice and/or that she is not worthy. There is considerable psychological literature which documents the depths of depression and other forms of stress, and even physical illness, felt by women who feel they have failed in a relationship (Cf., Freud, 1957; Brown and Harris, 1978; Scarf, 1981; papers in Kirkpatrick, 1982; Ayduk, Downey and Kim, 2001; Lepore and Greenberg, 2002; Fisher, 2004).

Males also have primordial sexual fears. Young boys entering adolescence, hearing stories from older boys, wonder if they will be able to get anything to come out of their emerging manhood, which they find themselves increasingly playing with; are they normal? Will it work they wonder as only a little bit of sticky fluid emerges? This is in addition to what they fear their mother will say if she finds out about what they are surreptitiously doing. Later they fear that they will be unattractive to females, that they will never have a girlfriend. When a girl does pay some attention to them they fear that it is only temporary, that the girl is looking out for the best looking or the best hunter or the best basketball player; his mother may very well tell him that this is the case with most women – except her – and not to trust girls. And even when a female settles down with him he has a recurring fear that she is disappointed and that she always has an eye out for someone else, someone she should have held out for. And can he satisfy her? If he fails on a particular occasion will that make her think he is not really up to it? And until recently, with the increasing availability of DNA testing, a male could never be sure that a baby was a result of his own sperm; can he really trust not only her but also his friends and alliance mates?

As we have seen, males evolved a number of emotional mechanisms and behavioural responses, which to some extent at least help to overcome the dangers of reproductive failure, but these have not completely eliminated male sexual and reproductive fears – indeed, a number of these fears play a major role in keeping the original behaviours in place. Males are motivated to spread their cheap sperm as widely as possible through the inheritance of a powerful sex drive. But attempting widespread sexual access usually only works for the most handsome, wealthy, high status or powerful individuals, such as warrior heroes, sports, film and music stars, and the politically powerful (who, it has been said, have the greatest sexual access of all in modern societies – Betzig and Weber, 1992). But there is danger even for these 'heroes': 'I risked my life for pussy' complained a former drug dealer in a modern city sub-culture in which accumulating ostentatious wealth and even killing another man became almost required to attract females (cited in Courtwright, 1998, p. 236). Even for powerful individuals with multiple wives/sexual partners, there is always the fear of being cuckolded, attacked and/or overthrown by testosterone-charged young males – even from among relatives and neighbours.

From principles of sexual selection we can argue that forming an emotional bond which includes exclusive sexual relations with a particular female may be the best way for the vast majority of human males to reduce exclusion and the dangers of taking on powerful individuals, and for fulfilling desires for having regular sex (and for having children). But this does not completely eliminate fear among males; males are acutely aware of their own lust (and secret desires for their friend's female), and can easily project both feelings onto their sexual competitors, and indeed onto their own alliance mates. Thus, any male, including friends, can be a threat who will try to take away a female. As a result, males fear the sexual 'drive' in other males as well as any sexual interest a female they desire might show others. That they do so is most vividly seen cross culturally in the extensive amount of fighting which males often indulge in over females (Symons, 1979; Wilson and Daly, 1992; Buss, 1994; also amply represented in mythology – Cf., Graves, 1960). Commonly, therefore, males fear their own, their friends' and their partner's potentially uncontrollable sexual desire, which leads them to create and support codes of alliance loyalty, honour and sexual morality which closely regulate their own, their mates' and general male and female sexuality.

However, these never completely eliminate male sexual fears; indeed, they can add to the transgressions that can be committed, generating additional areas of concern. In significant part, Paris' crime was considered to be so great, and Agamemnon and his followers were so incensed (and compelled to vengeance), because Paris was a guest – honorary household member – in Menelaus' house when he stole Helen. The legend of King Arthur is a tragic conflict involving the powerful force of raw sexual desire (the various temptations to adultery that Lancelot, Guinevere and Arthur face), a romantic fantasy (Lancelot and Guinevere's deep love) and the idealized vengeance brotherhood (the Round Table) where everyone is expected to respect each other's sexual claims upon females, indeed is *honour-bound* to jointly protect these claims against internal transgression and external aspirants.

In this context it is relatively easy for males (and females) to assign blame for transgressions onto 'the dangers of *female* sexual enticement': 'And I find more bitter than death the woman, whose heart *is* snares and nets, *and* her hands *as* bands: whoso pleaseth God shall escape from her; but the sinner shall be taken by her' (*Ecclesiastes* 7:26). A major part of Hamlet's depression was due to the fact that his mother not only appeared to reject him but, more importantly, that she did so for sexual desire for another male as much as for status or power – Hamlet could almost have understood the last reason (Carroll, 1985). So, Hamlet blamed his mother's sexual passion for her actions and consequently for the terrible dilemma he felt himself to be in. Hamlet turned a certain amount of his aggression inward into depression, but not so much so that he completely condemned male sexual passion; the cause of his loss was projected onto the dangers of female sexuality.

At the time that Shakespeare wrote *Hamlet* the fear of female sexuality was especially strong in Europe; at this time he also wrote *Troilus and Cressida* and in both these works moral decay is seen to be everywhere and is largely considered to be a result of the evil influences of female sexuality (Carroll, 1985). This view had smouldered in the thinking of a number of the early church fathers (Foucault, 1985), but really began to have force in Europe from about 1300 onward (Hays, 1966; Stone, 1977; Kelly, 1982; Carroll, 1985). By the mid-1500s women were believed to be 'fickle' and to be 'troubled with evil humours' (Scot, 1585, in O'Faolain and Martines, 1973). The cause of women's evil was most often said to be their uncontrollable sexual desires, which could castrate, make impotent, kill and otherwise mutilate men (Hays, 1966; Griffiths, 1976). Poets, playwrights and storytellers of the time created women who were over-sexed, fickle, foolish, quarrelsome and given to cheating on their husbands; henpecked husbands being common characters in drama (Notestein, 1955; Hays, 1966).

Preachers constantly railed against female fashions, saying they were signs of vanity and desires for 'glory' and 'empty honour', all designed to lead men into sin (Green, 1971, p. 177). Bishop Aylmer conceded that some women were good (because he was preaching in front of Queen Elizabeth), but others were useless and 'doltified with the dregs of the Devil's dunghill' (Stone, 1977, p. 196). Joseph Swetnam's *The Arraignment of Lewd, Idle, Forward and Unconstant Women* went through ten editions between 1616 and 1634 (Stone, 1977, p. 197). Shakespeare's Lady Macbeth and Cleopatra were portrayed as positively dangerous. Some women were 'willing to make war on chastity' (Green, 1971). Witchcraft was said to be a result of 'carnal lust, which is in a woman insatiable' (*Malleus*, 1486, pp. 122–123). Jean Bodin observed: 'there are fifty women witches to one man', not because of the 'weakness of this sex, because we can see untameable stubbornness in most of them' but because of 'bestial lust which pushes the woman to these extremes' (in Andreski, 1982).

These are extreme examples taken from a period of history in which both males and females were caught up in conditions of considerable sexual, kinship, economic and political ambiguity and conflict, circumstances which included significant demographic changes (see below under witchcraft). They nevertheless represent a basic male (and female) fear of female sexuality that has had considerable historical and cross-cultural expression, and continues in modern novels, plays, films and television soaps; the *Film Noir* genre made this a major theme. A fundamental fear of female sexuality may also derive from the previously discussed primordial fear of the controlling/engulfing power of mothers. Not uncommonly, mothers are uneasy with a young male's emerging, seemingly rampant, sexual urges and with all the females he shows an interest in – whom she fears are using their sexuality to lure him into danger. The result can be a negative picture in the son's mind of what many females may be up to with their sexual attractiveness. But at the same time he can be powerfully drawn to a sexually attractive female.

And so he is in a dilemma – not only is female sexuality a potential danger, so too is the stifling control of a mother who tries to deprive him of the 'pleasures of the flesh'. It is probably no accident that sons and mothers rarely talk to each other about the son's emerging sexuality (except for her to warn him against sexually enticing females). And it is not surprising that young males first turn outward to male companions when they are in the process of breaking the family bond. Donald Winnicott (1965) has gone so far as to argue that males have a tendency to accept, indeed seek, domination by other males because of fear generated by a powerful male fantasy of an overpowering woman (the mother).

This fear leads them to seek, and even welcome domination by a known human being, especially one who has taken on himself the burden of ... limiting the magical qualities of the all-powerful woman of fantasy ... The dictator can be overthrown, and must eventually die; but the woman figure of primitive unconscious fantasy has no limits to her existence or power (Winnicott, 1965, p. 165).

The ancient myth of the Amazons represents this fear. This is the story of powerful female warriors destroying men, including sons. In literature, females rarely actually kill husbands or sons but rather engulf/emasculate them through their 'overpowering mothering'. This fear of emasculating females is seen in the *Bible*, the *Canterbury Tales*, D.H. Lawrence's novels and in Dostoevsky's *Crime and Punishment*, for example (Carroll, 1985). In this last novel the protagonist hero, Raskolnikov, was made sexually impotent by the double-binding sexual and love behaviour of his females (Carroll, 1985; Wilkinson, 2009). In a sense he was hit by both extreme separation anxiety and engulfment fears at the same time; he almost represents the human dilemma in archetypical form. His separation anxiety was fuelled by an absent father and by his complete incompatibility with his mother and her lack of sensitivity to his feelings. However, his mother and his sister seemed to have only one aim and that was to sacrifice for him through drudgery and poverty so that he could achieve greatness. His sister was prepared to provide passionless sex in a loveless marriage to help him (and their mother); both these females considered that they were extending the ultimate in love to him.

To be worthy of this 'love' he had to totally dedicate himself to his studies and to reject his own passions and ambitions. His intense feelings of guilt – turned in on himself – for failing in these 'family obligations' was magnified by observing the life imposed by male lust on the downtrodden young prostitutes he encountered in the streets of St Petersburg, and by the position his prostitute friend Sonia had been put in by her drunken father and harsh step-mother. Raskolnikov came to detest both male and female sexual passion as being a generator of the selfishness which seemed to cause so much distress for humans. When male aggression generated by disgust and fear of sexuality turns inward it can lead to depression and impotence, as in the case of Raskolnikov, but all this might lead to activities like religious fervour with a strong puritan streak, or celibacy as part of an obsession to save humankind, or the obsessive mastering of an art or a science. It is hard to know all the motivations involved, or how widespread such rejection of sexuality is among males generally, but famous individuals such as St Paul, St Augustine (along with many other less famous male religious celibates, such as monks and priests), Mohandas Gandhi, Isaac Newton and J.M. Barrie gave up the pleasures of the flesh for higher contemplation for relatively long periods of their lives.

The dilemma of having energetic sex with a partner and having a loving mother at the same time can be resolved if a male leaves home and finds a sexual-mother substitute, thereby re-claiming mother love and a child-like state of play and dependency, but now with the added incentives of orgasm, sexual daydreaming and sexual fantasies included. But such females can be hard to find, raising the possibility of the loss (separation anxiety again) not only of a sexual partner but also of a substitute 'sexual mother', a double loss as it were. So, traditionally this fear has motivated males to desire for marriage females who are virgins not 'whores'. Males often do not respect females who have sex with them before marriage because they fear that if they will do it with them they will also do it (or may already have done it) with somebody else, particularly if they seem to enjoy it too much. Even after marriage, in the long run, it is less frightening to have 'bad' sex with a woman one can trust than 'good' sex with an un-trustworthy female.

Males, therefore, also hold prostitutes in low esteem; prostitutes epitomize complete untrustworthiness in that they blatantly and promiscuously use their sexuality to their own ends and without concern for morality – morality largely being defined as female loyalty and mothering tendencies. Prostitutes represent a complete withholding of mother-love. At the same time, for those males primarily interested in power and self-esteem through sexual conquest, prostitutes represent no victory at all. A male statement of contempt, and a claim for sexual self-control and sexual worthiness, is an often expressed refrain: 'I do not pay for sex'; prostitutes in their offering of sex also represent the dangers of sex in terms of diseases, lying and stealing. But, alas, it is hard for well-settled males to not sense that they may be missing out on something rather exciting which has no responsibilities attached to it: that they have given up on a variety of titillating sexual experiences.

From a consideration of our primordial reproductive fears we see that we can easily learn to fear loss: loss of attention, care, friendship, sex, significant others and love. We also fear a loss of freedom and independence, which can be in conflict with the first set of fears. And, regardless of which type of loss is predominant in a particular individual, or is most likely in a given culture, it seems universal that we incorporate a fear of potential loss of these things into our self-identity formation, which can manifest socially as a fear of losing self-control, self-esteem, respect, honour, status, wealth, property and authority. These fears help motivate the development of various rules, taboos, boundaries, social subordinations, allegiances and social hierarchies, which go a long way to minimize – if not in some circumstances eliminate – reproductive fears.

Additionally, long before we had the ability to philosophize about love, romance, sex, status and power, we evolved a multitude of means for determining if others were physically, sexually, mentally and/or socially dangerous or, indeed, compatible; and we spend considerable time in courtship to try to make sure we get it right. To start, we have inherited from our primate ancestors a relatively strong emotional sense of who to stay away from (to fear). Try as we might to not show it, some combination of anxiety, repulsion and general disgust exists when we encounter what is generally considered to be 'abnormal humanity'. For example, in a number of societies the notion of evil itself has been related to 'physically rotten, misshapen, and ugly' objects and individuals (Parkin, 1985a, p. 7). Strange humans are often considered to be either aliens or, more commonly, not quite fully human, beast-like or demon-like; for example, '"monster" is the oldest word in our tongue for human anomalies' (Fiedler, 1981, p. 16). During much of history cripples have caused repulsion, and for this (as well as for economic reasons) have often been either allowed to die young or have been instantly put to death (Daly and Wilson, 1981; Fiedler, 1981). At other times, however, they simply have been looked at in amazement or experimented on (Leroi, 2004); often they are laughed at (Wiseman, 2007). In some societies twins have been killed at birth (but in others they have been welcomed as portents of good things to come).

All of this, of course, makes evolutionary sense. Not only do most biological abnormalities generally represent potential genotypical incompatibility but also a phenotypical expression of possible genotypical 'inferiority'; additionally, disabled individuals, other things being equal, can be undesirable in the here and now as partners. So, along with fears of abnormality we have evolved desires to mate (and socialize) with like phenotypes – including with similar personality characteristics (see Chapter 3) – which has gone some way to forestalling the dangers of some of the potential sexual and psychological mismatches, and possibilities of loss, described above. But there is never an exact fit and a perfect mate is not always available. So, potential mates are more or less desirable, ranging from highly desirable along to 'will have to do', and so a particular *range* becomes established as a breeding pool separated from outsiders.

Cultural processes usually work to exaggerate the 'value' of insiders and the almost non-human nature of outsiders. Throughout human history, for example, outsiders have often been given stigmatizing names and descriptions which variously signify inferiority, stupidity, deviousness, cunning, scheming and general dangerousness (Cf., papers in Sanford and Comstock, 1971; Glover, 1999), while the name given to insiders has been the word for 'people' or 'humans', with the possibility in some cases of the heroes among them becoming gods when they die. In modern societies our fear of strangeness often manifests itself in anxieties about things such as different skin colours, different sexual orientations, perceived sexual deviants, radically different cultural presentations, such as 'weird' hair styles, different ways of dressing, different ways of speaking (dialects/accents), minority religions, opposing political views, particular status demarcations (untouchables, for instance), chemical substance users and the mentally ill, and in some cases fears about foreign ideologies and foreign states.

But, considering the emotional nature of human evolution, especially of human bonding and caring and sharing, it is not always easy to discard an offspring just because they are a little misshapen, or to end a partnership just because a previously vigorous, healthy reproductively worthy person has become crippled. And someone with a handicap might even be more worthy as a potential mate (Zahavi, 1975; Johnstone, 1995) because they have shown an extra ability in coping with their apparent handicap; a person with a deformed arm, for example, may be highly intelligent, very caring, extremely wise and have very good genes in other respects, all evidenced in part by having been able to overcome a disability all their lives. Or, a disabled individual may be slightly less productive than an able-bodied one, but be better than no partner at all given the potential scarcity of partners during the early stages of human evolution. And when a society has to merge with another for any reason, or is taken over by another, a certain amount of ability for accommodation and integration of cultural differences can be an aid to reproductive success.

So, evolution has resulted in a human ability to sensitize against the fear of differences/deformity, and if they possess enough force of personality, power, status and wealth, even a person with a physical abnormality can appear vigorous and attractive. One of the most famous past presidents of the USA, Franklin Delano Roosevelt, fits this category. And it is amazing how plural, multi-racial and multi-cultural societies can become in particular circumstances. Still, the primary fear of strangeness is sufficiently strong, and has to be actively overcome, sometimes with great effort, that its dynamic remains fundamental to our understanding of human nature. So, we have spent great efforts to try to help 'abnormal' individuals 'live normal lives'. But this act itself starts from the premise that they are not normal and require 'abnormal' attention. Whatever the exact feelings, the long-term disabled, even in modern societies, usually remain 'neither sick nor well', they are between death and health – they are undefined, ambiguous people (Murphy, 1987; Lappier, 2005).

The result of coping with strangeness/'abnormality' is that we are often ambivalent and sometimes fascinated by strangeness and strangers; we are afraid, yet we are curious; we are repulsed but we stare; we avoid but we want to help. A dramatic representation of this can be seen in our concept of 'freaks' and freakish behaviour. The notion of freaks, however, takes us a step beyond simple strangeness or deformity or abnormality; it takes us to questions concerning the very nature of what is required to be human; it raises the very issue of normal versus abnormal; it addresses our anxieties about our own – often seemingly feeble – struggles to be normal.

Throughout history the concept of freaks has largely been based on the idea of *unfinished* humanness, the feeling that something fundamental is missing; as such, apparent freaks cause an automatic and complicated mixture of: fixed staring, repulsion, pity, wonderment and awe; a condition 'of full-passioned fixity, of panic' (Twitchell, 1987, p. 43; see also, Otto, 1945); freaks are people who stir 'both supernatural terror and natural sympathy' (Fiedler, 1981, p. 24), and wonderment. Do they have special mental and sexual knowledge and techniques, enhanced desire, stamina and enthusiasm, or are they devoid of all sexual or romantic feelings; do they feel pain? We wonder at just how they feel about things generally and what images of themselves they have and what value they feel for themselves. Therefore, the curiosity, fear and ambivalence we feel concerning 'freaks' and 'freakish behaviour' illustrates a number of largely unconscious human fears about ourselves, especially fears concerning our sexuality and our self-identities.

Freaks

While Franklin Roosevelt's physical characteristics of deformity/disability were able to be concealed from the public at large, and among his inner circle they were overcome with presentations of self (charisma) and with social performance on the part of FDR, this is not so easily done with what Leslie Fiedler (1981) refers to as 'true freaks'. According to him, true 'Freaks' are not: 'strange people', 'oddities', 'very special people', 'curiosities', the dumb, blind, paraplegics, but 'humans more marginal than the poorest sharecropper or black convicts on a Mississippi chain gang' (p. 17). They are more marginal because they challenge the boundaries and measures of normality we have created to escape from four basic 'primordial fears': fears about 'scale, sexuality, our status as more than beast, and our tenuous individuality' (p. 17).

Only the true Freak challenges the conventional boundaries between male and female, sexed and sexless, animal and human, large and small, self and other, and consequently between reality and illusion, experience and fantasy, fact and myth (p. 24).

The true freaks are: The Dwarf, The Fat Lady, The Bearded Lady, The Human Skeleton, Giants, Hermaphrodites, Siamese Twins, Lionel-the Lion-faced man 'and Jo-Jo the Dog-faced boy, the greatest an-thro-po-log-i-cal mon-ster-os-i- ty in captivity' (p. 22).

For children, according to Fiedler, fears concerning scale are very important. The works of L. Frank Baum (the *Oz* books), James Barrie (*Peter Pan*) and Lewis Carroll (*Alice*) all play upon the relativity and precariousness of size. For children, dwarfs, giants and fat men and ladies are the most intriguing freaks. Children are born into a world of powerful giants. Just as they are about to grab something exciting a 'giant' comes along and sweeps them up, laughs at them and puts them out of its reach. Size for a child changes over time and the changes are largely out of the child's control. They want to grow up but it seems to be taking forever. Or, they do not want to grow up and it seems to be happening despite their wishes; the *Oz* books and *Peter Pan* suggest a fear of losing childhood and provide a magical fantasy world for perpetual children to be happy in (Byatt, 2004). Additionally, many adults believe that children have wonderful imaginations and powers of creativity which are lost with adulthood; and so some wonder if dwarfs retain certain special qualities, for example, such magical craftsmanship as in the German myth of the Ring of power to rule the world made by the Nibelung dwarf Alberich, and in ancient Norse mythology it was believed that dwarfs could make magic weapons (Redfield, 1994).

However, it is above all the sex lives of freaks which both threaten and intrigue us most. During and after puberty fears of possible sexual inadequacy begin to assert themselves. Sexual fears, however, start even before puberty, according to Fiedler, in that the child is confronted with the enormity, if not extreme 'strangeness', of the parental genitals. Do freaks have sex lives, and if so what monstrosities might emerge from such matings, and anyhow are not most freaks products of copulations with the beasts of the field, or at the very least, of very unnatural sex acts (Davidson, 1991; Ehrenreich, 1997)? Throughout history, sex with animals and with demons and witches has been condemned as wicked and twisted, and every sort of freakish result has been imagined and pictured. Yet we are often fascinated by the sexual adventures we fantasize that freaks might get up to and even by the secret potencies we imagine them to have (Fiedler, 1981): 'Is not a midget's "cum" essential to good sorcery'. Yes, freaks do have sex lives, sometimes very active ones. In the past male dwarfs were often kept as sexual pets by aristocratic ladies. Catherine the Great of Russia thought that giants (men of about seven feet tall used as guards to scare the opposition) were likely to have special sexual qualities, but was apparently disillusioned and so, it was said, turned to stallions.

It is, however, the 'real' sex freaks, the ones with both parts, with over-developed parts, a man with three balls, a woman with four breasts, the real boundary crossers, hermaphrodites and, not least, Siamese twins, which challenge the boundaries between 'normal' and 'abnormal', between 'proper' and 'perverted' sex, between the sexes themselves and between self and other. Take Myrtle Corbin 'called the "Four-legged Woman from Texas"', who had two vaginas. From one of these she allegedly bore three children and from the other, two. Or, the Cuban, 'Jean Baptista dos Santos who had, in addition to a pair of fused supernumerary legs, two large, fully operative male organs, and who, they report was possessed of extraordinary animal passion, the sight of a female alone being sufficient to excite him. He was said to use both penises, after finishing with one, continuing with the other' (Fiedler, 1981, p. 221). Then we have the story among the Macurap of Brazil about the man with three penises who made love to three females at the same time in his hammock (Mindlin, 2002).

Hermaphrodites and Siamese twins, especially, challenge boundaries between physical categories and also between self and other, between fantasy and reality. Siamese twins challenge our notions about sexual privacy and our 'normal' distinction between self and other by raising such questions as do they think as one, do they share mates, are they jealous and, above all, how do they organize it? Often the first question asked about Siamese twins is, 'Do Siamese Twins make love?'. Yes they do. Violet and Daisy Hilton, for example, both very attractive and sexually alluring (picture in Fiedler, 1981, p. 208) had very active and interesting sex lives. When Daisy was making love, said Violet, 'Sometimes I read and sometimes I just took a nap'. The most famous Siamese twins from whom the name comes – Chang and Eng – each with a wife (sisters), produced between themselves twenty-two children and it is estimated today at least one thousand living grandchildren and great-grandchildren.

Freaks, thus, embody a number of basic human fears. On the one hand, freaks appear revolting, potentially dangerous, even evil, possibly with strange powers, but we also feel 'there but by the grace of God go I'. As young people we wonder if we will grow up as other people seem to have done and really be adults; when will we shave; when will our breasts develop; when will our pubic hair grow? Dwarfs represent a fear that we will suffer the indignity of being treated as perpetual children, unfinished, never respected, pets, playthings. Giants, on the other hand, suggest that there are many out there that are extremely powerful, much stronger than us, and that we will never be able to match that. As we go into the world, we fret: do we appear 'strange', 'weird', inadequate, to others, especially to members of the opposite sex? Will our offspring be normal, or because of our own misbehaviour during their conception or their gestation have we given them a curse of freakishness that they will have to live with? How would I be looked at if I have a stroke, go blind, if I lost part of my face in an accident; will I turn into a freak as I grow older and hobble about all bent over, slobbering?

Our ambiguous fear and fascination with freaks represent fears about the precariousness of being a normal human (Fiedler, 1981) and of being sexually attractive, probably two of the greatest of human anxieties. The bearded lady reminds females of the precariousness of maintaining their femininity and sexual attractiveness. The man with two penises reminds men of their potential inadequacy, females of their unfulfilled libido as a result of the speed of their men's first efforts. The woman producing children from two vaginas reminds females of their potential barrenness, men of the sexual paradise they might have had. So, sexual freaks taunt people about their own unfulfilled sexual desires, or worse, of sexual desires never to be fulfilled, while at the same time suggesting that we may be repulsive in the eyes of others, that the thought of making love with us is revolting. The need to share between Siamese twins reminds us that, try as we might, sex may not always be private; our childlike passions have been observed; our 'perversions' have been noticed. The mother's warnings have been ignored, the father's little girl is wicked and the whole world knows about it. Siamese twins remind us that we are not alone, that we cannot do what we want; it frightens us that our individuality is at the mercy of others and there is no escape, no divorce.

Dwarfs, when we hear of their wild sexual escapades (there are dwarf porn sites), challenge our belief that 'little people' are innocent; they suggest that even the 'tiny' are already corrupted – does that include children, our children? And what about humanity itself; are we really beasts in disguise with a very thin veneer of civilization just barely keeping us in check? Is it true that 'man has no pre-eminence above the beasts: for all is vanity' (*Ecclesiastes* 3: 19)? Will our deepest secret feelings of envy, jealousy and hatred suddenly burst forth and will we be like wild animals, raging and uncontrollable? Are we really incomplete creatures, an evolutionary experiment not yet finished, just one stage in a much more ambitious programme of God's planned experimentation; or are we an experiment somehow gone wrong, soon to be replaced? Or, are we really God's special creation put on Earth to redeem it but needing to improve our efforts before it is too late? With the arrival of Enlightenment science – with its worship of rationality – we have progressively come to feel that we are indeed the rational gods of creation and that beasts are here to serve us and that freaks are here to be cured by our science. But even with enlightenment optimism and science, the personal question remains – even if most humans are god-like, what am I? Have I made it? Will I make it?

The existence of freaks thus reminds us of the precariousness of our own sense of self-worth; and so we try to avoid/hide them; and we try to subdue our guilt by insisting that they be cured. And to avoid their fate we ask that if we do our very best to be – or at least appear to be – normal, and to follow traditions and customs, and to dress properly, and to follow purity standards, and eat with our mouths closed, refrain from uncontrolled babbling, obey modesty taboos, keep our sexual desires in check and stick with relatively standard expressions of them (basically to keep sex private and not to brag about it), and if we do not openly indulge in animal-like outbursts of anger and violent destruction, we will not be treated with contempt, horror, disgust or pity – the plight of the freak; we hope that our (personally perceived) abnormalities, our (suspicions of) physical inadequacies, and our less than heroic persona will be ignored. We will, we hope, be considered fully human and be included in society.

Alas, physical abnormalities and related behaviours are not the only human attributes that can lead to exclusion from the category of being fully or properly human. Even 'normal' humans in a physical sense, ones with no visible unfinished aspects or animal-like characteristics, can be excluded on the basis of *behaviours* that are said to be non-human in origin and extremely dangerous in effects. The behaviours which have been attributed to witches, for example, are by most definitions behaviours which are said to be the result of individuals being under the control of external *powers*, which drive witches to *invert* just about every tenet of acceptable personal, social and moral behaviour imaginable.

Witchcraft

When humans conceptualize the universe in terms of abstract powers – such as the power of nature, spirits, ancestors, ritual, prayer or gods – these powers are not always good, safe, protecting powers but sometimes evil, dangerous ones. Nowhere in the human archive has there been a conceptualization of a power which causes more anti-social, dangerous and fundamentally evil behaviour than that of the power behind witchcraft as developed in parts of Europe between about 1450 and 1650. This picture of witches and witchcraft powers and practices was largely derived from accusations and trials during this period, but at the same time the content of these was theorized into wildly imaginative demonological treatises to 'guide' inquisitors, magistrates and judges in doing their work of rooting out witches. This conceptualization was then used by anthropologists when studying primitive societies (Cf., Malleus, 1486; Evans-Pritchard, 1937; Douglas, 1970a; Midelfort, 1972; Cohn, 1975; Kieckhefer, 1976; Anglo, 1977; Vannelli, 1977; Russell, 1988).

There is absolutely no reliable historical evidence that individuals practised witchcraft as an art or as a systematic religion during these years of persecutions in Europe (or in primitive societies as seen in the later anthropological literature). The activities of modern white witches – a contradiction of the term as understood in the above contexts – bears very little, if any, relationship to the historical and anthropological phenomenon of witchcraft beliefs and persecutions. In both primitive societies and historical Europe, witches were 'discovered' through accusations, arrests and trials, not by individuals having broken specific laws, committed heresy, having certain immediately identifiable physical characteristics which would clearly set them apart from other humans or of having gone around trying to recruit followers. There were a few confessions, but these were derived from feelings of complete social rejection – and probably behavioural depression – and from torture, virtually never as a result of individuals proudly proclaiming their powers of witchcraft or actively propagating a new religion.

From a standpoint of historical scholarship it is possible to identify two traditions of witchcraft beliefs – a 'common tradition' and a 'learned tradition' (Midelfort, 1972; Cohn, 1975; Kieckhefer, 1976). Witchcraft in the 'common tradition', both in Europe and in primitive societies, was most often generated from the bottom up and was part of a process in which common people projected their fears onto certain of their neighbours, kin and even supposed friends by claiming (through gossip and accusations) that the accused had become possessed by an evil and awesome force/power, and that this power was doing enormous harm (*maleficia*), often through the behaviour of these individuals, to other individuals and to a community. The nature of the force was rather ambiguous; it was felt to be enticed by, or almost generated and nurtured by, both the 'bad inner biology' and 'bad psychology' of the accused witch, but at the same time having vague external sources. The harm it was said to cause included storms, diseases, infections, impotence, premature births, premature deaths, crop failures and tragic accidents.

This power could intensify, become an end unto itself, making infected people do all sorts of very *anti-human* things; witches were generally said to be individuals who succumbed to the temptation to use any means: incest, adultery, lying, cheating, bestiality, infanticide, deceit, sexual attractiveness, 'abnormal' sexual techniques, cannibalism, or whatever else might come to hand, for the fulfilment of their own selfish, greedy, envious, jealous and 'perverted' desires. In a social sense, a suspected witch's badness usually implied some inability to play fair or to control self in seeking pleasure, material comforts and glory, or to restrain envy, jealousy and vindictiveness when dealing with others. They were said to meet at night in semi-conspiratorial groups in order to exchange ideas about how to succeed in their evil endeavours and to gossip about what fun they were having.

In Europe a 'learned tradition' was generated by propagandist demonologists – most usually during trials – and grafted on to the 'common tradition' (Cohn, 1975; Kieckhefer, 1976); however, the learned tradition only caught on in a minority of specific locations (Midelfort, 1972). In the learned tradition the Devil, as the anti-God/anti-authority agent, was identified as the source of the evil power, and it could be awesome; his witches were not just a number of individuals getting together at midnight for some perverted sex and gleeful discussions about how to cause trouble; in this tradition, witches did not practice their craft to simply seek pleasure and to gain material goods and to get even with a neighbour; they did it systematically with intent and forethought with the specific purpose of destroying human civilization, or at least God's control over human affairs. Recruits were not just highly untrustworthy individuals who would cut corners to get what they wanted; they were part of a real *conspiracy* to destroy god-loving individuals, the church, communities, crops and social order. The Devil (it was believed) at best fooled gullible people into becoming witches but, at worst, found only too many enthusiastic recruits. And the Devil gave them incredible powers of deception so that they were not easy to spot or to get to confess; one never knew who exactly was a witch.

In the common tradition the majority of witches were thought to be women, a belief greatly exaggerated in the European learned tradition. During this period of European history, as we have seen, female wills had come to be considered as extremely strong and their sexual appetites virtually uncontrollable. In the European learned tradition many females were said to be easy prey for the Devil, who came to them as a handsome young man, and who offered them fulfilling sex, money and other goods in exchange for their services as his agents in the grand conspiracy to destroy human decency and morality. They, in turn, were expected to use their sexual attractiveness and appetites to lure males into the conspiracy, and to use the powers given to them by the Devil to purposefully destroy crops, bring storms, break up families, steal land and castrate and otherwise destroy the sexual capabilities of males.

As is so often the case in the human story, theory and practice were not always in harmony. When we look at those accused in primitive societies, males were almost as common as females; even in Europe the percentage of men among the accused and brought to trial was greater than beliefs or gossip would have predicted. In both primitive societies and Europe those accused – males or females – were usually troublesome, somewhat antisocial individuals, but they could also be individuals caught up in personal and community quarrels; individuals seeming to succeed 'too easily' (such as beer sellers giving short measures or suspected chicken thieves); sometimes individuals seen as being too arrogant and disrespectful of authorities such as elders, shamans, magistrates or clergy; or they might be seen as unappreciative subordinates; or be midwives ('abortionists') and medicine women (poisoners); occasionally they were people with some local political authority; very rarely were they the poorest in a community.

In Europe, accused females generally tended to be old, on their own, widowed, wrinkled, ugly and believed to be quarrelsome, spiteful, envious, jealous or greedy; they were often defiant and arrogant; in certain locations they were beggars demanding alms or 'wisewomen' or (as noted) suspected poisoners and abortionists. This was the bulk of accusations despite the learned tradition's theoretical emphasis on the dangers of female sexuality and the use of seduction in the Devil's grand plot. Men accused had many of the same characteristics as the women; they were old, loners, cranky, quarrelsome, wrinkled and generally friendless. Well, so much for the witch as a sexually irresistible *femme fatale*, as a female who sold out God and men and families and civilization and was going to use sex to destroy them all for selfish, immediate gratification. And so much for the power of the Devil to recruit an elite force to take over the world and instigate evil and debauchery as a way of life.

In both traditions in Europe, and in primitive societies (common tradition), individuals were most often accused by their own kin, 'friends', neighbours, local enemies and competitors, not by outside institutions or authorities. One man in England, for example, compared the Devil of his vision with 'old Rice Williams of Newport' (Thomas, 1971, p. 475). The social contexts of accusations invariably included communities in which conflict, competition, envy and jealousy were rife. In primitive societies confessions were often a means of re-integrating an accused individual back into a community; this was less common in Europe where a confession did not eliminate the possibility of the death sentence. Although individuals sometimes confessed to being witches in Europe, it was almost always after isolation and torture.

Beliefs about witchcraft and accusations against alleged witches were never sufficient causes for trials to take place. The frequency and conduct of trials depended upon the attitudes and circumstances of political authorities and judges, secular and ecclesiastical; if they were keen, trials took place; if they were not, they did not; and often in a locality various authorities were not altogether in agreement; certainly authorities were by no means automatically in favour of witch hunts; and they did not go around making specific accusations themselves. In fact, the vast majority of people believed by common people to be witches, and even those accused of being witches, were never brought to trial; those accused who were simply old, wrinkled, grouchy and/or argumentative were usually not summoned, or sent home early if they were summoned. Punishments handed out by European courts ranged from public penance to death by burning, sometimes hanging. Although results varied greatly from location to location, and from circumstance to circumstance, the majority of those tried for witchcraft in Europe did not receive the death sentence. In primitive societies the death sentence was very rare indeed.

The common tradition, it can be argued, represents basic human fears and was found in both Europe and primitive societies; the elaborations of the learned tradition represented specific interests under extreme threat and existed only in some parts of Europe, albeit the two traditions often became fused and confused in various European locations. We learn from our consideration of witchcraft *beliefs* as seen in both traditions that humans fear an unknown, invisible invasion, a penetration of our bodies and minds, of our souls, by dangerous powers capable of doing great harm. The nature of witchcraft power was never clearly spelled out but an overriding characteristic of it was that it was almost impossible to detect – it certainly was not visible – and that the harm it could do was unpredictable and its extent not fully known. It was believed that, while some witches might not have actively sought this power, it had infected them nevertheless; the power might or might not do harm to the infected individual themselves (the witch), but it certainly would greatly harm other individuals and the community in general, often activated by the infected witch's spiteful anti-social behaviour, but also through any spells and incantations that the witch might perform out of sheer vindictiveness or for the joy of it.

The nature of this power, it can be argued, is very similar to modern fears of such things as nuclear energy, drugs, junk food, toxic waste, genetically modified food ('Frankenstein food'), food additives, mental illness, injections, heresies, advertising, television, social media and foreign ideologies. All of these have in common that they, like witchcraft power, suggest to some that an uninvited, deceitful, often invisible, but extremely powerful, force can enter our bodies and minds, and commandeer our physical and mental states, and consequently our behaviour, with possible harmful effects for ourselves and for other individuals, and even a community. These negative effects, it is feared, might not show up for several years but lie dormant, waiting for just the right moment or event to unleash them.

Generally, it is very clear in witchcraft beliefs that many aspects of sex and sexuality can be considered potentially dangerous. It is possible to suggest that fears of a contamination from a penetrating danger as seen in witchcraft power may have a degree of similarity with human fears of sexual penetration. A female may not only consider it sinful but also fear that it will hurt. And what if the baby from a sinful (regretted) or forced sexual penetration contains elements of a bad, vicious, untrustworthy, devious male? Films, such as *Rosemary's Baby*, in which a female is impregnated by the Devil and produces an evil child, have played upon this kind of fear. And males could fear that they will catch a devastating disease from penetration and could end up being forced to marry someone they do not like and provide for a child they do not want.

A fear of an uncontrollable female sexual urge and capacity, as represented in the common beliefs, was greatly amplified in the demonologies of the learned tradition.

There it was argued that an uncontrolled sexuality was a danger to children (paedophilia and infanticide were considered common witch practices), and men were warned against being lured by sexual promises which would greatly endanger them. At the same time a fear of sexual competition on the part of chaste women and wives was well represented in such beliefs because witches were said to use such unfair methods as wild, readily available sex and orgies, and seemingly asked little in return from those they seduced (after all they simply discarded any babies that resulted). And given witches' purported general vindictiveness, and their arrogance based on their newly gained witch-power, this scenario seemed by many very likely.

Beliefs further suggest a fear that some female witches would use their sexuality not only for their own selfish pleasures but would also seek ways and means of using their power to humiliate and take vengeance against all those who had crossed or otherwise irritated them, such as competitors in romantic and sexual pursuits, men who spurned them, men whom they had grown tired of, preachers who had lectured them and magistrates who had threatened them. In the learned tradition, religious leaders and magistrates argued that females were especially prone to be taken in by the Devil's seductive approaches when he was recruiting anti-civilization followers; he, it was argued, cynically played upon female romantic desires and sexual urges by appearing as handsome, dashing, charming and kind, with no indication whatsoever of wanting anything other than to fulfil their romantic wishes. But this did not absolve the potential witch; she had a choice; she could have, should have, repressed her sexual desire so that she was not vulnerable to this deviousness. But once under the spell of, or infected with, the power, wild sex and doing evil was inevitable and often outside her control. Additionally, it was feared, once she was infected, these activities may well start to give her great pleasure and she would begin to seek them out.

And so magistrates and religious leaders could fear a breakdown of families and child-care and social order in a free-for-all of sexual indulgence, paedophilia and abandoned children brought on by wild female hedonism and male susceptibility to it. There is also, however, a strong lesson here for females: do not trust the handsome and charming man (the Devil, the Warlock). He will only impregnate you with the seeds of evil, with a contamination which may well destroy your body and will take you over and make you do harm to all those who have loved you; then he will leave you to your fate, to your disgrace and condemnation by the community. In later years the Vampire myth (which has been very common in a variety of forms in Western civilization) conveyed similar fears concerning charming, handsome men, although here the penetration was by teeth and blood, not seminal fluid (Twitchell, 1987).

In fact, both males and females often fear the charismatic powers of certain individuals. People who lead other people into taking drugs, into prostitution, who make criminals of gullible admirers, who brainwash followers, who encourage others to violence and vandalism but themselves stand apart; individuals who it seems are never in trouble themselves – always two or three steps removed – are, it appears, true pupils of the Devil. And more often than not (especially when the suspected evil-doers are relatively highly placed in society) we fear that secret societies and conspiracies may be involved in aiding and abetting these anti-social efforts. We clearly see in witchcraft beliefs that humans fear individuals who demonstrate a desire to take short-cuts to pleasure, wealth, status and power. We fear individuals who have a tendency to lie, cheat, steal and sponge, who do not respect the feelings, property or vulnerabilities of others; we fear being cheated – in life as well as in personal dealings. Hard, honest effort with a definite deferring of gratification can seem to be for nothing where there are those who seek unfair short-cuts through the use of non-acceptable powers and methods.

Actual accusations, although not as dramatic as beliefs (especially beliefs as represented in the learned tradition), reinforce this last element of the witchcraft phenomenon. They clearly show that humans do not like individuals who are thought to be quarrelsome, greedy, spiteful, envious, jealous, arrogant, too successful, too knowledgeable (relative to their peers) and who are willing to violate taboos for their own 'selfish' ends. Thus, accusations turn intense fury towards those who seem not to want to control their passions, towards those who in their 'me-first' approach to life make demands on others and a community, and seemingly offer little in return, or who sometimes defy authority and blame others for their own misfortunes, or who let their arrogance and quarrelsomeness make social life difficult for others. Witchcraft beliefs and accusations represent an epic struggle between human passions and attempts to control them, and demonstrate a human tendency to powerfully condemn those who do not control them, especially those who do not even seem to be trying to control them; and they represent a fear that our own self-control will be cheated by the deviousness of others.

From a consideration of witchcraft *trials* and *punishments* we learn much more about social and political conflicts in specific groups, kinship networks, localities, regions and emerging kingdoms than we do about basic human fears. Nevertheless, such a study suggests social and political conditions in which the basic fears as represented by common witchcraft beliefs are likely to be intensified and extensively manifest. The social context of the witchcraft phenomenon in Europe, for example, was characterized by vast economic and demographic transformations. It was a period in European history when vast numbers of people moved from rural areas to cities and the cash nexus and rent were replacing feudal obligations regarding access to land and to a living; at the same time centralized judicial and political authorities had not yet fully evolved to regulate or compensate for the harsh results that these changes could generate. The context included considerable hardship, but at the same time new lucrative, commercial opportunities. Very significantly, the period also included both a dramatic rise in the age of marriage and a substantial increase in the number of single individuals in the population (Hajnal, 1965; Hill, 2001). This greatly disrupted sexual, family, economic and social relations in many localities (Cf., Vannelli, 1977).

What seems to have been the case is that deferred gratification based on the expectation of eventually achieving marriage, sex, family and children, and some land to live by or a position in service, facilitated and protected by kinship, neighbourhood and community networks (the Church/manors), no longer guaranteed reproductive, personal or social success. It was an age of increasing individualism and personal independence, including for females, but the Reformation, commercialization, industrialization and state legislators had not yet generated, clarified or been able to enforce new means of rewarding deferring gratification, or of providing safety or 'success' within a secure social/political framework. Individuals struggled with their unfulfilled desires, especially with sexual, romantic, status and economic desires. Unmarried females competed with men for labouring work and for running small enterprises (beer brewing, sewing and baking, for example), leaving men struggling to find land, a job or a profession in order to be able to claim a wife; females with their capacity to offer sex seemed to have a special advantage when it came to securing positions in service. And, of course, females could become prostitutes whereas the demand for males in this activity was almost non-existent.

The prostitutes that the vast majority of men could afford would possibly have been older women who were past their best for finding husbands (possibly widows), perhaps prone to stealing, lying and drunkenness, and to passing on diseases, fuelling the idea that not only was female sexuality dangerous but that those most prone to use it were somewhat polluted, highly untrustworthy, cunning women. Quite apart from the practical difficulties faced by females, and by the degrading activities they were often forced into, reports circulated about single women roaming the countryside at night causing mayhem and terror (Hill, 2001); widows and beggars were cut free from families, manors and alms; they wandered scavenging and made out as best they could. But the existence of these could be guilt-inducing because there were still vestiges of the idea that the giving of alms, plus care and respect for the old, were moral obligations; so it is possible that many elderly individuals were accused of being 'witches' so they could be refused (and ignored) – indeed, widows, lone individuals and beggars, for example, were especially prone to being accused of being witches in several locations (Macfarlane, 1970; Midelfort, 1972; Monter, 1972, 1976; Kieckhefer, 1976). From accusations and trials we clearly see that old age and old people can easily be feared, especially if they are cantankerous and demonstrate independence and arrogance, and make demands on us.

To conclude this section, it is to be noted that during most of human history witchcraft beliefs and fears have been either absent or only vaguely formulated; in most primitive societies people did not believe in, or fear, witchcraft; nor did they throughout the greatest part of the whole of European history; and in the relatively short time that beliefs, accusations and trials took place (approximately between 1450 and 1650) most locations in Europe were not affected, and no area continuously over that period. It seems that it was only for limited periods in particular circumstances (where desires and their control were well out of balance, and where there seemed little chance of fulfilling fundamental desires, let alone any further expectations, and political authorities were weak or non-existent), that human fear potentials led humans to turn on each other, especially on easy scapegoats, for explanations, blame and punishment, encouraged by witch finders and/or magistrates (Vannelli, 1977). This may well be because there were no obvious explanations or identifiable real enemy to blame for the vast social changes and community conflicts that existed in witchcraft locations. And, of course, once set in motion, fears, gossip, accusations, trials and fantastical beliefs fed on each other.

In most times and places in human history, however, more predictable species-typical 'threats' (such as those far removed from self on a social hierarchy, neighbouring societies, foreign ideas or 'world economic' conditions) have been available to blame for our anxieties. And, at the same time, some balance between desires and their control has been achieved through the identification of a number of 'evils', and the generation of measures ('thou-shalt-nots') as guidance to the avoidance of behaviours which activate such 'evils'. More firmly, these have been established as – more or less accepted and enforceable – taboos, rules and boundaries.

Struggling against evil

Human notions of evil range from 'strong' – for example, witchcraft – to 'weak' – for example, gluttony (papers in Parkin, 1985). In the strong sense, evil is fundamental; it is the essence of an object, social other, social 'system' or cosmic force; it is dangerous, terrifying, detestable, and must be eliminated – we have seen this, for example, in the case of witchcraft as described in the European learned tradition. In a weak sense evil is little more than 'bad' – drinking too much alcohol, for example; weak evil is usually redeemable: confession, ritual purification, atonement and/or abstinence can make amends for it; the death penalty is usually deemed appropriate for individuals engaged in strong evil.

Evil as perceived of in various social contexts will usually fall somewhere on a continuum between these poles rather than there being only two very distinct kinds, strong and weak; however, as was the case with witchcraft, certain 'evil finders' will try to push most seemingly weak evils towards the strong end. Objects, events, behaviours or people associated with evil have tended to be characterized as having at least one of four fundamental attributes. These are: (1) they are unfinished; (2) they are polluted – impure; (3) they cause great, inexplicable harm; and (4) there is an inversion of what is considered 'normal' in social living – sex with children, incest, cannibalism, infanticide, murder of old age pensioners, child murder, the wilful violating of trust and the breaking of oaths, for example (Cf., papers in Parkin, 1985; Russell, 1988; Baumeister 1997a; Hankiss, 2001; Cohen, 2004; Baron-Cohen, 2011).

In the case of the first two attributes, evil has roots in the human fear of freaks, freakish behaviour, strangers and a fear of one's body being infected, penetrated and/or taken over. Human languages still equate 'bad/evil' with filth, dirt, excreta, rubbish, rottenness, ugliness, bad smells, twistedness and brokenness, for example. The third attribute of evil has roots in the human fear of unknown forces of destruction, which cause, for example, a sudden devastating storm, a flood, a major accident, an incurable disease, a birth defect, mental illness or nuclear radiation. The forth attribute of evil – the causing of gratuitous harm, an inversion of normality, the violation of an oath – came into its own with the evolution of a human ability for a recognition of the dangers inherent in human social interactions.

The causes of evil often seem mysterious and largely unknown, but we like to look for them, gossip about them and accuse others of causing them. In this process we have tended to locate the sources (causes) of evil in awesome nature, specific individual characteristics/acts or in abstract forces. As time has passed, science has explained some of the unpredictable elements of nature (such as the causes of storms, various diseases or deformity), and at the same time the power of supernatural causes of evil (such as evil spirits or the Devil) have taken a back seat to more secular/scientific approaches; and so the notion of evil has tended to focus more on 'man's inhumanity to man', such as in trying to understand extreme exploitation or genocide. When evil has been attributed to abstract forces, such as to anti-gods and evil spirits, there often has been a dualism between the powers of good and evil – God opposing the Devil, or angels in deadly combat with demons, or workers struggling with capitalists, for example.

The particular nature of evil and its causes as conceptualized in various societies has much to do with the history and the political and social relations that have evolved in those societies. However, when we take a cross-cultural look at evil the thing which strikes one is the *universality* of the existence of a notion of evil and of the characteristics and behaviours which are considered to be evil. Besides the cases of physical deformity, physical disintegration/rotting (Davidson, 1991), strange diseases, destructive forces of nature, pollution, mental illness and the direct breaking of trust, as suggested above, evil seems to be universally attributed to murder, adultery, incest, jealousy, greed, coveting the property, status and privileges of others, lying, cheating, failing to live up to obligations, selfishness, arrogance, lust, cruelty, malice, envy and vindictiveness. Numerous cultures add such things as usury, disobedience to authority, parents and to God, dereliction of duty generally and unfairness. In all cultures there is a strong emphasis on the importance of controlling passions and of avoiding excess and striving for the middle, or moderation, in order to avoid evil.

This last gives us a major key to understanding the role of human desires and fears in human civilizations. In Chapters 3, 4 and 5 we could ascertain that humans desire (among other things) trust, love, respect, loyalty and honour. Human concepts of evil represent a fear that we might not receive them because we know that self-centred human passions can be so strong that they sometimes override these virtues. Our consideration of witchcraft drew our attention to fears that certain others might not always restrain themselves; that they might even secretly conspire to fulfil their wishes even if it means great harm to others. There was a message that deferred gratification by self can be undermined by the free or devious expression of passion in others; that the unrestrained or secret desires of others can do us harm; that any trust, love, respect, truth or loyalty others might show us could be false and often at our expense.

Besides a consideration of the emotional responses derived from processes of sexual selection during human evolution, a glance at psychoanalysis, mythology, various human stories and the musings of numerous philosophers, there is an additional implication that we may fear our *own* potentially uncontrolled desires as much as those of others. We have the capacity to imagine our uncontrolled desires being fulfilled; we get excited by this but the excitement also generates mental tension/anxiety when we imagine the violating of trust and the lying, cheating and deceiving we might have to do to fulfil these desires; we also fear that we will be found out, punished and humiliated; but we have these fears along with a lingering anxiety that others may cheat us, deceive us, make a fool of us if we totally restrain ourselves. The emotional-cognitive response is often considerable anxiety and confusion, and feelings of guilt, but also a conceptualization of potential evil forces, working away both inside ourselves and hovering around us in the zeitgeist (greedy, pushy individuals, witchcraft, conspiracies, for example), that are the *cause* of this painful state of ambiguity and ambivalence.

From this atmosphere we easily come to fear that our own desires might get out of hand and get us in trouble; more cognitively/analytically we may realize that to have sex with our friend's wife, or to physically hurt our enemies, or to abandon our children for lust, for example, will cause more trouble and all-round agony than the pleasure we could derive from doing these things; that unrestrained lust and anger will get us rejected and ignored, that violations of trust will get us outlawed, that achieving success/power at all costs will alienate, and hurt, many relatives, friends and fellow humans, so we consider desires that might cause these behaviours as being *evil* and set out to avoid such evil in our lives. The social/political response is, as noted, the erection of taboos, laws and boundaries designed to regulate desires, and avoid evil, and thus to generate what we consider to be civilization.

The mental mechanisms and motivators for this process have resulted from sexual and social selection in which repressors – fear images – evolved to override, or at least mediate, evolving desires during the evolution of a highly self-aware, social species. With developing cognitive abilities we have turned these into moralistically prescribed injunctions and codes, with a corollary of a species-typical sense of evils to be avoided. While, with the socio/political evolution of civilizations, various cultures have emphasized some aspects of evil at the expense of others, and have projected fears in slightly different ways, all have devised methods to teach, preach, persuade, bully and shame individuals into the 'truth' that moderation is the best policy. The penalty for uncontrolled desires/excess is the loss of love, the loss of respect, even from mothers and fathers. Mythology, theology, philosophy, ideology and science have all played their part in conceptualizing a physical and social universe in which desires are kept in check and evil thus held at bay.

A common theme in creation myths, for example, is that at the end of mythic time old gods and heroes (uncontrolled passion, conflict, violence) were replaced by a new supernatural order in which love, altruism and duty reign (Overing, 1985). In the *Code of Hammurabi* (about 1728–1686 BC) severe penalties are set out for making false accusations, stealing from family and friends, female adultery and entering another's house uninvited (Johns, 1904). In the Egyptian *Book of the Dead*, the dedication to Osiris highlights the dedicator's sense of social responsibility and, among other things, notes that he/she has not committed murder, sodomy or adultery, and has not oppressed members of his family or his servants and has not lied, cheated or stolen (Budge, 1934; Redfield, 1994). The early Hebrews were concerned with bad, impure spirits and with 'twisted' animals and people (Taylor, 1985). These could be dealt with through sacrifice, rituals of purification and, occasionally, confessions. When the Hebrews became the elect of God, however, laws were more clearly set down. Moses received a contract and it was not to be broken. So long as the Hebrews kept their trust with God, and with each other, they would be protected. If not, however:

> The Lord shall smite thee with a consumption and with a fever, and with an inflammation, and with an extreme burning, and with the sword, and with blasting, and with mildew; and they shall pursue thee until thou perish ... And thy carcass shall be meat unto all fowls of the air, and unto the beasts of the earth ... The Lord shall smite thee with ... the scab ... itch...madness ... blindness ... Thou shalt betroth a wife, and another man shall lie with her (*Deuteronomy* 28: 15–68).

This move in the direction of the 'strong' concept of evil was also a move from unfulfilment and pollution to the violation of trust as the major characterization of evil. 'Thou shalt love God above all things' is a statement that there is a higher order of obligation than to self and immediate others. 'Thou shalt not commit murder' meant, as it does everywhere else, that one should not willingly cause the death of one of *'us'*; causing the death of an outsider, an unfulfilled, unfinished, basically non-human 'monster', is fine – there are often prizes for doing so. 'Thou shalt not commit adultery' does not mean that men shall not have more than one wife but that alliance partners will respect the sexual property of their mates. 'Thou shalt not covet thy neighbour's ox, house, goods, wife,' and so on, means the same with regard to status, wealth and power. It means that one must keep one's envy and jealousy under control.

One could argue, however, that a crucial commandment is: 'thou shalt not bear false witness'. At some stage humans were able – to some degree – to imagine others thinking as self might be thinking. As a result, self's endeavours to deceive were perceived *by self* to be vulnerable to discovery, and the dangers of deception were brought directly and vividly into human consciousness. This, of course, does not mean that humans suddenly stopped bearing false witness or deceiving; indeed, too much reproductive success is gained by a certain amount of deception and in fact humans became quite good at it. As deception became more and more sophisticated, however, it was deemed necessary to keep it under some kind of observation and control; the result was to conceptualize certain deceptive behaviours into an abstract array of evils, and from there to oppose these with an abstract code of morality. At this level of awareness 'do unto others as you would have them do unto you' comes to have real meaning, and by accepting the commandment to not bear false witness one accepts – agrees – the 'need' to live with others in some sort of arena of trust. As an ancient Anglo-Saxon chronicle strongly advised: 'Man must... keep everything in balance,/keep faith with men,/and be pure in wisdom. Each of men must be even-handed/with their friends and their foes' ('Seafarer', Anglo-Saxons.net, 2013) and Dante (*Inferno*) saved his worst part of hell for those who were betrayers and traitors concerning God, friends, fellow citizens and comrades.

The establishment of commandments against bearing false witness and for being even handed, even with foes, represents a sort of human-nature-inspired social contract. It is a contract only partially lived up to, however, and so vigilance against its violation has become a major human concern. As the Christian Church evolved, for example, arenas of trust as established among individuals and within communities gradually became translated into 'thou shalt love the Church and obey its ordinances' because the Church and its priests represent God's moral guardians on earth and as such are curators of trust within a parish of people. Additionally, the Church also provided administrative and political processes which went some way towards the provision of guarantees of trust, and thus a degree of social order, within a community of worshippers (parish). According to St Thomas Aquinas, one of the Church's leading thinkers regarding administrative and political issues (among other things), individual slips against order were only venial sins and could be redeemed, but sins against the principles of order (or against nature as Aquinas conceived of it) were the much more serious mortal sins from which there was no salvation (Aquinas, cited by Davis, 1935 in Taylor, 1985, p. 48; Davidson, 1991).

The Church set up elaborate mechanisms for confessions and ritual purifications to deal with slips in individuals' struggles to control the evils within themselves that they faced in everyday life. The Church came to the view that humans were born with original sin (an innate tendency to excessive desires for sex, knowledge, god-like status and the like, inherited from Adam and Eve), but original sin could be temporarily mitigated with baptism and could be more or less contained with pure living. An individual who slipped a bit but had not sinned too badly could be continually re-integrated into the order of the universe and into their own personal community through confessions, communion, penance and prayer. Protestants were stricter; sin was deeply embedded and was an ever present danger; a ritualistic baptism or a bit of confessing was not going to alter that; the only way to keep original sin under wraps and to avoid evil generally was to seek to be as pure as the Christ (Taylor, 1985). This required a lot of individual praying and Bible-reading, and at times very austere living; theoretically it brought the struggle back to the level of self, reducing the role of The Church.

The other world religions, Hinduism, Buddhism and Islam, preach a similar story concerning the controlling of passions and the nature of sin – although there is less dualism between good and evil, and generally a weaker concept of evil than is found in Christianity or in a number of primitive societies (Lowie, 1924; Smith, 1958; O'Flaherty, 1976; papers in Parkin, 1985). Nevertheless, the injunction to control desires in order to live the good life is the basis of the concepts of evil as found in these religions. The *Ramayana* and the *Mahabharata*, Hindu epics which teach people how to live, emphasize truthfulness, loyalty, duty, affection, honesty and devotion to God. Vindictiveness, jealousy, envy, power-seeking with malicious intent and cheating are constantly cited as evil. The *Bhagavad Gita* (Trans. Zaehner, 1966), a long Hindu epic poem, strives towards a higher spirituality for giving purpose to life while clearly recognizing the dangers of uncontrolled passions.

Let a man think of things of sense,- /Attachment to them is born.
From attachment springs desire, /From desire anger is born.
From anger comes bewilderment, From that the wandering of the mind,/From this the destruction of the Soul (intellect):/With the soul destroyed the man is lost.
But he who moves among the things of sense, /His senses subdued to self,
From hate and passion free, /And is self-possessed,/Is not far off from calm serenity. (chap. 2, verses 62–64)

Islamic law especially condemns adultery, stealing, rejection of kin and intentional violations of the law. Buddhism makes a case that in order to reach Nirvana it is necessary to follow the eight-fold path of right: (1) view, (2) intention, (3) speech, (4) action, (5) livelihood, (6) effort, (7) mindfulness and (8) concentration. To do this one must purge self of excessive desire and instead practice restraint; knowledge is seen as a major means of achieving this. In all the above religions, leaders and teachers residing in temples and mosques play a similar role to that of priests in Christian churches as described above. And, as with the history of the Church in Christianity, the power and authority of these have waxed and waned throughout history.

The teaching and theorizing roles of religious leaders brings us to philosophers, many of whom played a similar preaching role in combating evil, but here from a secular standpoint. From Confucius and Socrates through to Marx, Durkheim and John Rawls, for example, the good life is the life of moderation and human reason is the means of controlling passions to this end. 'For it is in the contempt and disdain of pleasure that virtue is most conspicuous' (Cicero, *Laws*, in Knoles and Snyder, 1954). Earlier, and perhaps less concerned with completely avoiding pleasure, the Greek historian Herodotus had argued for the virtues of moderation, humility, modesty and common sense in daily life (Herodotus, *The Histories*, approx. 450–425 BC [Dewald and Waterfield, 1998]). Aristotle was especially concerned that humans use their intellect (reason) to control passion through discovering and living by 'the mean'.

Virtue then is concerned with feelings or passions and with outward acts, in which excess is wrong and deficiency is also blamed, but the mean is praised and is right; and being praised and being successful are both characteristics of virtue (*Nicomachean Ethics*, in Knoles and Snyder, 1954; see also, Aristotle, 2002).

This is a remedy for dealing with weak evil, but it pointed the way for Aristotle to a more absolute (strong) category of evil. Uncontrolled passion and acts that the above formula could not easily be applied to could be considered evil pure and simple: he cites shamelessness, envy, adultery, theft and murder as instances; a little bit of adultery or thieving is not better than no adultery or thieving, and it is hard to conceive of a desirable mean halfway between murder and not murder, for example. For Aristotle, a major means of keeping people on the straight and narrow was the use of 'righteous indignation' – re-discovered by sociobiologists as 'moral aggression' (Trivers, 1985) and by some games theorists as 'grudger' type behaviour; here Aristotle clearly recognizes the role of reason and community pressures in keeping evil at bay.

The philosophers' worship of reason was accompanied by their desire to locate the good life in social acts and social relationships. Confucius identified five fundamental relationships in which the control of desires and the importance of total trust were especially important: (1) ruler and minister, (2) father and son, (3) husband and wife, (4) elder brother and younger brother and (5) friend and friend. Like Confucius, other philosophers, such as Plato, Aristotle, Cicero, John Locke, Hegel, Rousseau and Karl Marx, have looked to their own particular version of 'rationality' to act as the basis for creating laws, rules, moral imperatives and – often relatively specific – legal and political systems. More generally, however, for many philosophers the value of reason itself was more important than any specific 'rational system'; as soon as humans got rid of superstition, greed, vanity and emotionality, and turned to reason alone, it would soon enough be obvious to them what was good and what was evil in human existence.

The nature of the fears that can be ascertained from our consideration of the human struggle against evil is relatively basic. Many are fears we have encountered elsewhere. Again, for example, we learn that humans fear uncontrolled and unregulated passions, 'excess' in behavioural terms often being taken as evidence of such a lack of control. In general, we can see that many, if not most, humans fear biology and emotionality; these are feared as being the enemies of reason and 'rationality'; their elimination, it is assumed, would lead to the 'good life'. Nevertheless, at the same time we continue to feel that both good and evil emanate from the biology or souls of particular individuals. The idea that people of a different race, group, religion or world view than self are biologically inferior, physically polluted or naturally devious, indeed evil, for instance, remains strong in many cultures. The idea that we are extremely vulnerable to developing 'false consciousness', or to advertising or 'false religions', suggests that as humans we are mentally weak of will and thus subject to evil, and to being turned into perpetuators of evil.

From a consideration of evil, we have additional evidence that humans fear those who lie, cheat, steal, covet the possessions and partners of others, violate trust, are greedy, envious, jealous, selfish, vindictive, arrogant and show malice. We learn from a human struggle against evil that we fear those who do not live up to contracts and those who do not follow rules and do not respect generally agreed upon boundaries. But we do much more than dislike such individuals; we express powerful *moral indignation* and frequent aggression towards those who openly practice – or are exposed as practicing – these behaviours. At the same time, however, we are often hypocritical in that we frequently violate rules, and when we violate some rule or code we often express extreme moral indignation at *other* individuals for violating some other codes. We mentally get away with this by justifying our own violations in counter-moral terms: for example, a lower paying tax-payer cheating on his taxes might find this 'morally' justifiable through his indignation at the use of tax loop-holes by the wealthy – he is standing up for poor people by standing up to a 'bad government' that has not imposed stricter rules against the rich or even properly enforced some of the rules that exist; often such rationalizing individuals become especially vigorous, or at least loud, in condemning rule breakers generally.

However, when we violate a code our rationalizing does not always protect us. We may have apprehensions and anxieties that we will be found out, or upon reflection might be a little ashamed of the hypocrisy we have engaged in (or about hypocritical acts we may be thinking about engaging in). Feelings of anxiety and shame often turn into feelings of guilt, generating a degree of self-control, but projection mechanisms redirect some of these feelings outward, against violators – or potential violators – of rules, regulations and boundaries generally. Self, thus, cleanses itself of at least some guilt through vigilant support for the vast majority of existing rules, regulations and boundaries and in strong support for the *principle* of establishing and obeying additional ones.

In this chapter I have attempted to locate human fears through a cross-cultural consideration and through identifying behaviours which generate repulsion, terror, weeping, wailing, anger and despair. Given the fundamental nature of fears as part of human nature, and the long theological and philosophical obsession with the emotions/ passions that underpin them for a self-aware, social species, it is surprising that the social sciences have been so little interested in systematically investigating fears or evil. Instead they have argued for their repression in favour of 'rationality' and assumed that once rationality takes a firm hold in human societies human fears will disappear; or, in a related fashion, have argued in favour of a social constructionist perspective, which assumes that when we get our society right fears will simply not exist.

And one might think it strange that humans seem to have made extremely little progress in overcoming what appear to be species-typical fears despite certain elements of folk wisdom, theology, philosophy, politics and social science downplaying their significance and having suggested moral, common sense, philosophical, ideological and rational means to avoid and/or overcome them. Famously, it has been said that: 'the only thing we have to fear is fear itself – nameless, unreasoning, unjustified terror which paralyzes needed efforts to convert retreat into advance' (FDR, first inaugural address, 1933). Yet it is doubtful that his message reduced human fears generated by the great American depression of the 1930s, let alone in other situations in which the same sentiment has been invoked.

Perhaps it is not so remarkable that fears persist because what we have learned so far about human nature suggests that 'evil' (albeit not necessarily in its strong sense) will remain an integral part of human life because passions (drives/desires/fears) are not easy to repress or control. In fact many are not necessarily considered undesirable and are actively sought – such as love, sexual pleasure, excitement or joy. But nevertheless, even these 'positive' drives and desires can cause fears. Our own feelings of love can be undermined by fears of loss and abandonment; pleasure in our own achievements can be tainted by fears of other peoples' envy and a constant fear of future failure. And positive emotions in others, even our friends, can be a threat to self: for example, a sense that a friend's new found joy will result in an abandoning of self, or that the attractiveness and vivacity on the part of another (even a friend) will lead to the stealing of a partner from self, or that excitement and joy in another may be as a result of great achievements which will greatly outshine self. In the words of the Mijikenda of East Africa: 'being a nuisance is not a necessary human trait; being evil is' (Parkin, 1985b).

Whether 'evil' is or is not an inherent part of the human condition, the games of civilization continually move on, driven by an interaction of powerful desires and fears – ultimately unfulfilable desires and unbanishable fears. It is time now to draw the work together in the form of a list of human desires and fears, in terms of how these may or may not work together to make up what we think of as human nature motivators of human behaviour, and in terms of what the next steps might be in developing a science of human behaviour and a workable approach to justice.

7 A human science, justice and politics

In a search for species-typical desires and fears the hope has been that knowledge of these will give us a better understanding of the motivations underpinning human social behaviour, including its evolution. The rationalistic approach, which seeks to argue that humans evolved a brain and mind designed to solve extremely difficult environmental problems, and in the process became a 'unique' rational animal with destiny in its own hands, was challenged. Rather, the emphasis has been on processes of sexual selection as they might have resulted in an emotional-cognitive human nature which is compatible with the universally observed patterns of human sexuality, mating, friendship and political behaviour we can currently observe cross-culturally and historically.

The argument has been that this emotional-cognitive human nature can usefully be described as species-typical emotions, desires and fears that evolved from a number of primate drives and instincts, which (along with the repression of some of our primate heritage, and with the subsequent evolution of species-typical patterns of cognition – see Vannelli, 2001) motivated relative *reproductive success* in the conditions which likely existed at the time of human differentiation, and for some time afterward. These motivators, the argument went on, have continued to operate well beyond original sexual and mating behaviour, resulting in, for example, current friendship, group, alliance, status and political behaviour. Because human fears are often more hidden (very often acting unconsciously) additional material – primarily from psychology, anthropology, history and a number of human stories – was added to identify some of the conditions that led to involuntary flight, repulsion, disgust, weeping, terror, aggression, despair and clinical depression as indicators of species-typical human fears. In this analysis, conditions in which loss, or a fear of potential loss, loomed large, as did cross-cultural definitions of 'evil' and struggles against evil.

The further argument has been that during human evolution many reproductive-oriented desires and fears have remained unconscious, and many have been sublimated into motivations for such things as playing, developing strong friendships, pair-bonding, love, romance, alliance forming, vengeance seeking, group conflict, competition, deal-making, codes of honour, status allocating and politics; that is, into human species-typical socio-political behaviour.

Human species-typical desires and fears

Suggested species-typical emotions, desires and fears as derived from the above analysis are presented in tabular form below in terms of the following useful analytical complexes: (1) the lust complex; (2) the evaluative complex; (3) the dependency complex; (4) the attention complex and (5) the power/control complex.

Table 7.1 Human species-typical desires and fears

Desires	Fears
LUST COMPLEX – OFFSPRING	
mother's attention, cuddling, eye-contact, mother's breast	physical separation
viewing and playing with infantile physical characteristics and behaviours (other children)	physical isolation loss of attention
LUST COMPLEX – MALES	
viewing and playing with semi-infantile physical characteristics and behaviours; soft skin, smooth hair, and especially the childlike, symmetrical, average-human faces of females	old age, deformity, ugliness being sexually ignored; sexual rejection; being considered sexually unattractive
viewing and playing with female secondary sexual characteristics; pubic hair; noticeable breasts and hips, round buttocks	loss of virility, loss of masculinity, having a small penis, castration; being sexually inadequate
to be attractive to females; to flirt with females; to show off to females; to be thought to have male characteristics of manliness/heroism	female promiscuity, outward display of female sexuality
to masturbate, sexual climaxes, a large penis	wife stealers/being cuckolded; unknowingly raising another man's child
to copulate with females (dedicated homosexuals excepted, who prefer males)	loss of a mother and/or mother substitute
to be reduced, and reduce a partner, to a childlike state of play	appearing effeminate; being henpecked 'abnormal' sexual practices (including bestiality, incest, paedophilia and/or homosexuality)
to return to the mother's breast; to be mothered	
to have a monopoly of a female's sexual expression	
LUST COMPLEX – FEMALES	
to be physically/sexually admired by males	uncontrolled male sexuality; male sexual desires; rape
to attract sexually attractive males (face, slender buttocks, athletic build), successful, powerful, or high status males	male aggression, male philandering; males giving the 'baby's' food away to a seductive female
to attract kind/caring males, especially romantic males	male control of female sexuality
to coyly present self sexually, to be modest; to flirt	own sexuality/uncontrollable sexual desire; immodesty
to transform the sexual drives of 'worthy' males into caring, protecting and providing males	
to have a monopoly of a male's sexual, caring and economic output	husband stealers; loss of a partner
to be played with like a child by a sexual hero	loss of beauty/femininity/youth infertility, the loss of an offspring
to have and care for offspring (and often a male partner)	'abnormal' sexual practices (including bestiality, incest, paedophilia and/or homosexuality)

Table 7.1 (cont.)

Desires	Fears
EVALUATIVE COMPLEX (OFFSPRING, MALES AND FEMALES)	
to assess levels of: trustworthiness, loyalty, reliability and helpfulness among friends, allies and sexual partners (and generalized others with whom a contract is formed)	strangeness, strangers – cripples, 'freaks', polluted (dirty, slovenly) individuals
to classify and stereotype humans in terms of degrees of physical and social likeness to self	that friends and acquaintances may be untrustworthy, lying, greedy, envious, spiteful, arrogant, shameless or vindictive
to attribute status demarcations to specific individuals and groupings of individuals	of choosing a 'wrong' sexual partner; of a loss of romance
to feel that self is positively evaluated by others	individuals too rich or too poor compared to self; 'overly' ambitious individuals
to judge (label) the world in terms of safe and dangerous – good and evil	sexual show-offs
to believe that humans are uniquely rational creatures and are above biology	being considered an unworthy self of showing envy; being embarrassed, shamed, humiliated
to believe that humans have a spiritual nature and that humans can be perfected	'evil' (unpredictable and unpreventable disasters/uncontrolled passions in others)
	that humans are really animals and that animal dangerousness can erupt at any time
DEPENDENCY COMPLEX (OFFSPRING, MALES AND FEMALES)	
to be with other humans in the form of parents, friends, lovers, allies and partners	isolation, rejection, separation, being shunned and/or humiliated by friends or allies
to be part of a small network of 5–12 people	
to have lifelong parents – or substitutes in the form of husbands and wives	loss of care givers, love objects, friends, allies and patrons, or sexual partners
to have safe nests	unattached individuals: hermits, loners, strangers
to submit to allies, friends, patrons	
to worship and submit to heroes and heroines and to submit to the virtue of honour	being let down by a friend or ally, violators of contracts
to adhere to group norms (males perhaps more than females)	cowards, dishonourable individuals
to formulate contracts	extremely self-interested individuals, non-conformists
to submit to a higher order and purpose in the universe	unreliable allies
	cynics
ATTENTION COMPLEX (OFFSPRING, MALES AND FEMALES)	
to be sexually attractive as in the lust complex above	to be sexually unattractive as in the lust complex above
to be noticed (flirting, joking, singing, dancing, acting, gossiping, storytelling)	loss of attention (from parents, friends, sexual partners, allies and patrons)
to compete with nature and other humans and to win competitions (sexual, hunting, war, sports, business)	losing (in sexual, battle, sports, or business competitions)
to display self through decorating self and home and accumulating wealth and status objects	being considered a 'show-off', a poser, tasteless or a fool
to ritually sacrifice (hold feasts, give to charity, exchange gifts, buy drinks)	to be considered selfish, 'mean', 'tight'
to be admired, to be respected; to be given prestige/a recognized status position	to be considered a 'suck up'; a 'sell-out'
to achieve acknowledged 'success'; to be given honours	to be considered a failure, to lose face

Table 7.1 (cont.)

Desires	Fears
POWER/CONTROL COMPLEX (OFFSPRING)	
to have attention and care from parents	loss of parental attention
to be able to move freely	physical restraint; being made to wait, forced to take turns
immediate gratification	that immediate desires will not be fulfilled
to have what siblings have; to have one's own things	to have things taken away; siblings seemingly having everything their way at the expense of self; to have to share with domineering siblings/friends
to feel safe among playmates	
to win in games	
to have one's own space; to know where the boundaries of that space are	engulfment (especially by parents)
to be in control of presentation of self (hair, clothing and body decoration style)	being bullied
	to lose in games
	restrictions on showing off, going out and presentation of self; being embarrassed by parents
POWER/CONTROL COMPLEX (MALES AND FEMALES)	
to have speed, strength, physical freedom	physical disability, old age
to control nature, gather food, build shelters, mould environments	youth in others, especially when old
	natural disasters, invasions, threatening groups
privacy – to have one's own space; to know where the boundaries of that space are	invasion/lack of privacy; negative gossip
to feel in control of self and self's circumstances	loss of self-control, uncontrolled desires in self and others; loss of self-esteem, status and control of circumstances; humiliation
to be able to defer gratification for 'higher ends'	
to have control over offspring; to escape overpowering parents	rebellious, uncontrollable offspring; engulfment (especially by parents)
to feel secure with sexual partners, friends, allies and colleagues	violators of codes, rules and taboos – especially those who do not defer gratification and who violate contracts
to establish rules, taboos, boundaries to protect self and self's friends and allies, and to have them obeyed and enforced	
to protect self and self's friends and allies, and to have them obeyed and enforced	losing in competitions, vengeance feuds or wars; humiliation
to take vengeance (often with violence) for being humiliated and/or harmed, and against violators of rules, taboos and boundaries	that our superiors will treat us badly; that they will show us no kindness or respect
to mitigate vengeance seeking into a 'higher calling' for justice as fairness	penetration and invasion of the body/soul by unknown, invisible forces/evil powers fanatics
to win in competitions (in battle, sports, status and/or sexual competitions)	that the universe, including human social life, has neither purpose nor direction, and that individuals have no control over its unfolding
to gain strength and wisdom (from prayer, sorcery, meditation, study, self-help, diet, dedication, sport, devotion)	
to feel in harmony with the powers and purposes of the universe; to believe that universal order and purpose can be understood and controlled	being bored, feeling purposeless; depression

It seems clear from the above juxtaposition of human desires and fears that they can be related. For most desires it remains possible to find words or concepts which represent fears that are, more or less, their 'opposites'. These are not linguistic opposites but rather evolved species-typical opposites which repress aspects of our primate heritage and at the same time motivate species-typical human behaviours. These emotionally charged opposites (desires and fears) generally direct us to certain parenting, sexual, friendship, alliance, patronage and general political behaviours and repel us from others. This does not mean, however, that human desires and fears work in perfect symmetry. There are, in fact, a number of inbuilt conflicts inherent in our desires and fears, conflicts originally derived from the sexual and child-caring competitions, contradictions and ambiguities which existed early in evolution and can be seen to continue to exist; indeed, many have even become magnified through cultural elaboration and spiritualization. For example, in sanctifying concepts such as 'love', 'romance', 'family' and 'duty' as being above animalness, a conflict between powerful sexual urges and honour through restraint can be generated; indeed, a conflict between sexual desires and 'morality' is not uncommon in human societies.

Human infants evolved an extensive capacity for developing strong emotional dependencies on parents which come into conflict with their own later reproductive interests when breaking away from parents becomes advantageous. Infant dependency evolved to include very strong anxieties/fears of being isolated and/or of being abandoned alongside desires for parental attention, love and support. But, even within the first year and increasing up to and through adolescence, the desire for increasing personal power ('I do it myself'), plus the anxieties and fears which motivate an avoidance of engulfment, make their appearance – infant anger when restricted through to adolescent *embarrassment* and *disgust* when being kissed and ordered about by one's mother in public, for example. These conflicting feelings unconsciously remain for life and affect friendships, marriage, alliances and group loyalty.

Males can be sexually excited by 'hour-glass' figured, youthful, facially child-like, smilingly enticing, provocative and energetic females, but simultaneously fear female sexuality – *femmes fatales* are especially threatening; 'mother-types' are a better bet; males thus desire, at the same time, both mothers and whores. Males also desire to appear strong, independent, 'grown up', not to be turned into 'children' by either whores or mothers – so they admire heroes, presented as the epitome of independence, despite glorifying 'mothering' on the part of mothers, sisters and wives. Males desire the company of male companions, and are attracted to masculine appearances and presentations of self in friends and alliance mates; but at the same time fear even their best friends and alliance mates as possible sexual competitors and cuckolders; so in stories a male bond of honour is presented as one of the highest bonds humans can achieve, a bond that epitomizes trust and loyalty; but in these stories the possibility of violation of trust, dishonour and adultery, often hover in the background.

In fact, the danger of sexuality to males is such that in stories of the honourable hero, such as we see in stories of knights and cowboys, the heroes are very often almost asexual to emphasize the honour their bond represents; detectives are most often loners to suggest that honour and seeking justice can only be achieved in the *absence* of most social entanglements, including those of sexual/love relationships or family. When any of these heroes stray it is very often a *femme fatale* who is responsible. However, these desires and fears conflict with male desires for females for sex, companionship and mothering, and for families.

Most females desire kind, loving, protecting, home-oriented males, but are put off by over-dependent adult male 'children', and so, at the same time, they both desire and fear powerful, vigorous, aggressive, independent males. Females want pre-masturbatory sons, protecting fathers/brothers and wild studs, all at the same time; their challenge is to turn the last into the first two by virtue of their own child-like attractiveness, inner 'virtues' and kind actions. There have been times in history when females have gained a degree of positive recognition and status from being very good at mothering (having numerous offspring and keeping them alive) and at nest building (being extremely creative in being able to use materials from the environment to make a safe place to rear offspring), and at being a wife – having attracted a nice looking man who is dedicated to her and her offspring. But this is not usually where a sense of achievement and status comes from in many human societies – it largely being considered what females should do as a minimum, and that 'any woman' can do it. Kudos generally comes from noticeable success at gathering, hunting, alliance behaviour, warfare, farming, careers and politics. But if a female wants to engage in these activities – and many desire to do so, at least in some of them – such activities (and desires) come into conflict with wishing to attract males, have children and nest build. Such activities can also generate a fear of losing femininity.

As a result of some of the conflicts between male and female sexual desires and fears, and of the internal conflicts inherent in both male and female desires and fears, adult humans have evolved motivations (often unconscious) to repress, hide and, above all, control and manipulate sexual drives, desires and fears. The widespread existence of sexual shame, embarrassment, joking, teasing and stigmatizing, and of stories in which tragedy is the result of uncontrolled sexuality, testify to this. As a result, control and repression of sexuality have come to be strongly reinforced by religious and moral teachings, and by laws. But this can cause confusions, ambiguities and feelings of guilt. Every culture develops a notion of 'normal sexuality'. This is culturally accepted sexuality which is controlled, directed and oriented to pair bonding for the purpose of having children within the confines of family life; it is to be indulged in secretly, quietly, without bragging. In most cases self seeks to achieve 'normal sexuality', and so apparent 'abnormalities' come to be avoided, if not considered abhorrent, disgusting, repulsive.

To be normal we do not go around naked in public, we do not masturbate in public, we do not have sex in public, nor at work, nor with children, not forcibly and many find the idea of oral and anal sex disgusting. Yet we can be titillated by images if not fantasies of all these things; so we masturbate in secret but have fantasies that we are having sex in a public park, in the back of a cinema, with someone watching, in a threesome; we sneak a look at pornography, and we read novels full of sex and watch films that depict explicit sex scenes, and we wonder what it would be like to have sex with our next door neighbour, with a football player, with our teacher (Cf., Friday, 1980, 1976; Scott, 1994; Leitenberg and Henning, 1995). But we feel a degree of shame and guilt for these feelings, and we wonder why they sometimes spontaneously jump into our heads; we cannot seem to stop them; sometimes we have dreams about the above; we keep these dreams to ourselves.

So we know/feel that normal sexuality is precarious, but that makes us even more vigilant to try to make sure we end up 'normal'. In these efforts we condemn masturbation, blatant sex, adultery, pornography, prostitution, paedophilia, swinging clubs and homosexuality (for example). These seem to be sex made too easy, sex without responsibility, sex without love, sex to no good end, sex leading to desecration and exposure as being weak, dirty, predatory. But we like sex, and we may even secretly indulge in some of these, and we fantasize about a number of them, joke about them, snigger at them, or viciously attack practitioners of them. We aggressively project our anxieties and fears about possibly not being normal onto the apparently abnormal in order to try to convince ourselves and others that we are normal – and possibly even to alleviate any guilt we might feel for having indulged in one or two of them, and/or had fantasies about some of them. As a result of all this mental processing we come to actually fear a number of them.

Just as males are confused as to which are best, mothers or whores, and females are looking for both protectors and independent studs, both genders can thus be confused as to which persona to show to the world. Desires and fears at work here – as seen in storytelling and psychology – include feelings such as desiring love fulfilment and submitting to its pleasures but also fearing committing self too strongly to specific others and/or appearing 'needy'. Male sexual behaviour, as a consequence, tends to manifest from a range of desire and fear motivators; a male, for example, may be a very loyal husband; or he may be a good-enough husband but not resist extra-marital sexual opportunities, and perhaps he may even seek them; or his strongest desire may be to be completely free of all family commitments, and so he has one-night sexual encounters with as many willing partners as he can find; or he partakes of sex with prostitutes; or he has only male friends and he may or may not have sex with these males; or he might dream of becoming a private detective, strong and independent, not really needing sex or females. It seems that evolution has not edited male sexuality with one specific focus but rather has generated a species-typical range of possibilities. But whatever the behaviour of a particular male, a happy outcome is not guaranteed. He may start out a happy, loyal family man but get bored with home life, and wish he had made other 'choices' earlier on, for example; or a promiscuous jet-setter may come to regret not having a stable family.

A female's securing of resources and help in caring for offspring has been sublimated into an enduring capacity and desire to provide 'mothering-ness' for children, courting males, husbands, pupils, patients and humankind. And humans have 'played and civilized' with this enough to make devoted caring an almost spiritual undertaking. But evolution has also provided females, particularly during adolescence, with desires to look glamorous and sexy, and to have fantasies of being taken away and kept by a powerful, intriguing 'stranger', one who is a real hero. She may not even mind becoming the second or third or fourth wife of a rich and powerful patriarch for the standing, power and other benefits this may provide. However, most will seek a male who is successful, witty and kind, and will treat her as an equal 'adventurer' in making family life a success; or, she may want to go it on her own and compete with males at their own game; to achieve in her own right; to have a sparkling career and earn money as a farmer, beer brewer, lawyer or politician; or she may not be bothered with males or offspring and instead devote herself to a spiritual quest, perhaps having sex with other women.

But female sexual/reproductive behaviour – at whatever point on the range – does not guarantee tranquillity and peace of mind. Even if she desires to find a man 'of her own', and does not mind marrying a 'normal' guy, she may be unhappy about the home and the amount of goods and social standing provided, given the sex and mothering she is contributing. And she may later become bored by the child-husband she has tamed and still harbour fantasies of meeting a romantic stranger, or of following a career; she may even go looking for this man in secret, or go back to college, while nevertheless keeping her husband bringing in the bacon and sharing in child-care activities.

Whatever the case with specific individuals, however, most parents (especially females) continue to have powerful desires to have and to nurture and protect offspring. But at the same time, there is often a fear that offspring responsibilities will greatly restrict freedom of choice and opportunities in a number of potentially exciting and enjoyable areas (including sexual expression) and that offspring may limit existing desires for material and status accumulation. These particular potential contradictions are now largely dealt with through modern methods of birth control and abortion rather than infanticide as in previous times (and still in some parts of the world), but the underlying contradictory motivations are similar; indeed, it is at first sight almost a paradox of selection theory that, with prosperity, humans have fewer rather than more offspring; however, it is not a paradox but rather a clear example of early motivators for status achievement and display, as mate-attracting mechanisms, remaining stronger than motivations for promiscuous reproductive output, even when this might be possible. As a result, we get such 'civilized' things as two kids per family and materialism and 'keeping up with the Joneses'.

We can also see that instead of strict repression and direct sublimation we often hide our feelings, divert them and make light of them. We try to disguise anger, envy and fits of jealousy; but this is in conflict with the fact that we do get angry, jealous and envious of others. We tease each other, and play, flirt, lie and joke as we excitedly talk about everything but sex and power when sex and control are what we are most interested in. We smile and small-talk, and even laugh at jokes, among our friends, colleagues and when interacting across delicate status and power positions (for example, boss and worker, teacher and pupil) or when entertaining members of the opposite sex, regardless of how we really feel about each other, and despite what our real intentions might be. The potential for mistakes in all this, the often exposed hypocrisy, the silliness and the tension springing from our capacity for anxiety born of empathetic worry that everyone can see through us, is seen over and over again in comedy, drama and romantic stories.

Early competition and conflict for mates and among siblings and between generations have left us with a heritage of competitive motivators – and often motivations to fight. But we have also evolved desires and fears to want friends, lovers and good colleagues; and we want to be liked. So we are motivated to try to hide, and even repress, our competitive impulses. We play and watch competitive sports, and many love them and get extremely excited by them, but then we say sports are recreation, 'not really important'. We compete for promotion at work but we call it dedication and conscientiousness; we compete with our neighbours and friends by expensively decorating our houses and calling it homemaking, and by landscaping our gardens and calling it a love of nature. Religious and political cults compete viciously for followers and call it 'enlightening (or saving) the unenlightened'; we are thrilled by heroic deeds of warriors in mythology, art, music, stories and films, and we decorate contemporary war heroes with medals, but we often reject war as being a dreadful, inhuman activity to be eliminated; in some societies we compete in politics almost as if it was a war between two opposing sides of righteousness unable to compromise, and we call it democracy based on 'rational' discussion.

And along with this – *despite this* – *in spite of this* – *because of this* – we claim a 'higher purpose' for humans, a non-competitive, cooperative purpose; it is a seeking for an 'above-biological' dimension – a rationality, a spirituality, which clearly separates us from animals. Nowhere is the denial of animalness more evident than in the arts, religions, philosophy and science. The arts are portrayed as lifting us above our dog-eat-dog animal nature, above our normal human nature, into states of heightened awareness; religion is seen as a linking-up with a higher spiritual order and that this implies *purpose* and *meaning* in the universe, including in our existence. Philosophy is presented as a reaching out to find a universal harmonious 'Truth', and science is thought of as the pinnacle of our use of reason applied through observations and experimentation in order to fully understand the world so that we can use this knowledge to make life 'better' and ourselves more 'perfect'.

Yet, we pay less attention to the 'high arts' than to soap operas; when probed, our meaning and purpose in life tend to be our own desires and avoidance of our specific fears; we ignore philosophy completely; being told that: 'now that you have been set free from sin and have become slaves of God, the fruit you get leads to sanctification and its end, eternal life' (*Romans* 6: 22) has less effect than drink, lying, sex and deviousness on our everyday behaviour. For example, in post-medieval Europe (when according to mythology everyone was a religious fanatic), many people treated the church and its prayers and rituals as magic for achieving immediate gratification, a cure or an end to their troubles, and they did not discriminate among prophecies, oracles, sorcery, portents, the mass, sacraments, prayer, elves, fairies, angels, saints, relics or witchcraft, as to where the best results might be obtained; certainly, 'Luck' competed with God for omnipotence (Thomas, 1971; see also, Green, 1971; O'Grady, 2012). Bishop Jewel's survey of the English clergy in the sixteenth century led him to conclude that it is not so much that 'they know not what the Scriptures are; they know not that there *are* any Scriptures' (cited in Thomas, 1971, p. 164). Science is feared as much as it is held to be a pinnacle of human rationality; what we want from it is a cure for our immediate, personal problems not the saving of humankind.

These observations are not meant as a criticism of superstitions or the supposed irrationality of 'some people' but rather it is a statement about some of the powerful ingredients of human nature, ingredients present in everyone, ingredients which, however, are in conflict with a spiritual (religious or secular) view of humans as the 'above animal' creature – which, of course, is also a major aspect of human nature. But, conflicting or not, the two are not as separate as it might seem when it comes to basic human biology. The arts are biology in action; we *'feel'* the arts stir our biology, especially our adrenaline and emotions; the arts extend our emotional imaginations, and give us a buzz like a drug might. Dancing and music are elaborated and disciplined stamping, marching, hooting and drumming; a great deal of visual art is depictions of war, glory, agony, heroes and naked women; religion is based on such things as daydreams, dreams, visions, fantasies, prayers and teleology, generating a primordial emotional *feeling* that there are above-human forces/dimensions, if not a higher order, in the universe, and that we need not fear, not even death.

These feelings in turn are derived from something like our feelings of guilt for disrespecting parents, for lying to parents and friends, letting friends down, having lustful and selfish desires, and from powerful fears of being rejected, leading to the idea that struggles for self-control, sacrifice and cooperation with the brotherhood of man are virtues greater than self's selfish thoughts. Mostly philosophy derives from the fact that sometimes people like to talk a lot (and listen) about teleologies, prejudices, stereotypes and rationalizations that seem to solve complex problems; and science, in essence, is a result of an innate curiosity and a desire to play with things and see how they work.

And despite the claim that we are rational beings in control of our destinies, certain fears persist; fears of abandonment, for example, are as powerful in modern societies as they have ever been in the human lineage, because with full self-awareness and powerful memories, fears of *anticipated* abandonment (and of loss and failure) join our mental mix, while at the same time fears of engulfment and a curtailment of freedom are increasingly visualized and theorized. Even in modern times, times of 'rationality', high arts and science, we cannot escape certain fundamental dilemmas – especially dilemmas concerning the relationship of self to others. Our desires and fears motivate us to want to feel that we are in charge of our circumstances, and to try to present ourselves as being 'in control'; but we also want to be considered loyal friends, faithful mates, reliable team members (followers). We desire to ingratiate self with specific others but seek to avoid the humiliation of many types of subservience.

This can be argued to be the basic human dilemma, attempting to balance control and dependency in regard to our relationship with others. We often worship and submit to heroes but avoid doing the same with our friends and enemies; we seek status but do not brag too much and we symbolically humble ourselves; we desire power but say we only wish to 'care' and serve. We give omnipotent power to gods while demanding individual responsibility for personal actions (including for ourselves). We advocate strong principles but try to understand and forgive others who transgress; we admire 'sticking to our beliefs', but encourage others (and ourselves) to 'learn from experience'. So, we want to control, or at least feel safe in, our own circumstances, but we cannot escape a nagging awareness of the potential dangers posed by unpredictable self and/or by other humans.

Running to stay ahead of the dangers we often generate *expectations* (a desire and fear-influenced mix of cognitively orientated 'objectives' in a particular cultural context – Vannelli, 2001). Fulfilling expectations can require controlling, manipulating and/or influencing, but also befriending, submitting to, integrating with and worshipping, others. It often comes to include a sense of a *right* to the fulfilment of our expectations as a reward for our efforts and devotions (and emotional energy invested), and consequently feelings of outrage when they seem obstructed. In the process, natural cognitive tendencies to constantly evaluate, classify, stereotype and teleologically explain our social environments (in the context of species-typical desires and fears) are at work. As a result, the grading of activities, tasks and humans, in terms of how dangerous, helpful or neutral to self (including to self-image and self-worth) they are perceived to be, is commonly undertaken. When socially shared, such hierarchical evaluations help an individual decide what and whom to approach or avoid and/or to what it might be possible to aspire. Grading humans in this way also gives them social worth.

If there was a *master behaviour(s)* in human evolution arguably it was the process of dependency and the related human nature processes of emotional-cognitive evaluating, classifying, stereotyping and deceiving – in other words, politics. This includes the development of desires to present self in a favourable light, and in fears that warn us of the potential dangers involved in encounters with others. These desires and fears, in turn, motivate us to develop social hierarchies and to construct protective social rules, taboos and boundaries, from the level of self to the level of the state.

Social valuations set up patterns of social relationships for us, and rules, taboos and boundaries provide a degree of predictability and protection. However, it is very difficult for humans to completely agree on these things, or for human life to remain static so that existing valuations and relationships remain relevant. In general, it is possible to argue that humans feel safest – most tranquil/relaxed – when their *expectations* are seemingly being met and potential dangers to them appear well under control, and are very upset when expectations seem blocked, and when dangers to them appear to be lurking. In many circumstances, however, humans can adjust expectations, defer gratification, fantasize about the future in order to accept the present; personal compromises are made.

All this may more or less work, but seldom satisfies everyone. Are equal opportunities sufficient or must equal outcomes be assured in economic and status competitions? Can a principle of 'from each according to their ability and to each according to their need' be made to work? Who is to decide on individuals' abilities and needs? And anyway these are relative concepts and subject to constant change – abilities increase with age and experience, then begin to decline; perceived 'needs' tend to escalate as they are satisfied; any agreement/compromise in human interactions requires *trust* to have any chance of permanency; but trust too can be a fickle element of human life. Fears of loss and rejection are ubiquitous, envy and jealousy common, arrogance difficult to conceal, cheating not unknown. So, individuals argue over these things, and sometimes, in particular circumstances, get violent and kill and torture each other over them.

This brings us to the importance of politics – as continuous politicking. We politic to present self in the best of all possible lights, to attempt to appear attractive to others, to gain respect and to make social environments safe. Our politics include seeking environments in which we feel our expectations have every chance of fulfilment, and, come what may, we will be *protected* against external dangers, disasters and our neighbours.

Through politics we construct 'protective' social rules, taboos and boundaries, (many of which regularize patterns of *expectations*, or status claims, reinforcing social hierarchies). But, again, human nature makes it inevitable that particular arrangements will not last forever, and in many cases not for very long; human civilization is a dynamic, ever-volatile process of constant politics. A number of primate drives, instincts and emotions – having been sublimated and spiritualized in the evolution of human nature – are the driving forces behind such politics.

So, if human personal and social behaviour is largely politics, driven by emotions, desires and fears, and a great deal of this operates unconsciously, what can be the role of reason, science and the quest for the perfection of humans? To start with, we must accept the importance of emotions, desires and fears and not treat them as enemies of reason or of human happiness, or as not part of our '*real*' nature. While environmental challenges during human evolution will have selected for some of the analytical skills (reason) as presented in the traditional scenarios of human evolution, as noted, the changes in mating patterns analyzed in this work did happen, and can be argued to underpin much of the human nature basic to an understanding of modern social/political life.

Humans did come to be able to survive in extremely harsh environments, and to apply clever technology to the task. As a result a number of analytical skills evolved, along with a desire to apply such skills. But these skills, and the desires that supported them, did not replace the sexual, social and political motivators highlighted in this work; selection would have held them in check; the same would have been the case for the evolution of over-extensive reflection and philosophy. This is because the emotional and political motivators discussed here have provided an amazing amount of reproductive success. As a species we have filled just about every environmental niche on the planet and have grown in population size beyond anything imaginable even for herd mammals; and it has been possible for humans to live and let live in population concentrations of individuals, families, groups and communities in densities well beyond anything any other primate species could manage.

To be sure, we have progressively gained knowledge about how many things in nature work – including our own minds and bodies – and we have invented more and more complex tools, shelters and material objects. Some of these have been very useful in survival and reproductive pursuits, but largely they have been toys with which to indulge our desires and/or to feel we are avoiding dangers through their employment. We have never been able to feel content with simply eating sufficient food, with living in shelters that simply keep us warm, dry and comfortable, or with making love instead of war; we have wanted toys, monuments, religions, status and glory. As a result we have developed a strong belief in the idea of progress to be achieved through philosophical and scientific understandings.

Yet, none of our knowledge, religion, philosophy or science has made it possible to predict the exact direction in which our toys will take us, or their specific consequences – and this is certainly not because no one has ever tried; it is just that those who have tried have very often been so extremely mundane and/or general in their predictions that the predictions have been meaningless (and usually wrong in just about all specifics when provided); and if someone does get something spectacularly right we usually only know in retrospect because at the time no one believed them because theirs was one among dozens of competing predictions (almost all seemingly credible at the time). This is because predictions are most often based on extrapolations in terms of idealized, 'rational', teleological models of 'where we *should* go', rather than on an acceptance of the vagaries of human nature with all its emotional, creative and political fickleness.

Living up to 'the human potential' formulae have ranged from advocating selflessly serving others to vigorously competing with others (both to get the 'best' out of us). A robust sex life is sought by many for health and happiness while others see sex as a distraction from philosophical or religious contemplation, if not as an impure activity. For many, getting married and nurturing a family is the most purposeful thing we can do; for others educating ourselves to the highest level possible and dedicating ourselves to a 'worthwhile' career represents the most exalted of human purposes. The integration of all of us into one 'family of humanity' is seen by some to be the highest human purpose, while others worship at the altar of respecting individual difference and local autonomy. Some feel that humans should be given maximum freedom through a minimum of restraints and others that we require very specific guidelines, taboos, rules and laws in order 'to be free'.

There is no evidence from our search for desires and fears that human existence is basically a rational process. Certainly we can reason, and we do it all the time. But it is clear that this powerful human capacity is guided by our emotions, desires, fears and species-typical patterns of cognition. It might be said that we rarely know why we feel or 'know' something, but we try to *reason* an explanation. In this sense emotions and reason work together, often generating a sense of purpose regarding most activities we undertake. Practical reason helps us solve a wide variety of everyday problems, but when it comes to our self-images, happiness, love, war, sex, romance, parenting, sports, status, 'success' and politics, emotions, desires and fears play the major motivational role.

Additionally, we will have to accept that there are no grounds to believe that our civilizing process has uncovered any discernible meaning or purpose for our lives, or the universe, other than the ones generated from our own desires and fears. No sooner has a spiritually oriented individual or group declared that they have discovered 'the human purpose' than rival claims emerge. Sometimes these can be decidedly in opposition to each other; for example, claiming that the human purpose is to love God and do God's bidding even at the expense of individual wellbeing, countered by a claim that there is no god and that the human purpose is to fulfil individual human potential and that a belief in god is a superstition that gets in the way of this.

The same lack of understanding can be said to accompany the concept of the evolution of human language as a 'rationalizing force'. While it can be argued that language has undeniably been extremely important for the evolution of what we think of as human civilizations, and that any type of hedonic-based social organization with a population much larger than a basic group of about five to fifteen individuals would be very difficult without it, the sublimation of sexuality into politics predates language. Nevertheless, language makes it possible to do politics on a scale larger than band or tribal societies. With all its 'creative' possibilities, as commonly observed in political discourse through the ages, language allows for the formation of relatively flexible social arrangements that are able to change as circumstances change. This, however, does not mean that it elucidates rational discourse but rather that it is able to negotiate a degree of practicality, to do deals, to make compromises, to form patron/client relationships and to interact with relative strangers; language makes it possible to conduct tribal battles without the need for violence or combat. As a result we have been able to link bands into federations of sorts, and federations into even larger political units.

We can see that the sublimated reproductive and sexual motivators which generate such things as alliances, male brotherhoods, kinship loyalties, friendship networks, welfare provision, patronage networks and political administrations are rooted in emotional processes and desires and fears much more than in rational or analytical cognitive processes. During hominin evolution these motivators took on a life of their own and the behaviours they generated were indulged in for their own sake given the high degree of reproductive success they provided. For instance, as noted in Chapter 5, 'becoming a dedicated hunter, an eager competitor, a person with friends and support, a good team player, one trusted by allies and powerful patrons, one who appears to love children and who is looking to improve his rank ... has made many males very attractive to females'. In being attracted to these males, female motivators too selected for the evolution of what we think of as civilized behaviour; as did female desire and fear propensities that motivated a female to become good at attracting help in child rearing and in managing campfire gossip and neighbourhood politics, and in controlling husbands and their friends and in making safe nests. Male attractions to these females also, of course, in feedback, selected for what has become our idea of human civilization.

So, does all this mean that we can never scientifically understand humans or human behaviour? Not necessarily. Like all of nature, human biology follows certain laws of physics and chemistry that we can understand, and even make certain predictions from; equally, it should be possible to use the consistency of human nature in the form of 'species-typical desires and fears' to gain a deeper understanding of human behaviour and to formulate a workable approach to justice.

A human nature basis for a social (human) science

In Chapter 1 it was argued that it is possible to have a social science even if we take the full implications of Darwin's epistemological legacy on board, a legacy that implies that there is no grand design or purpose in human existence; and we can do so even if we question the post-enlightenment assumption that humans are supremely rational creatures. If we consider human species-typical desires and fears (derived from evolution) as motivational forces – forces made up of an integration of species-typical drives, instincts, emotions, learning abilities and emotionally charged patterns of cognition – we can analyse human behaviour and history as being far from completely random, even if we have rejected a designer or some natural or human force working for an inevitable outcome.

This approach allows us to analyse human patterns of sexuality, mating practices, offspring care and parent/offspring relationships, settlement patterns, taboo, rule and law creation, exchange arrangements, status hierarchies, social boundary building, politics and social evolution in terms of their degree of *compatibility* with species-typical desires and fears, even though human nature and resultant behaviour may never be completely harmonious. While this means that we can stay in line with the Darwinian elimination of teleology, inevitable progress and supernatural powers in our explanations of human events, is it worth *understanding* and studying a creature that has no hope of achieving moral salvation, rationality, social perfection or progress? Yes. Yes, because most humans, most of the time, *like* being alive and enjoy seeking to fulfil their desires and interacting with other people; this is despite the fact that this can be problematic and difficult, and that great emotional dangers can be encountered in our search for satisfaction. A scientific understanding of human nature might, just might, be able to help us achieve this enjoyment of being alive more often, more regularly.

But for this to be at all possible we have to feel that we are living in relationships and social conditions that are relatively safe; that our desires have a real chance of fulfilment; that we are protected from the dangers highlighted by our fears. For this to happen we need to be able to feel that we can *trust* others to live up to their emotional (and other) obligations; above all, we want to feel that others act *fairly* in their dealings with us. This brings us to a dimension of human nature that will be basic to our science of humans, one not often linked to our biological or psychological 'being', but arguably one that is very firmly rooted in both; that is an apparently innate sense of fairness (justice) regarding human relations. Here I will argue that this sense is closely related to our species-typical desires and fears.

In search of justice

To start, our study of desires and fears suggests that it will always be impossible to achieve a perfect harmony among human desires and fears, even in the mind of one individual, let alone among several individuals. This means that there will always be a potential for disagreements and conflicts in a society. A strong human nature propensity to designate certain events, people or behaviours as being *evil* just because we do not like their behaviour, or have been harmed by them, or are in conflict with them, does not solve the problem – indeed, calling for the eradication of evil usually makes it worse if it leads to such things as torture, persecution and genocide – generally followed by eventual retaliation; in both cases innocent people often suffer more than the perpetrators of some wrong or harm.

While some things seem clearly evil (for example an inexplicable event such as an earthquake that destroys a multitude of innocent people or a sadistic serial killer or a torturing dictator), and others seem clearly good (an apparently miraculous recovery from an illness or a stranger sacrificing themselves for a child), it is impossible for us to draw a line between 'clearly' evil and good in the vast majority of the events, activities, encounters and people that we face each and every day. More importantly, however, designating something evil will not prevent it in the future; for example, considering a destructive hurricane *evil* will not prevent another one from happening; attributing evil to a serial killer or a dictator will not stop them – it may encourage them.

In general, we have learned that it is best to give inexplicable events of nature a *neutral* moral value and let science explain them as best it can; and so we attribute the common cold to bacteria and viruses rather than evil spirits; and in many cases we have realized that it is all too easy for individuals who dislike something or someone to designate them as being evil, or as causing evil, and that too many innocent people get hurt when vengeance is sought against evil deeds and people, and some very unhelpful theories of cause and effect can be generated – for example that a strong desire for sex among certain females makes them accept powers from the Devil to cause 'evil' such as storms, crop failure, female barrenness and male infertility. So, in most cases, political authorities have built dams against floods, financed vaccinations, and have taken over the vengeance business and the arresting and trying of individuals, and for the most part they have had little time for what 'seekers of evil' or local gossip might think appropriate in these matters.

More controversially, history and a study of human nature suggest that it is also not helpful to try to establish higher *moral codes* and force people to live by them because moral codes are relative to cultures, ethnic groups, religious dogmas, special interests, fanatical cultists, dogmatic bigots and visionaries. The last thing most moralists can accept is that human nature is conflict-prone, ambiguous, ambivalent and subject to emotionally driven behaviour and occasional emotional outbursts. Moralists do not like to compromise; they rarely imagine that they might be wrong: 'How can I argue with a man who thinks he will go to heaven for slitting my throat' asks Voltaire (1972). Many witches, sinners, heretics, accused criminals, deviants and political opponents have suffered greatly at the hands of moral outrage.

Unfortunately, a striving for *rationality* has not served much better in achieving justice. A number of rationalistic theories have been advanced to this end, but these approaches have generated extremely abstract formulations which present idealized pictures of humans and human societies rather than 'what humans actually are or do'; they have been abstract theories in which abstract humans find abstract justice in abstract societies (Cf., Plato, 1970 [348 BC], 1974 [432–408 BC]; Rawls, 1971; Kant, 1974 [1797]; Locke, 1980 [1662]). And when attempting to implement rational justice in the form of policies and institutional arrangements one person's rationality is very often another's irrationality. For example, for a free marketeer, socialism is an irrational 'system' that encourages laziness, lethargy, skiving, a destruction of creativity, political tyranny and a complete abdication of individual responsibility; this they oppose to the maximizing of individual reason, creativity, effort and freedom in a context of healthy competition and a free exchange of ideas and materials, as they see free markets doing; for a socialist, a free market approach represents selfishness, cut-throat competition, greed, a complete disregard for the welfare of others and letting the losers fall where they may – poverty and early death if that is their fate; which they oppose to a rational *system* based on *planning* and the satisfaction of *needs* (rather than wants) by dedicated, non-selfish, cooperative workers – which is how they see socialism working once selfishness and greed are no longer encouraged (or have been eliminated).

Both camps claim that they represent a 'modern' approach based on human reason, rationality and science – in contrast to traditional, superstitious, emotional ways of life as presupposed to have existed in pre-industrial/pre-enlightenment societies. And both think their approach is not only the most rational but also the most moral because of the assumptions behind the characteristics listed above; and it is rare that they even start to comprehend the stance of the other – indeed, they often consider the other not only irrational and immoral but also evil. History clearly suggests that, when taken to extremes, neither approach lasts very long because they are abstract conceptualizations based on abstract notions of rationality rather than a rounded consideration of humans and human nature with its inbuilt conflicting emotions and ambivalences.

The most dramatic examples of the dangers of extreme 'rationality' as a road to human advancement and superior social justice were seen in Chapter 1 where supposedly purely rational systems were seriously tried in practice. In these examples, we could see that as many, if not more, humans have been annihilated by other humans attempting to impose 'rational systems' as by those defending 'good' against evil or upholding moral absolutes (as devastating as these can be). And the 'progress' of science has not been able to prevent this; indeed, it has not only been invoked but also employed. The failed Nazi regime used science and technology to effectively kill at least six million Jews and other 'lesser' humans to make room for a scientifically bred and nurtured race to rule the world.

The Soviet Union and the Chinese 'cultural revolution' killed considerably more people to clear the way to create a utopian paradise in which all bourgeois superstitions, prejudices and exploitive tendencies (injustices) would be eliminated through the use of a 'true materialist science' ('scientific socialism'). Mao, as quoted in Chapter 1, declared the 'new' system to be 'the most complete, progressive, revolutionary and rational system in human history' (Mao, 1966, p. 23). In the Soviet Union 'scientific' psychology was dedicated to creating a 'New Soviet Man' to run and appreciate the justice of this 'system' (Bauer, 1952), and to identifying and incarcerating the 'mentally ill' individuals who did not see the rationality, justness and clear logic of the 'new system' – leaving the traditional elements of the secret police to send 'evil' obstructionists to Siberian labour camps.

Considering the damage done by witchcraft persecutions, moral outrage, heresy hunting, religious wars, crusades, suicide bombers and attempts to implement rational utopias, many theorists of justice (from relatively early on and continuing through modern times) have distrusted 'the search for the Good', morality and 'rationality' as a basis for politics and for justice (Cf., Voltaire, 1759, 1972 [1764]; Morgenthau, 1946; Machiavelli, 1950 [1513, 1531]; Hume, 1958 [1739–40], 1963 [1742]; Hobbes, 1968 [1651]; Glover, 1999; Nietzsche, 2010 [1872–88]). One of the oldest suggested alternatives has been that justice can best be served by attempting to continuously apply an innate human sense of *fairness* to human affairs, whatever the socio-political context. This idea goes back to the ancients and remains strong even today, as in the golden rule – 'do unto others as you would have them do unto you' (Cf., Aeschylus, *The Oresteia*; Rawls, 1971, 2001; Nietzsche, 1974 [1882], 1990 [1886]; Plant, 1991; Glover, 1999; Honick and Orians, 2012; Sloane, *et al.,* 2012; Wright, *et al.,* 2012). The idea that we have an innate sense of fairness to rely on in our endeavours has gained some support from modern primate studies in that higher primates seem to have a rudimentary sense of it (De Waal, 1996, 2006; Hauser, 1997, 2007; Brosnan and de Waal, 2003; Khamsi, 2007; Proctor, *et al.*, 2013). For humans the argument goes something like this: through self-awareness we know that other humans are different from self, but a capacity for sympathy and empathy allows us to 'understand' how others might feel in a number of circumstances. So, we understand how various actions, including our own, can affect them, and so we can *empathetically* suffer negative effects with them, if only by imagining how we would feel if it were happening to us.

How, then, does this translate into a concept of justice? If two children of the same age and apparently similar behaviour are about to have a piece of cake, but the one who cuts the cake cuts a much bigger piece for his or herself than for the friend, our automatic response is that this behaviour is unfair to the one getting the smaller piece. We also empathize that this will hurt the feelings of the one receiving the smaller piece. This can be solved by imposing a rule that one child cuts the cake and the other has first choice of pieces. This is only just if both children want (*desire*) as much cake as possible, and if both have a *fear* of being cheated. If one child wanted only a small piece, and the other as large a piece as possible, cutting two equal pieces as a result of this rule satisfies neither, whereas both could have been. So, we ask each what they want; if they both say they want a bigger piece than the other we cut equal pieces, and while neither may be satisfied neither can claim that they have been unfairly treated.

To apply a degree of functional rationality to this problem we might say that the cake should always be cut equally to avoid confusion, to save time and to be able to introduce efficiency into cake cutting. But this assumes that all individuals desire cake to the same degree; that an individual's desire for cake will remain constant throughout their lifetime. A rationalistic/scientific approach might suggest that neither should have any cake because cake is bad for them and that cake-making be prohibited because of its harm to health. But most humans *love* cake; most accept that they will not get as much as they desire on some occasions, but that if they cannot agree on different sizes they will have to do with equal-sized pieces of cake and feel that this is fair. They know that it probably is not as good for them as vegetables but they love it as a treat and, more importantly, they begin to fear a loss of freedom and control over their lives if someone is going to be able to force them to stop eating cake (and force them to eat vegetables) however rational not eating cake might be: what will be next? Smoking, alcohol, sex, religion, travel, debate?

We emphatically know in this example that even a small child has a sense of fairness (Cf., Honick and Orians, 2012; Sloane, *et al.*, 2012). We can also see that a dialogue concerning fair (just) behaviour can be carried on (with an outcome which no fair person could object to) without any recourse whatever to a concept of 'good' and 'evil', morality or a superior form of rationality. To say that the greedy cutter was *evil* would be to say that a child's tendency to self-interest and a sweet-tooth represented something fundamentally wrong with them; to say that the cutter was immoral would be to condemn their love of cake and childish expressions of a natural will to power as being a violation of a higher absolute code that designated them as being in need of punishment and/or redemption. Applying the concept of fairness avoids all this and leaves most people feeling that both children have been treated justly. We can see from this example that the degree of desire for cake and a desire not to be cheated (and a fear of being cheated) are crucial to our understanding of fairness in this situation; so, we can argue that justice as fairness requires an understanding of species-typical desires and fears.

One might feel that dividing cake is not a serious matter; that it is an issue which does not involve the moral depth of the things with which we should be concerned. But the same principle applies in much more serious issues. Take capital punishment for premeditated, vicious murder for example. Some feel that individuals convicted of especially vicious murders are evil pure and simple and should be executed to rid the world of that evil. Some consider the commandment 'thou shalt not kill' to be a moral obligation for which a violator must pay with what they took – a life. Others, however, read this injunction to mean that 'not kill' also means not killing in revenge, and so judicial killing is equally immoral. Others take the view that even the most vicious of individuals who kill someone are under the control of psychological forces barely understood and that such individuals are spiritually and/or scientifically redeemable, and execution is a cruel and overly harsh punishment for an act so little understood, and over which the killer had such little control.

We have powerful human nature desires and fears at work in all this: we have a desire for vengeance against the perpetrator who activates our sense of danger, and great fears of loss if we empathize with the victim and imagine a similar thing happening to one of our loved ones; we fear the killer might kill again, and if their act was especially sadistic, humiliating and degrading of the victim, we fear a sense of helpless vulnerability and shame, almost as if it could have been us, humiliated, hurt and killed. We feel that we are not safe on the streets or even in our homes; our fear turns to powerful resentment and hatred for the killer. We want vengeance. But we do not know if we, ourselves, would be able to personally kill the killer; and we can empathize with a person waiting to be executed the next morning, having a last meal; what must be going through their mind we wonder, and we shudder at the thought of it and we feel frightened; we may even feel a little bit guilty that somehow we cannot make it alright, start over, undo the act of murder.

So, might not forgiveness be a better approach? We have spent a great deal of our lives forgiving our parents, partners, children and even friends for transgressions, so should not the principles of forgiveness be applied here as well; is it not a higher *moral* principle than killing even a killer? Well, not necessarily. We have developed strong love and respect, and even bonds based on fear and guilt, with parents, children and friends, and so we forgive them many transgressions because we emotionally cannot do much else; however, we do not automatically feel that the *concept* of forgiveness has a spiritual power for good in the larger society, one that we should be compelled to follow. And anyway, it is very unlikely that we would forgive even one of our own who did a vicious, sadistic killing, even if they turned against strangers rather than the likelihood that they would turn against us; after all, we remind ourselves, the condemned person has been convicted of cold-blooded murder; they demonstrated no mercy, no forgiveness. And there is the Confucian question: 'if you pay back evil with good how are you to pay back good?' (Kapuscinski, 2008).

I know of no way of generating agreement for formulating policies concerning what should be done about the vicious killing of one human by another. We get little help from metaphysical ideas about good and evil, moral arguments or science; the psychology of reconstructing – redeeming – an individual is in its infancy at best, and we have little more than moralistic or rationalistic dogmas that it works, and a gut feeling that it does not. But let us see where an approach from the point of view of justice in terms of fairness might take us. It seems greatly unfair to a victim of murder that they were murdered and to any others who have been dependent on that individual for material, moral or emotional support, but executing the killer after the fact will not bring back the victim or affect the victim in any way, or restore any of the victim's previous support for dependents. It may satisfy our and the relatives' desire for vengeance, but it may also generate feelings of guilt when we think of the executed person's last hours.

Not executing the killer might be unfair to citizens because the killer might escape and kill again, and/or it will be expensive for tax-payers to keep them in prison; and if there is any deterrent effect of execution it will be lost. However, murderers very rarely escape, and the deterrent effect of execution over life in prison is very far from clear (Sellin, 1959, 1967; Passell and Taylor, 1977; Radelet and Lacock, 2009; Fagan, 2012), and incarceration for life in a secure facility can be cheaper than the cost of prosecution, numerous appeals and eventual execution (Fagan, 2012). But above all there is a major overpowering potential unfairness in the use of executions: the executed person may have been innocent of the murder. Recent uses of DNA testing, along with the uncovering of other 'new evidence' in 'watertight' cases, shows just how possible it is for mistakes to be made, including in a significant number of capital cases (Findley, 2002; Innocence Project, 2012). Even when it seems that there is *absolutely clear* evidence that a particular person is guilty, research shows that this is not always the case. Of the people found innocent through DNA and other evidence for a variety of crimes, including murder, a large percentage (72 per cent) were convicted on the basis of being misidentified by witnesses (often cross-culturally); or were the result of false confessions, flawed forensic evidence, spiteful informants, bad lawyering or jury misunderstanding (Findley, 2002). And we can never know how many not found to be innocent later really were.

And, with wrongful convictions leading to executions, of course, there is the possible unfairness to a community that the real perpetrator remains at large, and possibly relatively free, because the authorities have closed the case and there is no one in jail continually protesting innocence, with lawyers working on the case. While the fact of discovering innocence after the execution will have no effect on the executed person, it can have a powerful negative effect on their relatives and supporters who can easily feel that, given innocence, they did not do enough to get the person off; but above all they will empathize with the horror felt by the innocent person going to their execution for a crime they did not commit. And this type of guilt could also apply to the friends and relatives of the victim of the original crime who may feel extremely guilty about their feelings of hate directed at an innocent person, especially when they imagine the condemned person's last moments and the grief of their family. As a general rule, then, the greatest unfairness here seems to the vast majority of humans, regardless of views on good and evil or morality, to be the execution of an *innocent* person.

This is probably due to a deep-seated fear of death looming as we imagine ourselves trapped, no one listening, no escape possible and time running out – it is almost like being stuck in a cave able to hear rescuers talking about giving up, or like being still alive when put in a coffin and hearing the dirt being piled on. We do not feel these fears most of the time, but learning of the execution of an innocent person brings them dramatically to the front of our consciousness. The possibility of executing an innocent person also generates a fear that powerful authorities can be arbitrary and as individuals we can be extremely vulnerable, our pleas laughed at, sneered at; that justice is for others, not us. So, a theory of justice based on fairness would do away with the death penalty to avoid the potentiality of this unfairness. One might add that this fairness should include that every legitimate possibility of proving innocence after conviction be available to the incarcerated person; but also, if incarceration is to replace execution, any release on the grounds of redemption (rather than discovered innocence) should face extremely severe tests that redemption has been achieved beyond any shadow of a doubt.

It is clear from these examples that justice as fairness depends not only on an understanding of human desires and fears but also on some form of administration. If we say the one who cuts the cake has second choice there has to be someone to impose that rule, and if that rule does not seem fair, to ask each person what size they want. Someone has to catch a murderer and someone has to try them and someone has to try to make sure that the right person has been convicted. And sometimes a community has to make difficult collective decisions as to what is fair. Take the issue of abortion, for example. If successful, those who argue that abortion should not be allowed would be restricting the behaviour of those who want to have an abortion, thus being unfair to them. And because those who want the option of abortion on request do not insist that those against it use it, they are not being unfair to them by restricting their actions or making them do something they do not want to do. On this basis abortion on request should be allowed.

However, what about fairness to the foetus, to the unborn child? Abortion terminates the life of the foetus and so is unfair to the foetus, it is argued. But is it possible to be unfair to a foetus? At what point does a bunch of cells become vulnerable to being unfairly treated; we certainly cannot ask them. Some will argue that at the moment of conception the union of sperm and egg have 'a human essence' and should be treated with all the rights to fairness of adult humans. But this is based purely on faith with no scientific support whatever. Science can tell us that an early bunch of cells is biochemically 'less' than a wide variety of creatures we kill as pests, let alone those we kill to eat. Nevertheless, it cannot tell us at what point a group of cells moving in the direction of becoming a human baby should be considered human, other than to demonstrate that at about 24 weeks' gestation a foetus may possibly survive outside the womb. But it is not clear as to how viable they might be (which, of course, is true of all newborns, although chances are much greater for those that go full term).

However, some would argue that faith and morality are more important as means of conducting our lives than science. Additionally, because anti-abortion views are often allied to a strict code of sexual morality which conflates abortion with irresponsible sexual promiscuity, anti-abortionists argue that the immorality of abortion goes even further than simply killing babies, it also encourages sexual immorality. But this tends to be a moralistic position imposing faith and its moral imperatives onto other humans who take a different practical, moral or scientific view than self (and self's group). And, as we have seen, imposing moralities on others has caused significant harm and untold unfairness to millions of humans throughout history, and so we have learned to be wary of moral/religious imperialism.

Can an understanding of human desires and fears help us on the abortion issue? We clearly have very powerful protective feelings for babies. We have seen that human attractiveness itself has many infant-like characteristics that make us sexually excited and extremely protective, and that many of these characteristics remain in offspring at least until puberty. And we know that very powerful bonds of love can be formed with offspring, and friendship and confirmation of self can be part of the interaction with them. And we can fantasize about offspring bringing status to us with imagined successes and comforting us in our old age – and in providing us with grandchildren to coo over. So, we have powerful desires to protect children at all costs, and we have even spiritualized 'childness' to some extent.

But we also have desires and fears which push us in the opposite direction. The evolution of female choice has motivated females to want to choose the time, place and 'right' partner for the purpose of becoming pregnant; so young human females have developed a strong fear of unwanted pregnancies, not only for the potential restricting effects on future romantic or career possibilities, but also for practical and emotional reasons – it seems like their whole future has suddenly been determined. Nevertheless, powerful sexual and romantic desires/images can lead to unwanted pregnancies, generating for females a sense of despondency and depression. For unmarried females, for example, an 'illegitimate' child usually brings shame, but also can be a great burden, not just on her but on her family (who may also feel guilty that they have failed to 'protect' her and/or to provide the foundation for her to find a 'worthy' mate). If her family abandons her, an unsupported mother may be unable to provide sufficient care for the child, who suffers greatly as a consequence, and this can cause guilt-suffering on the part of the 'failed' mother.

Our fear of all this has been sufficiently strong throughout history that we have directly shamed pregnant unmarried females, and have generally imposed a life of restricted options for them, as a lesson to try to prevent it happening among young unmarried girls. And we have put pressure on males who might impregnate a female to be responsible for that female and child to give them some support, but also as a general warning, again to try to minimize it happening. But the imagined future for such an arrangement is not necessarily a happy one, envisioned as potentially being an unhappy marriage, possible divorce and children raised in a broken home. But, as noted, these injunctions are up against very powerful sexual, love and romantic emotions that do not usually include these fears when in full flow; these fears only take effect after the fact of pregnancy.

Newly married couples carried away with love and passion can end up with a pregnant female long before such is desired, causing a sense of disappointment and depression concerning the loss of the anticipated joys of early marriage and planned joint adventures together. Even longer-term pair-bonded/married mothers and fathers with existing offspring can feel the practical and emotional demands of existing infants 'mounting up' and becoming increasingly burdensome. Many of us feel how depressing it can be to just make ends meet, and to do nothing else but child-care, day after day after day; we remember the crying, screaming, not sleeping and constant demands of the last child that seemed to go on for such a long time. And we fear that we simply do not have the energy (or time) to give to the new born that which we believe any child deserves. Or, we fear that an unwanted additional child will take away resources and attention from existing offspring whom we are desperately struggling to protect and to give a good start in life. If parents have rebellious teenagers they may well feel that they have their hands full, that it will not be possible to do more, that all the offspring will suffer if another one is added. In the context of these memories (and maybe even memories of post-natal depression and very difficult births) many females (and males) feel that they could not face another pregnancy, let alone another potentially traumatic birth. Additionally, some females (and their partners) may greatly appreciate and want to keep the trim, well looked-after, sexy body that attracts attention and makes them feel desirable – perhaps only barely maintained after two children.

With all these feelings and thoughts uppermost, the individuals in the circumstances described above – from unmarried young girls to older married pairs – do not fantasize about having a lovely baby, and there is nothing about an embryo, hidden away inside them that biologically or emotionally causes them to be attracted to it; desires to hug, cuddle, protect, care for and sacrifice for a child have long faded, or at least have been submerged. Anthropology and history do not suggest that there is such a powerful love of babies in all humans that we would do everything we can to have as many as possible. Indeed, it is questionable as to just how much emotional pressure individuals are under to sacrifice everything to make sure all babies survive.

Most parents are very sad – indeed deeply depressed – when a baby dies, but this does not usually generate sufficient long - term feelings of guilt and depression to make them sexually, personally and socially immobilized. Indeed, given the high rate of infant mortality during much of history it would have seemed reproductively counter-productive for a human to become emotionally immobilized at the death of a child. In fact, the suspected frequency of infanticide clearly indicates that such an emotional disability did not generally exist. As noted, even among those who want and are ready for children, there clearly is no evidence of a human desire to produce children one after another for as long as possible.

In fact, humans have sought means of birth control through a variety of approaches throughout human history – at times by instituting periods of sexual abstinence, or by encouraging withdrawal before ejaculation, and sometimes by backstreet/wise woman abortions, or by practicing infanticide (often through simple neglect). In many cultures infanticide and/or high infant mortality were made more palatable by delaying giving offspring full human status until their teen years (until they menstruated or went through often painful initiation rituals); sometimes excess offspring have been farmed out to wealthier families as servants. Historically, however, where economic conditions have permitted, the protective nurturing emotions have tended to eliminate infanticide and the sending away of children, but a desire to avoid unwanted pregnancies and/or to restrict family size has never gone away.

With the achievement of a relatively good standard of living, and with scientific knowledge, the most common way to restrict births is the use of mechanical and chemical means of birth control, but abortion as a last ditch means of birth control has remained a feature of modern life, most usually put in the hands of medical specialists. Science has, thus, provided means of 'family planning' that are more reliable and economically and emotionally less stressful than sexual restraints or infanticide, and are more acceptable for many than abortion, levelling average family size out at something between one and two children. At the same time, most people have come to feel that the existence of an unwanted offspring (with reduced care to existing offspring, a lack of emotional commitment on the part of tired parents, the burden on resources) are unfair to all concerned: the child, the parents and society that might have to care for unwanted offspring. But most people also feel that infanticide is just too uncaring and unfair to an aware creature and that backstreet abortions are unfair to the woman who has to risk life in having one. So, the beginning of 'authentic' humanness is no longer delayed until puberty or after a specific ritual but rather begins at birth. We have made backstreet abortions, infanticide and child neglect illegal – often considering these 'murder' or at least unlawful killing.

So, the issue of abortion comes down to: at what point is medically performed abortion, by choice and consent of a pregnant woman, murder (or unlawful killing)? Medically approved abortion, then, exists in the twilight zone between generally acceptable methods of birth control and perceived murder. Following the above historical developments, in the context of human desires and fears, it seems that, currently, only after the point an embryo or foetus has been deemed to be a potential human being is abortion considered an unlawful killing. We have seen that science can tell us that an early bunch of cells is bio-chemically 'less' than a wide variety of creatures we kill as pests, let alone those we kill to eat. But, as noted, it cannot tell us when these cells become human; science does suggest that after about 22–24 weeks a foetus might survive outside the womb.

In the end, then, a community will have to decide at what point a foetus becomes a human. Modern economically developed communities have tended to place the definition of 'human' somewhere past the mid-point of gestation, and have deemed it *unfair* for moral campaigners to prevent abortion before this point, and so have made it legal (often, however, attaching restrictions and specific medical procedures to be followed), but have made it illegal after this point (in most circumstances). It is almost as if a workable balance between the two human nature tendencies regarding adult propensities to child-care/love versus restricting the number of offspring has been reached through trying to be as *fair* as possible to both tendencies.

Absolute moral claims – such as the 'right to life' or unrestricted 'freedom of choice' – have not led us to this point; they have only led to conflict, intolerance, violence and even arson and murder. There are no grounds – in terms of desires and fears, history, science or justice as fairness – for claiming that abortion is immoral other than a strong faith held by some that it is so. Equally, those who argue that 'freedom of choice' for women is the moral imperative at stake have no grounds in human nature, science, history or justice as fairness to argue this. In the first place, neither human desires and fears nor justice drive us towards complete freedom of action for individuals; indeed our desires and fears work to repress and restrict a great deal of behaviour – like having largely repressed primate male tendencies to kill the previous offspring of a female they are interested in, or eventually greatly reducing complete and widespread bonobo-like promiscuity. Indeed, for human society to have evolved there are a number of things that communities have not allowed freedom of action to apply to – like murder or stealing or constantly breaking contracts, for example.

But those who argue for 'freedom of choice' regarding abortion usually mean freedom of choice by a female over *her own* body, not freedom to act against others. This is a stronger argument, but we do not allow people to have complete freedom over their own bodies. In many societies we prohibit public nudity; we say, for example, that individuals cannot inject themselves with certain drugs; we discourage them from using too much alcohol or tobacco; we say that they should not self-harm or commit suicide. If they do too much of any of these we take children away from them, we socially reject them and sometimes lock them up in prisons or mental institutions. On an absolute moral ground of freedom of choice over one's own body we should not condemn the above or lock them up, but in terms of fairness to dependents and social others who may be in danger from these activities, and perhaps to the 'offenders' themselves, we restrict them.

In the case of abortion we have decided that it is unfair for a female to not be allowed an abortion before mid-term, but unfair to the potential human child thereafter. So, the principle of justice as fairness can be applied, despite the many ambiguities and conflicts of desires and fears surrounding an issue like abortion. It is, however, very clear from this example that a community has continually to be deciding what is fair as scientific knowledge and environmental conditions constantly evolve. Fairness is not a once and for all thing; it is an aspect of human nature (a sense of fairness) that allows for making judgments and rules, but also for changing them as new understandings and circumstances emerge. But it is not a free for all of total relativity; judgments of fairness take place within the confines of basic human desires and fears, and in light of scientific and technological understanding at a particular time.

Fairness based on desires and fears is not to be based on a particular political or cultural ideology or a particular moral code. If these violate an analysis based on an innate sense of fairness informed by species-typical desires and fears, it is the culture, ideology, social relationships or moral codes that require critical examination, not the approach to justice. But these ideologies or cultural precepts are not to be declared evil or immoral, but rather as being unfair. Anti-abortionists have every right to argue in a community forum that the point of being considered a human should be lowered (as do pro-abortionists that it should be raised) – both can use science if it helps, but what neither has a right to do (under justice as fairness) is to define abortion or anti-abortion as a moral issue for which there can be no compromises; nor do they have the right to label opponents as being evil, not worthy of existence, or to put irrelevant obstacles in the way of legally accepted abortion, or, conversely, of arguments against it.

Justice as fairness holds that both *tolerance* and *acceptance* of things (and others) we may not personally like is the price we often have to pay for desiring to live with other humans. But it does not mean that we will have to tolerate and accept any and everything (as shown in the examples above). Additionally, the complexities of human social life can result in great unfairnesses mounting up in sections of a community. This can be as a result of chance, unintended consequences, conflicts of interest, defensive reactions, exaggerated fears, competitions, clashes of moral fervour, different views as to the road to utopia/perfection and the bullying of fanatical extremists. While, as a result, extreme unfairness has happened in human history on many occasions, humans learned early on that it is desirable to establish procedures for deliberating and coming to conclusions concerning fairness, and for setting out rules and laws to try to prevent unfairness, if not provide fairness. And, above all, communities have looked to provide for means of enforcing judgments – which also, of course, can themselves be examined in terms of being just or unjust in terms of fairness.

Human nature and social/political life

It will require further research in line with the suggested social science above to uncover the specific nature of the most successful societies in providing justice. But what has been illuminated so far suggests that successful ones will be those that have been able to utilize our innate sense of fairness while recognizing and accounting for an imperfect human nature, a nature not always full of harmonious desires and fears. This approach does not presume that even the most successful set ups will make all the people feel happy or satisfied all the time; but it does mean that, consciously and unconsciously, most people will likely trust most other humans most of the time, be relatively tolerant of differences, generally have faith in humankind, feel that their society is more or less working, and that the future 'doesn't look too bad'.

And when we are considering justice it will be well to remember that, while witch hunting, the tyrannies of social utopias and suicide bombers tend to get all the headlines, most humans throughout history have not been involved in these things. They have lived in small wandering bands, on homesteads, gone to their jobs and raised families. They have had a fair share of successes, but also of emotional upsets and neighbourhood, personal and sexual conflicts, bad days and good days – life conditions which lend themselves to a notion of justice as fairness, usually dealt with by some form of informal and/or semi-formal enforcement by a community.

So the basis for justice as fairness is already there; we just have to learn to apply it to very large-scale societies, in a global network of societies. However, this likely means – considering, again, the vagaries of human nature and social interactions – that decision-making will often be messy and full of compromises, hypocrisy, u-turns, spin, deals and confusion. It is here that a science of humans based on desires and fears can help. Human customs, laws and political patterns can be examined in terms of the degree to which they are or are not congenial with species-typical desires and fears, and in terms of the degree of fairness they seem to provide. This can range from customs and laws regarding human sexual relations and family patterns, through those related to the institutions of civil society, with such things as economic behaviour and social hierarchies considered along the way.

It is to be noted, however, that applying justice as fairness and scrutinizing institutions in terms of species-typical desires and fears will not mean that there will be an end to individuals making moral claims in a community, or an end to individuals who think that they have a completely correct rational answer to issues of concern, or even an end to individuals viewing the world in terms of good and evil. Moral and rationalistic claims are based on very specific aspects of human nature (spiritualization, a strict categorizing and stereotyping of limited clusters of perceptions, teleological thinking, a powerful sense of righteousness and often an obsession with exposing evil, for example), and are held to with great tenacity against non-believers and against evidence; moralists and rationalists tend not to have much empathy with those who take a different stance from themselves. They usually find concepts such as 'compromise' or 'live and let live', 'tolerance' and 'muddle through' to be an anathema, the worst kind of laziness, naiveté, immorality and evil. To them their own claims appear so self-evidently 'true' that opposition to them seems to be the result of delusion, brainwashing, corruption, having sold out, stupidity and even malicious intent.

While individuals with extreme moralistic/rationalistic mental tendencies usually do not make up a majority of humans in a normal community, they can make a lot of noise and do a lot of bullying, and because of their obsession and tenacity they often have relatively influential, if not powerful, roles in a community. Furthermore, the vast majority of humans have at least some degree of propensities to see the world in terms of good and evil, or to make moral condemnations and to think that there must be rational solutions to a whole variety of problems. So, we all, sometimes, share moral indignation, and often follow those who seem to have an answer.

Therefore, it will be extremely important that some forms of *balances* of *power* and means to maintain them are put into the equation in order to balance conflicting moral claims and to prevent particular moral or rational solutions becoming 'final solutions'. To paraphrase Abraham Lincoln once again: such an approach will not satisfy all the people all the time; it will not satisfy some of the people any of the time; but hopefully it will satisfy most of the people most of the time. There will be disagreements; there will remain loud, bullying moral fundamentalists and purists; there will remain dangers – personal, social and political. Even at the best of times such political/judicial arrangements will appear messy and often be frustrating, but if they provide the greatest amount of fairness possible for humans it is no bad thing.

So, if morality is a danger to humans, and the concept of 'good and evil' is both a nonsense and a danger, and a search for justice is a constant struggle, and if political life is messy, is life worth living? Yes, because, as noted, for most people most of the time life is enjoyable, with fantasies of a decent, if not exciting, future when the everyday seems a bit boring. People like to have sex and fall in love and to make things, accumulate things, and to play with each other and to sing and dance; to gossip and joke about these things; to play and watch sports; to work hard; to build and decorate nests, and to watch TV and read books (and sometimes write them). And we plant our gardens; all because it is in our nature to do so and we get pleasure, or at least a sense of positive achievement, from it; and we get pleasure in thinking about doing these things in the future. We may often want things *not* to change because we are comfortable. However, our natural curiosity, our wanting to show-off, our feeling that our desires have never been fully satisfied, along with our often exaggerated fears of potential dangers – not to mention the tyranny and/or bounty of luck – will always make sure that nothing stands still. All the more reason that we like to play the games of life in a just environment, in a fair environment, because we are never sure what is going to happen next, as we plant our gardens.

References

Abbott, J., 1991, *In the Belly of the Beast: Letters from Prison*, New York: Vintage Books.
ACSF Investigators, 1992, 'AIDS and Sexual Behaviour in France', *Nature*, Vol. 360, pp. 407–409.
Adams, J., Maslin, M. and Thomas, E., 1998, 'Sudden Climate Transitions during the Quaternary', www.esd.ornl.gov/projects/qen/transit.html.
Adams, J., 2004, 'An Island of Hobbits', *Newsweek International* (8 November), p. 57.
Adams, J., 2005, 'A Quick Background to the Pliocene', Environmental Science Division, Oak Ridge National Lab., www.esd.ornl.gov/projects/gen/pliocene.
Aiello, L., 1996, 'Territoriality, Bipedalism and the Origin of Language', in Runciman, Maynard Smith and Dunbar, 1996.
Aiello, L. and Wheeler, P., 1995, 'The Expensive-Tissue Hypothesis: The Brain and the Digestive System in Human and Primate Evolution', *Current Anthropology*, Vol. 36, pp. 199–221.
Ainsworth, C., 2003, 'Running Repair', *New Scientist* (15 March), pp. 40–43.
Aitkenhead, D., 2012, 'The Buzz: How the Vibrator Came to Be', *The Guardian* (Friday 7 September).
Alden, A., 2013, 'Earthquake Prediction in the 21st Century', http://geology.about.com/od/eq_prediction/a/aa_EQprediction_2.htm.
Aldington, R. (ed. and Trans.), 1928, *Candide and Other Romances*, London: Murray's Book Sales.
Alexander, R., 1971, 'The Search for an Evolutionary Philosophy of Man', *Proceedings of the Royal Society of Victoria*, Vol. 84, pp. 99–120.
Alexander, R., 1974, 'The Evolution of Social Behavior', *Annual Review of Ecology and Systematics*, *Vol.* 5, pp. 325–383.
Alexander, R. and Noonan, K., 1979, 'Concealment of Ovulation, Parental Care and Human Social Evolution', in Chagnon and Irons, 1979.
Alexander, R. and Tinkle, D. (eds.), 1981, *Natural Selection and Social Behavior: Recent Research and New Theory*, New York: Chiron Press.
Ambrose, S., 1975, *Crazy Horse and Custer: The Parallel Lives of Two American Warriors*, New York: Meridian.
Andreski, S., 1982, 'The Syphilitic Shock: A New Explanation of the Witch Burnings', *Encounter*, Vol. LVIII: 5 (May), pp. 7–26.
Andrews, P., 1989, 'Review of, M.D. Leakey and J.M. Harris, 1987, *Laetoli – A Pliocene Site in Northern Tanzania*, Oxford: Clarendon Press', in *Jr. Human Evolution*, Vol. 18, pp. 173–181.
Andrews, P., 1992, 'Reconstructing Past Environments', in Jones, Martin and Pilbeam, 1992.
Andrews, P., 1995, 'Ecological Apes and Ancestors', *Nature*, Vol. 376: 6541, pp. 555–556.
Anglo, S., 1977, *The Damned Art: Essays in the Literature of Witchcraft*, London: Routledge and Kegan Paul.

Anitei, S., 2007, 'Why Do Women Have Breasts?', Softpedia (9 February), http://news.softpedia.com/news/.
Aquinas, T. (St.), 1966, *The Political Ideas of St. Thomas Aquinas* (ed. D. Bigongiari), Hafner Library of Classics, Kindle ed. Amazon.
Ardrey, R., 1961, *African Genesis: A Personal Investigation into the Animal Origins and Nature of Man*, New York: Macmillan.
Aries, P., 1985, 'Thoughts on the History of Homosexuality', in Aries and Bejin, 1985.
Aries, P. and Bejin, A. (eds.), 1985, *Western Sexuality: Practice and Precept in Past and Present Times*, Oxford: Basil Blackwell.
Aristotle, 1885, *The Politics* (Trans. Benjamin Jowett), Oxford: Clarendon Press.
Aristotle, 2002 [350 BC], *Nichomachean Ethics* (Trans. Joe Sachs), Newbury, MA: Focus Publishing.
Armstrong, L., 1978, *Kiss Daddy Goodnight: A Speakout on Incest*, New York: Hawthorn.
Ash, T., 2002, 'Truth Is Another Country', *The Guardian Review* (16 November), pp. 4–6.
Ashenburg, K., 2008, *Clean: An Unsanitised History of Washing*, London: Profile Books.
Ashton, R., 1996, *George Eliot: A Life*, London: Allen Lane.
Attah, R., 2014, 'A Review of Twin Deliveries in Aminu Kano Teaching Hospital, North-West Nigeria', www.njbcs.net/article.asp?issn=0331-8540;year=2014
Augustine, St., 1950 [413–426 AD], *The City of God* (Trans. M. Dods), New York: Random House.
Augustine, St., 2012, *On the Predestination of Saints*, [Kindle Edition], Fig, Amazon Digital Services Inc.
Axelrod, R., 1984, *The Evolution of Cooperation*, New York: Basic Books.
Ayala, F., 1978, 'The Mechanisms of Evolution', *Scientific American*, Vol. 239: 3, pp. 56–69.
Ayduk, O., Downey, G. and Kim, M, 2001, 'An Expectancy-Value Model of Personality Diathesis for Depression: Rejection Sensitivity and Depression in Women', *Personality and Social Psychology Bulletin*, Vol. 27, pp. 868–877.
Babchuk, W., Hames, R. and Thompson, R., 1985, 'Sex Differences in the Recognition of Infant Caretaker Hypothesis', *Ethology and Sociobiology*, Vol. 6: 2, pp. 89–101.
Badcock, C., 1986, *The Problem of Altruism*, Oxford: Basil Blackwell.
Badcock, C., 1994, *PsychoDarwinism: The New Synthesis of Darwin and Freud*, London: HarperCollins.
Bahceci, M., *et al.*, 2005, 'A Novel Missense Mutation of 5-alpha Reductase Type 2 Gene (SRD5A2) Leads to Severe Male Pseudohermaphroditism in a Turkish Family', *Urology*, Vol. 66: 2, pp. 407–410, Doi:10.1016/j.urology.2005.02.021.
Barash, D., 2001, *Revolutionary Biology: The New, Gene-Centered View of Life*, New Brunswick: Transaction Pub.
Barkow, J., Cosmides, L. and Tooby, J., 1992, *The Adapted Mind: Evolutionary Psychology and the Generation of Culture*, Oxford: Oxford University Press.
Baron-Cohen, S., 1995, *Mindblindness*, Cambridge, MA: MIT Press.
Baron-Cohen, S., 2004, *The Essential Difference: Men, Women and the Extreme Male Brain*, London: Penguin.
Baron-Cohen, S., 2011, *Zero Degrees of Empathy*, London: Allen Lane.
Bateman, R. and DiMichele, W., 2002, 'Generating and Filtering Major Phenotypic Novelties, Neogoldschmidtian Saltation Revisited', in Cronk *et al.*, 2002.
Bateson, G., Jackson, D., Haley, J. and Weakland, J., 1956, 'Towards a Theory of Schizophrenia', *Behavioral Science*, Vol. I, pp. 251–264.

Bauer, H., 1979, 'Agonistic and Grooming Behavior in the Reunion Context in Gombe Stream Chimpanzees', in Hamburg and McCown, 1979.

Bauer, R., 1952, *The New Man in Soviet Psychology*, Cambridge, MA: Harvard University Press.

Baum, M., Everitt, B., Herbert, M. and Keverne, E., 1977, 'Hormonal Basis of Proceptivity and Receptivity in Female Primates', *Archives of Sexual Behavior*, Vol. 6, pp. 173–192.

Baumeister, R., 1997, *Evil: Inside Human Violence and Cruelty*, New York: W.H. Freeman and Co.

Baumeister, R., 1997a, 'The Myth of Pure Evil', in Baumeister, 1997.

Beach, F., 1956, 'Characteristics of Masculine Sex Drive', *The Nebraska Symposium on Motivation*, Vol. 4, Lincoln, NE: University of Nebraska Press.

Beach, F. (ed.), 1965, *Sex and Behavior*, New York: John Wiley and Sons.

Beach, F., 1976, 'Sexual Attractivity, Proceptivity, and Receptivity in Female Mammals', *Hormones and Behavior*, Vol. 7, pp. 105–138.

Beach, F. (ed.), 1977, *Human Sexuality in Four Perspectives*, Baltimore: The Johns Hopkins University Press.

Beck, B., 1975, 'Primate Tool Behavior', in Tuttle, 1975.

Belinda, J., *et al.*, 2004, 'A Direct Brainstem–Amygdala–Cortical "alarm" System for Subliminal Signals of Fear', *NeuroImage*, Vol. 24: 2005, pp. 235–243.

Bell, N. and Vogel, E., 1960, *The Family*, New York: The Free Press.

Belle, D. (ed.), 1982, *Lives in Stress: Women and Depression*, London: Sage.

Benedict, R., 1946, *Patterns of Culture*, New York: Mentor Books.

Benjafield, J., 1992, *Cognition*, Englewood Cliffs, NJ: Prentice-Hall.

Benshoof, L. and Thornhill, R., 1979, 'The Evolution of Monogamy and Concealed Ovulation in Humans', *Journal of Social and Biological Structures*, Vol. 2, pp. 95–106.

Betzig, L., 1986, *Despotism and Differential Reproduction: A Darwinian View of History*, Hawthorne, New York: Aldine.

Betzig, L., 1992, 'Medieval Monogamy', in Mithen and Maschner, 1992.

Betzig, L, Borgerhoff, M. and Turke, P. (eds.), 1988, *Human Reproductive Behaviour: a Darwinian Perspective*, Cambridge: Cambridge University Press.

Betzig, L. and Weber, S., 1992, 'Polygyny in American Politics', *Politics and Life Science*, Vol. 12: 1, pp. 45–52.

Bhattacharya, S., *et al.*, 2010, 'Inherited Predisposition to Spontaneous Preterm Delivery', *Obstetrics & Gynecology*, Vol. 115: 6, p. 1125, doi:10.1097/AOG.0b013e3181dffcdb.

Bickerton, D., 2002, 'Foraging Versus Social Intelligence in the Evolution of Protolanguage', in Wray, 2002.

Biegert, J., 1963, 'The Evaluation of Characteristics of the Skull, Hands and Feet for Primate Taxonomy', in Washburn, 1963.

Binford, L. and Binford, S. 1966, 'A Preliminary Analysis of Functional Variability in the Mousterian of Levallois Facies', *American Anthropologist*, Vol. 68, pp. 238–295.

Binford, S. and Binford, L. (eds.), 1968, *New Perspectives in Archaeology*, Chicago: Aldine Pub.

Binford, S. and Binford, L., 1968a, 'Post-Pleistocene Adaptations', in Binford and Binford, 1968.

Birdsell, J., 1953, 'Some Environmental and Cultural Factors Influencing the Structuring of Australian Aboriginal Populations', *American Naturalist*, Vol. 87, pp. 171–207.

Birdsell, J., 1968, 'Some Predictions from the Pleistocene Based on Equilibrium Systems among Recent Hunter-Gatherers', in Lee and DeVore, 1968.

Birdsell, J., 1973, 'A Basic Demographic Unit', *Current Anthropology*, Vol. 14, pp. 337–356.

Bischoff, N., 1975, 'Comparative Ethology of Incest Avoidance', in Fox, 1975.

Bjarklund, D. and Kipp, K., 2002, 'Social Cognition, Inhibition, and Theory of Mind: The Evolution of Human Intelligence', in Sternberg and Kaufman, 2002.
BJS, 2010, 'Crocodile and Hippopotamus Served as Brain Food for Early Human Ancestors', *Science Blog*, 9 June.
Black, A., 1984, 'St. Thomas Aquinas: the State and Morality', in Redhead, 1984.
Blackmore, S., 1999, *The Meme Machine*, Oxford: Oxford University Press.
Bloch, M., 1961/1965, *Feudal Society*, Vols. I and II, London: Routledge and Kegan Paul.
Block, S. and Reddaway, P., 1977, *Russia's Political Hospitals*, London: Gollancz.
Blueher, H., 1912, *The German Youth Movement as an Erotic Phenomenon*, in Dollimore, 1998.
Blueher, H., 1917, *The Role of the Erotic in Male Communities*, in Dollimore, 1998.
Boas, F., 1965, *The Mind of Primitive Man*, New York: The Free Press.
Boaz, N. and Ciochon, R., 2004, 'Brute of Dragon Bone Hill', *New Scientist* (17 April), pp. 32–35.
Bobe, R., 2006, 'The Evolution of Arid Ecosystems in Eastern Africa', *Journal of Arid Environments*, Vol. 66:3, pp. 564–584.
Boehm, C., 1997, 'Egalitarian Behaviour and the Evolution of Political Intelligence', in Whiten and Byrne, 1997.
Booyens, K., 2008, *The Sexual Assault and Rape of Male Offenders and Awaiting-Trial Detainees*, Doctoral Thesis, South Africa: University of Pretoria.
Boudon, C., et al., 1995, 'Molecular Study of the 5 Alpha-Reductase Type 2 Gene in Three European Families with 5 Alpha-Reductase Deficiency', *The Journal of Clinical Endocrinology & Metabolism*, Vol. 80: 7, pp. 2149–2153.
Bowlby, J., 1969, *Attachment. Attachment and Loss* Vol. 1, London: Hogarth.
Bowlby, J., 1973, *Separation, Anxiety and Anger. Attachment and Loss*, Vol. 2, London: Hogarth.
Bowlby, J., 1980, *Loss: Sadness and Depression. Attachment and Loss*, Vol. 3, London: Hogarth.
Brady, G., 2000, 'Hox Genes: The Molecular Architects.' Irish Scientist Year Book, www.ireland.com/newspaper/science/2000/065/sci.htm
Brady, S., 2012, *John Addington Symonds (1840–1893) and Homosexuality: A Critical Edition of Sources*, Basingstoke: Palgrave.
Bragg, M., 2006, 'Greek Comedy', In Our Time, BBC Radio 4 (13 July).
Brain, C., 1981, *Hunters or the Hunted?*, Chicago: University of Chicago Press.
Braun, D., et al., 2010, 'Early Hominin Diet Included Diverse Terrestrial and Aquatic Animals 1.95 Ma in East Turkana, Kenya', *Proceedings of the National Academy of Sciences of the United States of America (107)*, National Academy of Sciences.
Brennan, M. and Antonyshyn, O., 1996, 'The Effects of Temporalis Muscle Manipulation on Skull Growth: an experimental study', *Plastic and Reconstructive Surgery*, Vol. 97, pp. 13–24.
Brin, D., 1995, 'Neoteny and Two-Way Sexual Selection in Human Evolution: Paleo-Anthropological Speculation on the Origins of Secondary-Sexual Traits, Male Nurturing and the Child as a Sexual Image', at, www.kithrup.com/brin/neoteny.html
Broadhurst, T., Cunnane, S. and Crawford, M., 1998, 'Rift Valley Lakefish and Shellfish Provide Brain Specific Nutrition for Early Homo', *British Journal of Nutrition*, Vol. 79, pp. 3–21.
Brosnan, S. and de Waal, F., 2003, 'Monkeys Reject Unequal Pay', *Nature*, Vol. 425, pp. 297–299.
Brown, G. and Harris, T., 1978, *Social Origins of Depression*, New York: Free Press.
Brown, P., 2003, 'In the Shadow of Fear', *New Scientist* (6 September), pp. 30–35.
Brunet, M. et al., 2002, 'A New Hominid from the Upper Miocene of Chad, Central Africa', *Nature*, Vol. 418, pp. 145–151.
Buck, R., 1984, *The Communication of Emotions*, London: The Guilford Press.

Budge, E., 1934, *From Fetish to God in Ancient Egypt*, London: Oxford University Press.
Buhle, P. and Wagner, D., 2002, *Radical Hollywood*, New York: The New Press.
Bunney, S., 1993, 'Science: On the Origins of the Midwife', *New Scientist*, Issue 1874 (22 May).
Bürglin, T., 2005, 'The Homebox Page', www.biosci.ki.se/groups/tbu/homeo.html.
Buss, D., 1989, 'Sex Differences in Human Mate Preferences: Evolutionary Hypotheses Tested in 37 Cultures', *Behavioral and Brain Sciences*, Vol. 12, pp. 1–49.
Buss, D., 1992, 'Mate Preference Mechanisms: consequences for Partner Choice and Intersexual Competition', in Barkow, Cosmides and Tooby, 1992.
Buss, D., 1994, *The Evolution of Desire*, New York: Basic Books.
Buss, D., 2000, *The Dangerous Passion*, New York: Bloomsbury Publishing.
Buys, C., 1992, 'Human Sympathy-Groups: Cross-Cultural Data', *Psychological Reports*, Vol. 71:3, p. 786.
Buys, C. and Larson, K., 1979, 'Human Sympathy Groups', *Psychological Reports*, Vol. 45, pp. 547–553.
Byatt, A. S., 2004, 'A Child in Time', *The Guardian* (24 October), pp. 16–17.
Bygott, J., 1979, 'Agonistic Behaviour, Dominance and Social Structure in Wild Chimpanzees of the Gombe National Park', in Hamburg and McCown, 1979.
Byrne, R., 2002, 'Social and Technical Forms of Primate Intelligence', in De Waal, 2002.
Byrne, W. and Whiten, A. (eds.), 1988, *Machiavellian Intelligence*, Oxford: Clarendon Press.
Campbell, A., 2013 (2002), *A Mind of Her Own*, Oxford: Oxford University Press.
Campbell, B. (ed.), 1972, *Sexual Selection and the Descent of Man*, Chicago: Aldine.
Caplan, P. (ed.), 1987, *The Cultural Construction of Sexuality*, London: Tavistock.
Carnegie, A., 1901, *The Gospel of Wealth and other Timely Essays*, New York: Century.
Carroll, J., 1985, *Guilt*, London: Routledge and Kegan Paul.
Carter, N. (ed.), 1980, *Development, Growth and Ageing*, London: Croom Helm.
Chagnon, N., 1968, *Yanomamo: The Fierce People*, New York: Holt Rinehart and Winston.
Chagnon, N., 1979, 'Is Reproductive Success "equal" in Egalitarian Societies?', in Chagnon and Irons, 1979.
Chagnon, N., 1981, 'Doing Fieldwork among the Yanomamo', *Contemporary Anthropology*, pp. 11–24.
Chagnon, N., 1988, 'Life Histories, Blood Revenge, and Warfare in a Tribal Population', *Science*, Vol. 239, pp. 985–992.
Chagnon, N. and Irons, W. (eds.), 1979, *Evolutionary Behaviour and Human Social Behaviour*, North Scituate, MA: Duxbury.
Chance, M., 1962, 'An Interpretation of some Agonistic Postures: The Role of "cut-off" Gaze and Postures', *Symposia of the Zoological Society of London*, Vol. 8, pp. 71–89.
Chance, M., 1988, 'Introduction', in Chance and Omark, 1988.
Chance, M., 1988a, 'A Systems Synthesis of Mentality', in Chance and Omark, 1988.
Chance, M., and Omark, D. (eds.), 1988, *Social Fabrics of the Mind*, London: Lawrence Erlbaum Associates.
Chang, J. and Halliday, J., 2005, *Mao: The Unknown Story*, New York: Random House.
Changeux, J-P., 1980, 'Properties of the Neural Network', in Piattelli-Palmarini, 1980.
CIA, 2013, *The World Factbook*, Central Intelligence Agency – USA, www.cia.gov/library/publications
Clancy, K., Hinde K. and Rutherford, J. (eds.), 2012, *Building Babies: Primate Development in Proximate and Ultimate Perspective*, New York: Springer.

Clark, D., 1968, 'Studies of Hunter-Gatherers as an Aid to the Interpretation of Prehistoric Societies', in Lee and DeVore, 1968.
Clarke, R. and Tobias, P., 1995, 'Sterkfontein Member 2 Foot Bones of the Oldest South African Hominid', *Science*, Vol. 269, pp. 521–524.
Clutton-Brock, T. and Harvey, T. (eds.), 1978, *Readings in Sociobiology*, San Francisco: W. H. Freeman.
Clutton-Brock, T. and Parker, G., 1995, 'Punishment in Animal Societies', *Nature*, Vol. 373, pp. 209–216, doi:10.1038/373209a0.
Coe, M., 2012, *Breaking the Maya Code*, London: Thames and Hudson.
Cohen, D., 1979, *Sleep and Dreaming: Origins, Nature and Functions*, New York: Pergamon.
Cohen, N., 2004, 'Where Be Monsters?', *The Observer* (18 January), p. 3.
Cohen, P., 2001, 'All Hung Up about Size', *New Scientist* (12 May), Vol. 170: 2290, pp. 34–35.
Cohen, P., 2002, 'The Bodyguard', *New Scientist* (14 September), pp. 28–33.
Cohn, N., 1975, *Europe's Inner Demons*, London: Chatto-Heinemann.
Colburn, R. (ed.), 1956, *Feudalism in History*, Princeton NJ: Princeton University Press.
Colson, E., 1974, *Tradition and Contract: The Problem of Order*, Chicago: Aldine.
Comte, A., 1974, *The Positive Philosophy*, New York: AMS Press.
Connor, R. and Krützen, M., 2003, 'Levels and Patterns in Dolphin Alliance Formation', in de Waal and Tyack, 2003.
Conquest, R., 1971, *The Great Terror*, Harmondsworth: Penguin.
Cooley. C. 1902, *Human Nature and the Social Other*, New York: Charles Scribners.
Coppens, Y., Howell, C., Isaac, G. and Leakey, R. (eds.), 1976, *Earliest Man and Environments in the Lake Rudolf Basin: Stratigraphy, Paleontology and Evolution*, Chicago: University of Chicago Press.
Copley, J., 2001, 'Here's Looking at You', *New Scientist* (12 May), p. 17.
Corballis, M., 2002, 'Evolution of the General Mind', in Sternberg and Kaufman, 2002.
Cosmides, L., Tooby, J. and Barkow, H., 1992, 'Introduction: Evolutionary Psychology and Conceptual Integration', in Barkow, Cosmides and Tooby, 1992.
Courtwright, D., 1998, *Violent Land*, Cambridge, MA: Harvard University Press.
Cowley, G., 2003, 'Where Living is Lethal', *Newsweek International* (22 September), pp. 52–55.
Cronk, Q., Bateman, R. and Hawkins, J. (eds.), 2002, *Developmental Genetics and Plant Evolution*, London: Taylor & Francis.
Cronk, L., Chagnon, N. and Irons, W. (eds.), 2000, *Adaptation and Human Behavior*, Hawthorne, New York: Aldine de Gruyter.
Crook, J., 1980, *The Evolution of Human Consciousness*, London: Oxford University Press.
Crystal, D., 2005, *How Language Works*, London: Penguin.
Cunnane, S., 2006, 'Survival of the Fattest: The Key to Human Brain Evolution', *médecine/sciences* (Paris), (Jun-Jul) Vol. 22: 6–7, pp. 659–663.
Cunnane, S., Harbige, I. and Crawford, M., 1993, 'The Importance of Energy and Nutrient Supply in Human Brain Evolution', *Nutrition and Health*, Vol. 9, pp. 219–235.
Curtis, V., 2007, 'Dirt, Disgust and Disease: A Natural History of Hygiene', *Journal of Epidemiology and Community Health*, Vol. 61: 8, pp. 660–664.
Dahlberg. F., 1981, *Women the Gatherer*, New Haven: Yale University Press.
Dailey, T., 2013, 'Comparing the Lifestyles of Homosexual Couples to Married Couples', Center for Marriage and Family Studies: www.frc.org/get.cfm?i=IS04C02
Dallos, S. and Dallos, R., 1997, *Couples, Sex and Power*, Buckingham: Open University Press.

Daly, M. and Wilson, M., 1981, 'Abuse and Neglect of Children in Evolutionary Perspective', in Alexander and Tinkle, 1981.

Damasio, A., 2000, *The Feeling of What Happens*, London: Vintage.

Dart, R., 1925, 'Australopithecus Africanus: The man-Ape of South Africa', *Nature*, Vol. 115, pp. 195–199.

Darwin, C., 1859 [1996], *On the Origin of Species by Means of Natural Selection, or the Preservation of Favoured Races in the Struggle for Life*, Oxford: Oxford University Press.

Darwin, C., 1872 [1998], *The Expression of the Emotions in Man and Animals*, London: HarperCollins.

Darwin, C., 1879 [1974/2004], *The Descent of Man, and Selection in Relation to Sex*, London: Penguin.

Daughaday, W., 1968, 'The Adenohypophysis', in Williams, 1968.

Davenport, W., 1977, 'Sex in Cross-Cultural Perspectives', in Beach, 1977.

Davidson, A., 1991, 'The Horror of Monsters', in Sheehan and Sosna, 1991.

Davis, H. (SJ), 1935, *Moral and Pastoral Theology* (Vol. 1.), London: Sheed and Ward.

Dawkins, R., 1976, *The Selfish Gene*, London: Oxford University Press.

Dawkins, R., 1986, *The Blind Watchmaker*, London: Longmans.

Dawkins, R., 1989, *The Selfish Gene* (New Edition), Oxford: Oxford University Press.

Dawkins, R., 2004, 'The Big Questions', Channel 5 Television (7 January).

Dawkins, R., 2004a, *The Ancestor's Tale: A Pilgrimage to the Dawn of Life*, London: Weidenfeld and Nicolson.

Dawkins, R., 2009, 'The Purpose of Purpose', Lecture Tour of US Universities, for example, www.centerforinquiry.net/.../richard_dawkins_in_minnesota_020409

de Tocqueville, A., 1956, *Democracy in America*, New York: Mentor Books.

DeVore, I., 1963, 'A Comparison of the Ecology and Behaviour of Monkeys and Apes', in Washburn, 1963.

de Waal, F., 1982, *Chimpanzee Politics: Sex and Power among Apes*, New York: Harper and Row.

de Waal, F., 1988, 'The Communications Repertoire of Captive Bonobos Pan paniscus compared to Chimpanzees', *Behaviour*, Vol. 106, pp. 183–251.

de Waal, F., 1989, *Peacemaking among Primates*, Cambridge, MA: Harvard University Press.

de Waal, F., 1996, *Good Natured: The Origins of Right and Wrong in Humans and Other Animals*, Cambridge, MA: Harvard University Press.

de Waal, F. (ed.), 2002, *Tree of Origin*, Cambridge, MA: Harvard University Press.

de Waal, F., 2002a, 'Apes from Venus: Bonobos and Human Evolution', in de Waal, 2002.

de Waal, F., 2005, *Our Inner Ape*, New York: Riverhead Books.

de Waal, F., 2006, *Primates and Philosophers: How Morality Evolved*, Princeton NJ: Princeton University Press.

de Waal, F. and Lanting, L., 1998, *Bonobo: The Forgotten Ape*, Berkeley: University of California Press.

de Waal, F. and Luttrell, F., 1988, 'Mechanisms of Social Reciprocity in Three Primate Species: Symmetrical Relationship Characteristics or Cognition?', *Ethology & Sociobiology*, Vol. 9, pp. 101–118.

de Waal, F. and Tyack, P. (eds.), 2003, *Animal Social Complexity: Intelligence, Culture, and Individualized Societies*, Cambridge, MA: Harvard University Press.

Deag, J., 1974, *A Study of the Social Behaviour and Ecology of the Wild Barbary Macaque, Macaca sylvanus*, Unpublished PhD dissertation, University of Bristol.

de Menocal, P., 2003/4, 'African Climate Change and Faunal Evolution during the Pliocene–Pleistocene', *Earth and Planetary Science Letters*, Vol. 220, pp. 3–24.

Deschenes, E. and Esbensen, F., 1999, 'Violence in Gangs: Gender Differences in Perceptions and Behavior', *Journal of Quantitative Criminology*, Vol, 15, pp. 63–96.

Dewald, C. (ed.) and Waterfield R. (trans.), 1998, *The Histories by Herodotus*, Oxford: University of Oxford Press.

Diamond, J., 1991, *The Rise and Fall of the Third Chimpanzee*, London: Hutchinson Radius.

Dickemann, M., 1979, 'Female Infanticide, Reproductive Strategies and Social Stratification: A Preliminary Model', in Chagnon and Irons, 1979.

Dickie, J., 2004, *Cosa Nostra: A History of the Sicilian Mafia*, London: Hodder.

Dimont, M., 1962, *Jews, God, and History*, New York: Simon and Shuster.

Dixson, A., 1998, *Primate Sexuality: Comparative Studies of Prosimians, Monkeys, Apes and Human Beings*, Oxford: Oxford University Press.

Dollimore J., 1998, *Death, Desire and Loss in Western Culture*, New York: Routledge.

Dorner, G., 1976, *Hormones and Brain Differentiation*, Amsterdam: Elsevier Scientific Publishing.

Dougals, M., 1970, *Purity and Danger*, Harmondsworth: Penguin.

Douglas, M. (ed.), 1970a, *Witchcraft Confessions and Accusations*, London: Tavistock.

Doward, J., 2004, 'Boys Aged 10 for Sale', *The Observer* (10 October), pp. 8–9.

Drake, S., 1994, *Galileo: Pioneer Scientist*, Toronto: University of Toronto Press

Drea, C. and Frank, L., 2003, 'The Social Complexity of Spotted Hyenas', in de Waal and Tyack, 2003.

Dumont, L., 1980, *Homo Hierarchicus*, Chicago: University of Chicago Press.

Dunbar, R., 1992, 'Neocortex Size as a Constraint on Group Size in Primates', *Journal of Human Evolution*, Vol. 20, pp. 469–493.

Dunbar, R. (ed.), 1993, *Human Reproductive Decisions: Biological and Social Perspectives*, Galton Institute symposium (13th), London: Macmillan.

Dunbar, R., 1993a, 'Coevolution of Neocortical Size, Group Size and Language in Humans', *Behavioral and Brain Sciences*, Vol. 16:4, pp. 681–735.

Dunbar, R., 1996, *Grooming, Gossip and the Evolution of Language*, London: Faber and Faber.

Dunbar, R., 1998, 'Your Cheatin' Heart', *New Scientist* (November), 2161, pp. 28–32.

Dunbar, R., 2002, 'Brains on Two Legs: Group Size and the Evolution of Intelligence', in de Waal, 2002.

Dunbar, R., 2004, *The Human Story*, London: Faber and Faber.

Duncombe, J. and Marsden, D., 1993, 'Love and Intimacy: The Gender Division of Emotion and "Emotion Work"', *Sociology*, Vol. 27:2, pp. 221–241.

Duncombe, J. and Marsden, D., 1994, 'Can Men Love? – "Reading", "Staging" and "Resisting" the Romance', *ESRC Research Report*, London: ESRC.

Durkheim, E., 1965, *The Elementary Forms of the Religious Life*, New York: The Free Press.

Eagleman, D., 2011, *Incognito*, London: Canongate.

Ehrenfeld, T., 2005, 'How Do You Feel Now? Emotional States: Technology Gives Autistic Kids a Hand', *Newsweek International* (28 February).

Ehrenreich, B., 1997, *Blood Rights: Origins and History of The Passions of War*, New York: Henry Holt.

Eibl-Eibesfeldt, I., 1971, *Love and Hate*, London: Methuen.

Ellickson, R., 1991, *Order Without Law: How Neighbors Settle Disputes*, Cambridge, MA: Harvard University Press.

Elliot Smith, G., 1927, *The Evolution of Man* (2nd ed.), Oxford: Oxford University Press.

Elsberry, W., 2009, 'Richard Dawkins and "The Purpose of Purpose"', */03/richard-dawkins-2.html#comment-panels* (10 March), http://pandasthumb.org/archives/2009

Enard W, *et al.*, 2002, 'Molecular Evolution of FOXP2, a Gene Involved in Speech and Language', *Nature*, Vol. 418: 6900, pp. 869–872.

Engels, F., 1940, *Dialectics of Nature*, London: Lawrence & Wishart.

Engels, F., 1940a, 'The Part Played by Tools in the Transition from Ape to Man', in Engels, 1940.

Epic of G., 1964, *The Epic of Gilgamesh*, Harmondsworth: Penguin.

Erikson, E., 1965, *Childhood and Society*, New York: Vintage Books.

Estronaut, 1999, 'Spontaneous Abortion', www.estronaut.com/a/spontaneous_abortion_common.htm

Evans-Pritchard, E., 1937, *Witchcraft, Oracles and Magic among the azande*, Oxford: Clarendon Press.

Evans-Pritchard, E., 1940, *The Nuer*, Oxford: Clarendon Press.

Evans-Pritchard, E., 1965, *The Position of Women in Primitive Societies*, London: Faber and Faber.

Eveleth, P. and Tanner, J., 1976, *Worldwide Variation in Human Growth*, Cambridge: Cambridge University Press.

Excerpts, 2013, 'Progenesis and Human Puberty: Pubertal Timing as a Heterochronic Variable', www.neoteny.org/neoteny/a/progenesispuberty.html

Eysenck, M., 1992, *Anxiety: The Cognitive Perspective*, Hillsdale, NJ: Lawrence Erlbaum.

Eysenck, M. and Keane, M., 1990, *Cognitive Psychology*, Hillsdale, NJ: Lawrence Erlbaum Associates.

Facts, 2013, 'Information about the Sexual Development of Youth', Facts about Youth, American College of Pediatricians, http://factsaboutyouth.com/posts/promiscuity/

Fagan, B., 1986, *People of the Earth*, Boston: Little Brown and Co.

Fagan, J., 2012, 'Capital Punishment: Deterrent Effects & Capital Costs', www.law.columbia.edu/law_school/communications/reports/summer06/capitalpunish

Fagles, R., 1984, *Sophocles: The Three Theban Plays*, New York: Penguin Books.

Fedigan, L., 1986, 'The Changing Role of Women in Models of Human Evolution', *Annual Review of Anthropology*, Vol. 15, pp. 25–66.

Feldman, *et al.*, 2000, 'Material Social Support Predicts Birth Weight and Fetal Growth in Human Pregnancy', *Psychosomatic Medicine*, Vol. 62, pp. 715–725.

Fiedler, L., 1981, *Freaks: Myths and Images of the Secret Self*, Harmondsworth: Penguin.

Findley, K., 2002, 'Learning from Our Mistakes: A Criminal Justice Commission to Study Wrongful Convictions', 'Introduction', *California Western Law Review*, Vol. 38: 2, pp. 333–353.

Finkelhor, D., 1980, 'Sex among Siblings: A survey on Prevalence, Variety, and Effects', *Archives of Sexual Behavior*, Vol. 9:3, pp. 171–194.

Fisher, H., 1982, *The Sex Contract: The Evolution of Human Behaviour*, London: Granada.

Fisher, H., 2000, 'Lust, Attraction, Attachment: Biology and the Evolution of the Three Primary Emotional Systems for Mating, Reproduction and Parenting', *Journal of Sex Education and Therapy*, Vol. 25:1, pp. 96–104.

Fisher, H., 2004, *Why We Love: The Nature and Chemistry of Romantic Love*, New York: Henry Holt.

Fisher, H., Aron, A, Mashek D. and Brown L., 2002, 'Defining the Brain Systems of Lust, Romantic Attraction, and Attachment', *Archives of Sexual Behavior*, Vol. 5, pp. 413–419.

Fisher, R., 1930, *The Genetical Theory of Natural Selection*, Oxford: Clarendon Press.

Foley, J., 2011, 'Fossil Hominids: The Evidence for Human Evolution', The Talk Origins Archive, www.talkorigins.org/faqs/homs/

Foley, R., 1994, 'Speciation, Extinction and Climate Change in Hominid Evolution', *Journal of Human Evolution*, Vol. 26:4, pp. 275–289.
Ford, C. and Beach, F., 1951, *Patterns of Sexual Behavior*, New York: Harper and Row.
Forgas, J. (ed.), 1981, *Social Cognition*, London: Academic Press.
Fortey, R., 2001, *Trilobites: Eyewitness to Evolution*, London: Flamingo.
Forward, S. and Buck, C., 1981, *Betrayal of Innocence: Incest and its Devastation*, Harmondsworth: Penguin Books.
Fossey, D., 1979, 'Development of the Mountain Gorilla (*Gorilla gorilla beringei*) through the First Thirty-Six Months', in Hamburg and McCown, 1979.
Fossey, D., 1983, *Gorillas in the Mist*, Boston: Houghton Mifflin.
Foucault, M., 1980, *The History of Western Sexuality: An Introduction*, Vol. I, New York: Vintage Books.
Foucault, M., 1985, 'The Battle for Chastity', in Aries and Bejin, 1985.
Foucault, M., 1990, *The History of Sexuality Vol. III: The Case of the Self*, London: Penguin.
Foucault, M., 1992 [1987], *The Use of Pleasure. The History of Sexuality*, Vol. II, Harmondsworth: Penguin.
Fox, D., 2002, 'What Came First, Bigger Brains or Lots of Sex', *New Scientist* (23 November), p. 24.
Fox, R., 1962, 'Sibling Incest', *British Journal of Sociology*, Vol. 13:1, pp. 28–50.
Fox, R., 1967, *Kinship and Marriage: An Anthropological Perspective*, Harmondsworth: Penguin Books.
Fox, R. (ed.), 1975, *New Directions in Anthropology: Biosocial Anthropology*, London: Malaby Press.
Fox, R., 1980, *The Red Lamp of Incest*, London: Hutchinson.
Freud, S., 1905, *Jokes and their Relationship to the Unconscious*, in Freud, 1974.
Freud, S., 1917, 'A Metapsychological Supplement to the Theory of Dreams', in Freud, 1974.
Freud, S., 1921, *Group Psychology and the Analysis of the Ego*, in Freud, 1974.
Freud, S., 1936, *The Problem of Anxiety*, New York: Norton.
Freud, S., 1952, *Totem and Taboo*, New York: Norton.
Freud, S., 1957, *Civilization and Its Discontents*, London: The Hogarth Press.
Freud, S., 1961 (1927), *The Future of an Illusion*, New York: Norton.
Freud, S., 1962 [1905], *Three Essays on the Theory of Sexuality*, New York: Basic Books.
Freud, S., 1971 [1916–1917], *Complete Introductory Lecturers on Psychoanalysis*, London: George Allen and Unwin.
Freud, S., 1973, *Introductory Lectures on Psycho-Analysis*, James Strachey and Angela Richards (eds.), Harmondsworth: Pelican.
Freud, S., 1974 [1953–1974], *The Standard Edition of the Complete Works of Sigmund Freud* (trans. James Strachey), London: Hogarth.
Freud, S., 1976 [1930], *The Interpretation of Dreams* (trans. James Strachey), London: Penguin.
Friday, N., 1976, *My Secret Garden: Women's Sexual Fantasies*, London: Arrow Books.
Friday, N., 1980, *Men in Love*, London: Arrow Books.
Friedlander, N. and Jordan, D., 1995, 'Obstetric implications of Neanderthal robusticity and bone density', *Human Evolution*, (Florence) Vol. 9, pp. 331–342.
Fry, P., 1976, *Spirits of Protest: Spirit-Mediums and the Articulation of Consensus among the Zezuru of Southern Rhodesia*, Cambridge: Cambridge University Press.
Furuhashi, Y., 2005, 'Sexism and Science: The History of Female Orgasm', *Seven Oaks* (14 June), www.sevenoaksmag.com/features/66_feat1.htmlCachedSimilar

Galbraith, J., 1975, *Economics, Peace and Laughter*, Harmondsworth: Penguin.
Galdikas, B., 1979, 'Orangutan adaptation at Tanjung Puting Reserve: mating and ecology', in Hamburg, and McCown, 1979.
Galdikas, B., 1995, 'Social and Reproductive Behavior of Wild Adolescent Female Orangutans', in Nadler, *et al.*, 1995.
Galik *et al.*, 2004, 'External and Internal Morphology of the BAR 1002'00 Orrorin Tugenensis Femur', *Science*, Vol. 305: 5689, pp. 1450–1453.
Gallup, G., Boren, J., Gagliardi, G. and Wallnau, L., 1977, 'A Mirror for the Mind of Man, or Will the Chimpanzee Create an Identity Crisis for *Homo Sapiens*', *Journal of Human Evolution*, Vol. 6:3, pp. 303–313.
Gay, P., 1989, *Freud: A Life for Our Times*, London: Papermac.
Genetic Home Reference, 2008, '5-Alpha Reductase Deficiency', http://ghr.nlm.nih.gov/about
Gerth, H. and Mills, C. W. (eds.), 1958, *From Max Weber: Essays in Sociology*, New York: Oxford University Press.
Gibson, K. and Ingold, T. (eds.), 1993, *Tools, Language and Cognition in Human Evolution*, Cambridge: Cambridge University Press.
Gilbert, P., 1984, *Depression: From Psychology to Brain State*, London: Lawrence Erlbaum.
Gilbert, P., 1989, *Human Nature and Suffering*, Hove and London: Lawrence Erlbaum.
Giles, J., 1984, 'Naked love: The evolution of Human Hairlessness', *Biological Theory: Integrating Development, Evolution, and Cognition*, Vol. 5, pp. 326–336.
Glass, D. and Eversley, D. (eds.), 1965, *Populations in History*, London: Edward Arnold.
Glover, J., 1999, *Humanity: A Moral History of the Twentieth Century*, London: Jonathan Cape.
Gluckman, M., 1963, *Order and Rebellion in Tribal Africa*, London: Cohen and West.
Goffman, E., 1971, *The Presentation of Self in Everyday Life*, Harmondsworth: Penguin.
Gonyo, B., 2006, 'The Forbidden Love', www.truthseekersinadaption.com
Goodall, J., 1964, 'Tool-using and aimed throwing in a community of free- living chimpanzees', *Nature*, Vol. 201, pp. 1264–1266.
Goodall, J., 1968, 'The Behaviour of Free-Living Chimpanzees in the Gombe Stream Reserve', *Animal Behavior Monographs*, Vol 1:3, pp. 163–311.
Goodall, J., 1986, *The Chimpanzees of Gombe*, Cambridge, MA: Belknap.
Goodall, J., 1986a, 'Social rejection, exclusion, and shunning among the Gombe chimpanzees', *Ethnology and Sociobiology*, Vol. 7: 3–4, pp. 227–236.
Goodall, J., *et al.*, 1979, 'Intercommunity interactions in the chimpanzee population of the Gombe National Park', in Hamburg and McCown, 1979.
Goodall [Lawick-Goodall], J., 1977, 'Infant Killing and Cannibalism in Free-Living Chimpanzees', *Folio Primatologica*, Vol. 28, pp. 259–282.
Goodman, M., 1963, 'Man's Place in the Phylogeny of the Primates as Reflected by Serum Proteins', in Washburn, 1963.
Goody, J., 1983, *The Development of the Family and Marriage in Europe*, Cambridge: Cambridge University Press.
Gould, S., 1977, *Ontogeny and Phylogeny*, Cambridge, MA: Harvard University Press.
Goy, R., and McEwen, B. (eds.), 1980, *Sexual Differentiation of the Brain*, Cambridge, MA: M.I.T. Press.
Grandin, T. (and Catherine Johnson), 2005, *Animals in Transition: Using the Mysteries of Autism to Decode Animal Behaviour*, London: Bloomsbury.
Graves, R., 1960, *The Greek Myths*, Vols. I and II., Harmondsworth: Penguin.
Gray, J., 2007, 'On Top of the World?', *The Guardian Review* (20 January), p. 8.

Gray, P. and Garcia, J., 2013, *Evolution and Human Sexual Behavior*, Cambridge, MA: Harvard University Press.

Green, V., 1971, *Medieval Civilization in Western Europe*, London: Edward Arnold.

Greenberg, G. and Tobach, E., 1988, *Evolution and Social Behavior and Integrative Levels*, Hillsdale, NJ: Lawrence Erlbaum.

Greenwald, E. and Leitenberg, H., 1989, 'Long-Term Effects of Sexual Experiences with Siblings and Nonsiblings during Childhood', *Archives of Sexual Behavior*, Vol. 18:1, pp. 389–399.

Grene, D. and Lattimore, R. (eds.), 1953, *Aeschylus I: Oresteia – Agamemnon, The Libation Bearers, and The Eumenides*, Chicago: University of Chicago Press.

Gribbin, J., 2002, *Science: A History 1543–2001*, London: Allen Lane.

Gribbin, J. and Gribbin, M., 1997, *Richard Feynman: A Life in Science*, London: Viking.

Griffin, D., 1999, 'Late Miocene Climate of Northeastern Africa: Unraveling the Signals of the Sedimentary Succession', *Journal of the Geological Society*, (July), pp. 817–826.

Griffiths, N., 1976, *Penelope's Web: Some Perceptions of Women in European and Canadian Society*, Toronto: Oxford University Press.

Gusfield, J., 1963, *Symbolic Crusade: Status Politics and the American Temperance Movement*, Glencoe: University of Illinois Press.

Haile-Selassie, et al., 2012, 'A New Hominin Foot from Ethiopia Shows Multiple Pliocene Bipedal Adaptations', *Nature*, Vol. 483: 7391, pp. 565–569, DOI:10.1038/nature10922.

Hajnal, J., 1965, 'European Marriage Patterns in Perspective', in Glass and Eversley, 1965.

Halberstam, D., 1969, *The Best and the Brightest*, Greenwich, CT: A Fawcett Crest Book.

Hall, J., 1986, *Powers and Liberties*, Harmondsworth: Penguin.

Hall, K., 1963, 'Some Problems in the Analysis and Comparison of Monkey and Ape Behavior', in Washburn, 1963.

Hallberg, K., Nelson, D. and Boysen, S., 2003, 'Representational Vocal Signaling in the Chimpanzee', in de Waal and Tack, 2003.

Halperin, M., 1974, *Bureaucratic Politics and Foreign Policy*, Washington, D C: The Brookings Institute.

Hamburg, B. and McCown, E. (eds.), 1979, *The Great Apes*, Menlo Park, CA: Benjamin/Cummings.

Hamilton, M., 2012, 'The Concepts of Implicit and Non-Institutional Religion', *I R (Implicit Religion)*, Vol. 15:4, pp. 511–520, doi:10.1558/imre.v15i4.511.520.

Hamilton, W., 1975, 'Innate Social Aptitudes of Man: An Approach from Evolutionary Genetics', in Fox, 1975.

Hamilton, W., 2001, *Narrow Roads of Gene Land*, Vol. 2, Oxford: Oxford University Press.

Hankiss, E., 2001, *Fears and Symbols*, Budapest: Central European University Press.

Hansell, et al., 2004, 'Myosin Gene Mutation Correlates with Anatomical Changes in the Human Lineage', *Nature*, Vol. 428, pp. 415–418.

Harari, N., 2014, *Sapiens: A Brief History of Humankind*, Amazon Media: Vintage Digital.

Haraway, D., 1989, *Primate Visions: Gender, Race, and Nature in the World of Modern Science*, London: Routledge.

Harcourt A. and de Waal, F., 1992, *Coalitions and Alliances in Humans and other Animals.*, New York: Oxford University Press.

Harding, S. and Teleki, G. (eds.), 1981, *Omnivorous Primates: Gathering and Hunting in Human Evolution*, New York: Columbia University Press.

Harman, L., 1988, *The Modern Stranger: On Language and Membership*, Berlin: de Gruyter.

Harris, H., 1975, *The Principles of Human Biochemical Genetics*, New York: Elsevier Pub.

Harris, M., 1968, *The Rise of Anthropological Theory*, London: Routledge and Kegan Paul.

Harris, M., 1977, *Cows, Pigs, Wars and Witches*, Glasgow: Fontana.

Harris, M., 1978, *Cannibals and Kings*, Glasgow: Fontana.

Hashimoto, C. and Furuichi T., 1994, 'Social Role and Development of Non Copulatory Sexual Behavior of Wild Bonobos', in Wrangham, *et al.*, 1994.

Hauser, M., 1997, 'Minding the Behaviour of Deception', in Whiten and Byrne, 1997.

Hauser, M., 2007, *Moral Minds: How Nature Designed Our Universal Sense of Right and Wrong*, London: Little Brown.

Hawking, S., 1995, *A Brief History of Time*, London: Bantam Books.

Hawley, J., 2004, 'Leg Count', *The Guardian Review* (22 May), p. 9.

Hay, R., 1976, 'Environmental Setting of Hominid Activities in Bed I, Olduvai Gorge', in Isaac and McCown, 1976.

Hays, H., 1966, *The Dangerous Sex: The Myth of Feminine Evil*, London: Methuen.

Hays, J., 2011, 'Health Care and Health Problems in the Developing World', Facts and Details, http://factsanddetails.com/world.php?itemid=1200#

Hays, R. and. Oxley, D., 1986, 'Social Network Development and Functioning during a Lifetime Transition', *Journal of Personality and Social Psychology*, Vol. 50, pp. 305–313.

Hegel, G., 1807 [1977], *Phenomenology of Spirit* (trans. A. V. Miller), Oxford: Clarendon Press.

Henderson, K., 2005, 'After Hours', *The Guardian Weekend* (23 July), pp. 42–43.

Hendrickson, K., 2012, 'Cesarean Sections in the U.S.: The Trouble with Assembling Evidence from Data', *Scientific American*, (28 March), http://blogs.scientificamerican.com/guest-blog

Henry, *et al.*, 2012, 'The Diet of Australopithecus Sediba', *Nature*, Vol. 487, pp. 90–93, doi:10.1038/ nature11185.

Herbenick, D., *et al.*, 2009, 'Prevalence and Characteristics of Vibrator Use by Women in the United States: Results from a Nationally Representative Study', *Journal of Sexual Medicine*, Vol. 6, pp. 1857–1866.

Hewlett, B. (ed.), 1992, *Father-Child Relations: Cultural and Biosocial Contexts*, New York: Walter de Gruyter.

Hill, A., 2003, 'Love at First Sight – For My Brother', *The Observer* (4 May), p. 15.

Hill, B., 2001, *Women Alone*, New Haven: Yale University Press.

Hill, J., 1984, 'Human Altruism and Sociocultural Fitness', *Journal of Sociological and Biological Structures*, Vol. 7, pp. 17–35.

Hill, K. and Kaplan, H., 1988, 'Trade offs in Male and Female Reproductive Strategies among the Ache', in Betzig, Borgerhoff and Turke, 1988.

Hobbes, T., 1968, *Leviathan* (C. Macpherson ed.), Harmondsworth: Penguin.

Hockett, C., and Ascher, R., 1964, 'The Human Revolution', *Current Anthropology*, Vol. 5: 3, pp. 135–166.

Holmes, A., *et al.*, 2003, 'TheGene Cassette Metagenome Is a Basic Resource for Bacterial Genome Evolution', *Environmental Microbiology*, Vol. 5:5, pp. 383–394.

Hooper, R., 2005, 'Genes Blamed for Fickle Female Orgasm', *New Scientist*, (11 June).

Horick, A. and Orians, G., 2012, 'Are We Born with a Sense of Fairness?' *Pacific Standard*, www.psmag.com/culture/

Horney, K., 1932 (1967), 'The Dread of Woman: Observations on a Specific Difference in the Dread Felt by Men and by Women for the Opposite Sex', *International Journal of Psycho-Analysis*, Vol. 13, pp. 348–360.

Hrdy, S., 1979, 'Infanticide among Animals: A Review, Classification and Examination of the Implications for the Reproductive Strategies of Females', *Ethology and Sociobiology*, Vol. I, pp. 13–40.

Hrdy, S., 1981, *The Woman That Never Evolved*, Cambridge, MA: Harvard University Press.

Hrdy, S., 1999, *Mother Nature: A History of Mothers, Infants and Natural Selection*, New York: Pantheon.

Hrdy, S., 2011, *Mothers and Others: The Evolutionary Origins of Mutual Understanding*, Cambridge, MA: Harvard University Press.

Huck, M. and Fernandez-Duque, E., 2012, 'When Dads Help: Male Behavioral Care during Primate Infant Development', in Clancy, Hinde and Rutherford, 2012.

Human Rights Watch, 2001, 'No Escape: Male Rape in U.S. Prisons', www.hrw.org/reports/2001/prison

Human Rights Watch, 2001a, 'Rape Crisis in U. S. Prisons', www.hrw.org/news/2001/04/18/rape-crisis-us-prisons

Hume, D., 1958, *A Treatise of Human Nature*, Oxford: Oxford University Press.

Hume, D., 1963, *Essays: Moral, Political and Literary*, Oxford: Oxford University Press.

Humphrey, N., 2012, 'This chimp Will Kick Your Ass at Memory Games – But How the Hell Does He Do It?', *Trends in Cognitive Science*, Vol. 16:7, pp. 353–355.

Hutchinson, G., 1959, 'Speculative Consideration of Certain Possible Forms of Sexual Selection in Man', *American Naturalist*, Vol. 93, pp. 81–91.

Ilyasova, A., 2006, 'Dykes on Bikes and the Regulation of Vulgarity', *International Journal Of Motorcycle Studies*, http://ijms.nova.edu/November2006/IJMS_Artcl.Ilyasova.html

Imperato-McGinley, J, *et al.*, 1974, '5alpha-Reductase Deficiency in Man: An Inherited form of Male Pseudohermaphroditism', *Science*, Vol. 186: 4170, pp. 1213–1215.

Ingold, T., Riches, D. and Woodburn, J. 1991, *Hunters and Gatherers (2 Vols.)*, Oxford: Berg.

Innocence Project, 2012, www.innocenceproject.org/Content/Facts_on_ PostConviction_DNA_Exonerations.php, Benjamin N. Cardozo School of Law, Yeshiva University.

Inoue, S. and Matsuzawa T., 2007, 'Working Memory of Numerals in Chimpanzees', *Current Biology*, Vol. 17:23, R1004-R1005.

Insideprison.com, 2006, 'Prison Rape: The Challenge of Prevention and Enforcement', www.insideprison.com/prison-rape.asp

INTJ Forum, 2013, 'Persistence in Courting ', http://intjforum.com/showthread.php?t=99111

Isaac, G., 1976, 'East Africa as a Source of Fossil Evidence for Human Evolution', in Isaac and McCown, 1976.

Isaac, G., 1976a, 'The Activities of Early African Hominids: A Review of Archaeological Evidence from the Time Span Two and a Half to One Million Years Ago', in Isaac and McCown, 1976.

Isaac, G., 1978, 'Food Sharing and Human Evolution: Archaeological Evidence from the Plio-Pleistocent of East Africa', *Journal Of Anthropological Research*, Vol. 34:3, pp. 311–325.

Isaac, G. and McCown, E. (eds.), 1976, *Human Origins: Louis Leakey and the East African Evidence*, Menlo Park, CA: W.A. Benjamin.

Isaac, G., Harris, W. and Crader, D., 1976, 'Archeological Evidence from the Koobi Fora Formation', in Coppens, *et al.*, 1976.

Isbell, L., 2006, 'Snakes as Agents of Evolutionary Change in Primate Brains', *Journal Of Human Evolution*, Vol. 51:1, pp. 1–35.

Isler, K. and Thorpe, S., 2003, 'Characteristics of Vertical Climbing in Wild and Captive Sumatran Orang-Utans (*Pongo abelii*)', *Journal of Experimental Biology*, Vol. 206, pp. 4081–4096.

Itzhaki, J., 2003, 'The FOXP2 Story', The Wellcome Trust, genome.wellcome.ac.uk/doc_wtd207979/

Izar, P., *et al.*, 2006, 'Cross-Genus Adoption of a Marmoset by Wild Capuchin Monkeys', *American Journal of Primatology*, Vol. 68, pp. 692–700.

Izard, C. (ed.), 1982, *Measuring Emotions in Infants and Children*, Cambridge: Cambridge University Press.

Izard, C. and Dougherty, L., 1982, 'Two Complementary Systems for Measuring Facial Expressions in Infants and Children', in Izard, 1982.

Jablonski, N., 2010, 'The Naked Truth: Why Humans Have No Fur', *Scientific American*, 20 January.

James, W., 1884, 'What Is Emotion', *Mind*, Vol. 9, pp. 188–205.

James, W., 1892, *Psychology: A Briefer Course*, London: Macmillan.

Janis, I., 1972, *Victims of Groupthink*, Cambridge, MA: Houghton Mifflin.

Jarvie, I., 1984, *Rationality and Relativism: In Search of a Philosophy and History of Anthropology*, London: Routledge and Kegan Paul.

Jencks, C., 1975, *Inequality: A Reassessment of the Effects of Family and Schooling in America*, Harmondsworth: Penguin.

Johanson, D. and Edey, M., 1981, *Lucy: The Beginnings of Humankind*, London: Granada.

Johns, C., 1904, *Babylonian and Assyrian Laws, Contracts and Letters*, New York: Charles Scribner's and Sons.

Johnson, A. and Earle, T., 2000, *The Evolution of Human Societies: From Foraging Group to Agrarian State* (2nd ed.), Redwood City, CA: Stanford University Press.

Johnson, A., Wadsworth, J., Wellings, K. and Field, J., 1994, *Sexual Attitudes and Life Styles*, Oxford: Blackwell.

Johnson, G., 2012, 'Why Straight Women and Bisexual Men May Not Really Exist', *Gay Star News*, 25 September.

Johnstone, R., 1995, 'Sexual Selection, Honest Advertisement and the Handicap Principle: Reviewing the Evidence', *Biological Reviews*, Vol. 70:1, pp. 1–65.

Jolly, C., 1970, 'The Seed Eaters: A New Model of Hominid Differentiation Based on a Baboon Analogy', *Man (NS)*, Vol. 5:1, pp. 1–26.

Jones, D., 1995, 'Sexual Selection, Physical Attractiveness, and Facial Neoteny: Cross-Cultural Evidence and Implications', *Current Anthropology*, Vol. 36:5, pp. 723–748.

Jones, D., 2000, 'Physical Attractiveness, Race and Somatic Prejudice in Bahia Brazil', in Cronk, Chagnon and Irons, 2000.

Jones, S., 2002, *Y: The Descent of Man*, London: Little Brown.

Jones, S., 2013, *The Serpent's Promise: The Bible as Science*, London: Hachette Digital.

Jones, S., Martin, R. and Pilbeam, D. (eds.), 1992, *The Cambridge Encyclopedia of Human Evolution*, Cambridge: Cambridge University Press.

Joravasky, D., 1971, 'A Hero in the USSR', *New York Review of Books* (23 September).

Jordan, D., 2013, 'Human Birth & Bipedalism: An Overview for College Students', http://anthro.ucsd.edu/~dkjordanresources/clarifications/HumanBirth.html

Kahneman, D., 2011, *Thinking, Fast and Slow*, New York: Farrar, Straus and Giroux.

Kant, I., 1974, *Groundwork for the Metaphysic of Morals* (trans. H. J. Paton), London: Hutchinson.

Kaplan, H. and Hill, K., 1985, 'Hunting Ability and Reproductive Success among Male Ache Foragers', *Current Anthropology*, Vol. 26, pp. 131–133.

Kaplan, M., 2002, 'Let the River Run', *New Scientist*, (28 September), pp. 32–35.

Kapuscinski, R., 2008, *Travels with Herodotus*, London: Penguin.

Kauth, M., 2000, *True Nature: A Theory of Sexual Attraction*, New York: Kluwer Academic/Plenum Pub.

Keeley, L., 1996, *War Before Civilization*, New York Oxford University Press.

Keith, A., 1931, *New Discoveries Relating to the Antiquity of Man*, New York: Norton.
Kelly, J., 1982, 'Early Feminist Theory and the *Querelle des Femmes, 1400–1789*', *Signs*, Vol. 8:1, pp. 4–28.
Kennedy, J., 1962, *To Turn the Tide* (J. Gardner ed.), New York: Harper Brothers.
Kenny, T. and Knott, L., 2012, 'congenital adrenal hyperplasia', patient.co.uk/health/congenital-adrenal-hyperplasia
Kerrigan, J., 1996, *Revenge Tragedies: Aeschylus to Armageddon*, Oxford: Oxford University Press.
Khamsi, R., 2007, 'Envious Monkeys Can Spot a Fair Deal ', *New Scientist*, Vol. 15.
Khrushchev, N., 1977, *Khrushchev Remembers* (Trans. and ed. by S. Talbot), London: Sphere Books.
Kieckhefer, R., 1976, *European Witch Trials*, London: Routledge and Kegan Paul.
Kinsey, A., Pomeroy, W. and Martin, C., 1948, *Sexual Behavior in the Human Male*, Philadelphia: W.B. Saunders.
Kinsey, A., Pomeroy, W. and Martin, C., 1953, *Sexual Behavior in the Human Female*, Philadelphia: W.B. Saunders.
Kirkpatrick, M. (ed.), 1982, *Women's Sexual Experience*, New York: Plenum.
Klein, M., 1952, *Some Theoretical Conclusions Regarding the Emotional Life of Infants*, London: Hogarth.
Kloor, K., 2002, 'Powder Keg', *Audubon Mag.*, http://archive.audubonmagazine.org/features0212/dispatch.html
Kluckhohn, C. and Leighton, D., 1946, *The Navaho*, Boston: Beacon Press.
Knoles, G. and Snyder, R., 1954, *Readings in Western Civilization*, New York: J.B. Lippincott.
Kohn, M. 2004, *A Reason for Everything*, London: Faber and Faber.
Kondo, S., Kawai, M. and Ehara, A. (eds.), 1975, Contemporary Primatology: Fifth International Congress of Primatology, Nagoya, Basel: Karger.
Koops, K., McGrew, W., Matsuzawa, T. and Knapp, L., 2012, 'Terrestrial nest-building by wild chimpanzees (Pan troglodytes): Implications for the tree-to-ground sleep transition in early hominins', *American Journal of Physical Anthropology*, DOI:10.1002/ajpa.22056.
Kortlandt, A., 1985, 'Comment on Quiatt and Kelso', *Current Anthropology*, Vol. 26:2, pp. 207–222.
Kramer, H. and Sprenger, J., 1928, [1486], *Malleus Maleficarum (trans. Montague Summers)*, London: Arrow Books Ltd. (1971).
Kroeber, A., 1917, 'The Superorganic', *American Anthropologist*, Vol. 21, pp. 235–263.
Krogman, W., 1951, 'The Scars of Evolution', *Scientific American*, Vol. 185:6 (December), pp. 54–57.
Kuchment, A., 2004, 'The More Social Sex', Newsweek International, (17 May), pp. 56–57.
Kuper, A., 1983, *Anthropology and Anthropologists: The Modern British School*, London: Routledge and Kegan Paul.
La Freniere, P., 1988, 'The Ontogeny of Tactical Deception in Humans', in Byrne and Whitten, 1988.
Lafont, H., 1985, 'Changing Sexual Behaviour in French Youth Gangs', in Aries and Bejin, 1985.
Laing, R., 1965, *The Divided Self*, Harmondsworth: Penguin.
Laing, R., 1971, *Self and Other*, Harmondsworth: Penguin.
Lapper, A., 2005, 'Beauty Unseen, Unsung', *The Guardian Weekend* (September), pp. 12–22.
Larson, R., 1997, *Red Cloud: Warrior-Statesman of the Lakota Sioux*, Norman: University of Oklahoma Press.
Laslett, P., 1983, *The World We Have Lost*, London: Methuen.

Lawick-Goodall, J., 1971, *In the Shadow of Man*, London: Collins.
Leach, E., 1966, 'Liberty, Equality, Fraternity', *New Statesman* (8 July).
Leakey, M. D. *et al.*, 1976, 'Fossil hominids from the Laetoli Bed, Tanzania', *Nature*, Vol. 262, pp. 460–465.
Leakey, M. G. and Walker, A., 1997, 'Early Hominid Fossils from Africa', *Scientific American*, Vol. 276:6 (June), pp. 60–65.
Leakey, M. G., Feibel, C., McDougall, I. and Walker, A., 1995, 'New 4-Million Year-Old Hominid Species from Kanapoi and Allia Bay, Kenya', *Nature*, Vol. 376:6541, pp. 565–571.
Leakey, R., 1981, *The Making of Mankind*, London: Book Club Associates.
Leakey, R. and Lewin, R., 1979, *Origins*, New York: Dutton.
Leakey, R. and Lewin, R., 1979a, *People of the Lake*, London: Collins.
Leakey, R. and Lewin, R., 1992, *Origins Reconsidered*, New York: Doubleday.
Leblanc, M., *et al.*, 2005, 'Evidence for Megalake Chad, North-Central Africa, during the late Quaternary from Satellite Data', www.elsevier.com/locate/palaeo/
LeDoux, J., 1998, *The Emotional Brain*, London: Weidenfeld and Nicolson.
Lee, E., 2014, 'Flirting 201: More than Meets the Eye', eHarmony: Get Matches, www.eharmony.com/dating-advice/
Lee, R., 1968, 'What Hunters Do for a Living, or, How to Make Out on Scarce Resources', in Lee and DeVore, 1968.
Lee, R. and DeVore, I., 1968, *Man the Hunter*, Chicago: Aldine Pub. Co.
Lehman, A., 2010, 'The Causes of Autism', www.originsofautism.com/
Leitenberg, H. and Henning, K., 1995, 'Sexual Fantasy', *Psychological Bulletin*, Vol. 117:3, pp. 469–496.
Leland, J., 2000, 'The Science of Women & Sex', *Newsweek International*, (5 June), pp. 66–78.
Lepore, S. and Greenberg, M., 2002, 'Mending Broken Hearts: Effects of Expressive Writing on Mood, Cognitive Processing, Social Adjustment, and Health Following a Relationship Breakup', *Psychology and Health*, Vol. 17, pp. 547–560.
Lerner, M., 1997, 'What Does the Belief in a Just World Protect Us from: The Dread of Death or the Fear of Undeserved Suffering?', *Psychological Inquiry*, Vol. 8, pp. 29–32.
Leroi, A., 2004, *Mutants: On the Form, Varities and Errors of the Human Body*, London: HarperCollins.
Leslie, I., 2011, *Born Liars: Why We Can't Live without Deceit*, London: Quercus.
Levi-Strauss, C., 1963, *Totemism*, Boston: The Beacon Press.
Levi-Strauss, C., 1966, *The Savage Mind*, London: Weidenfeld and Nicolson.
Levi-Strauss, C., 1969, *The Elementary Structures of Kinship*, Boston: Beacon Press.
Levy, O. (ed.), *The Complete Works of Friedrich Nietzsche*, New York: Russell and Russell.
Lewin, R., 1995, 'Little Foot Stumbles into the Crossfire', *New Scientist*, Vol 147, p. 14.
Lewin, R., 2003, 'When, Where and How?', *New Scientist*, (4 June), p. 5.
Lewis, J., 1998, *The West: The Making of the American West*, London: Robinson Publishing.
Lim, M. and Young, L., 2006, 'Neuropeptidergic Regulation of Affiliative Behavior and Social Bonding in Animals', *Hormones and Behavior*, Vol. 50, pp. 506–517.
Lippa, R., 2006, 'Is High Sex Drive Associated with Increased Sexual Attraction to both Sexes? It Depends on Whether You Are Male or Female', *Psychological Science*, Vol. 17, pp. 46–52.
Lippa, R., 2007, 'The Relation Between Sex Drive and Sexual Attraction to Men and Women', *Archives of Sexual Behavior*, Vol. 36: 2 pp. 209–222.
Lively, S., 2005, 'Homosexuality and the Nazi Party', Life and Liberty, www.lifeandlibertyministries.com/archives/000278.php

Lively, S. and Abrams, K., 2013, *The Pink Swastika: Homosexuality in the Nazi Party*, (5th Internet Edition): www.thepinkswastika.com/5201.html

Lloyd, E., 2005, *The Case of the Female Orgasm: Bias in the Science of Evolution*, Cambridge, MA: Harvard University Press.

Locke, J., 1980, *Second Treatise of Government*, Cambridge: Hackett Publishing Company.

Lodish,. H., *et al.*, 2003, *Molecular Cell Biology*, New York: W.H. Freeman.

Lopreato, J., 1984, *Human Nature and Biocultural Evolution*, London: George Allen and Unwin.

Lordkipanidze, D., *et al.*, 2013, 'A Complete Skull from Dmanisi, Georgia, and the Evolutionary Biology of Early Homo', *Science*, Vol. 342: 6156, pp. 326–331.

Lorenz, K., 1971, *Studies in Animal and Human Behavior, Vol. II*, Cambridge, MA: Harvard University Press.

Lovejoy, O., 1980, 'Hominid Origins: The Role of Bipedalism', *American Journal of Physical Anthropology*, Vol. 52 (February): pp. 397–398.

Lovejoy, O., 1981, 'The Origins of Man', *Science*, Vol. 211, pp. 341–50.

Lovejoy, O., 1981a, 'Is It a Matter of Sex', chapter 16 in Johanson and Edey, 1981.

Lovejoy, O., 1988, 'Evolution of Human Walking', *Scientific American* (November), pp. 118–125.

Lovejoy, O., 2009, Reexamining Human Origins in Light of *Ardipithecus ramidus*, *Science*, Vol. 326: 5949, pp. 74, 74e1–74e8.

Lowie, R., 1924, *Primitive Religion*, New York: Horace Liveright, Inc.

Lynn, R., 2012, 'An Examination of Rushton's Theory of Differences in Penis Length and Circumference and r-K Life History Theory in 113 Populations', doi:10.1016/j. paid.2012.02.016.

Maccoby, E. and Jacklin, C., 1974, *The Psychology of Sex Differences*, Redwood City, CA: Stanford University Press.

MacDonald, K., 1995, 'The Establishment and Maintenance of Socially Imposed Monogamy in Western Europe', *Politics and the Life Sciences*, Vol. 14, pp. 3–23.

Macfarlane, A., 1970, *Witchcraft in Tudor and Stuart England*, London: Routledge and Kegan Paul.

Machiavelli, N., 1950, *The Prince and the Discourses*, New York: Random House.

MacLean, P., 1978, 'The Evolution of Three Mentalities', in Washburn and McCown, 1978.

Maestripieri, D., 2001, 'Biological Bases of Maternal Attachment', *Current Directions in Psychological Science*, Vol. 10, pp. 79–83.

Mail Online, 2009, 'Twin Town', (15 July), www.dailymail.co.uk/home/index.html

Mail Online, 2014, Inside the Brazilian town known as 'Twin Land', www.dailymail.co.uk/news/article-2604223/Inside-Brazils-Twin-Land

Mair, L., 1962, *Primitive Government*, Harmondsworth: Penguin Books.

Mair, L., 1974, *African Societies*, Cambridge: Cambridge University Press.

Malleus, 1486, *Malleus Maleficarum*, in Kramer and Sprenger, 1928.

Mao, 1966, *Quotations From Chairman Mao Tse-Tung*, Peking: Foreign Language Press.

Marks, A. (ed.), 2010, *Nietzsche, Ultimate Collection*, Houston, TX: Everlasting Flames Pub.

Marks, L., 2004, '5α-Reductase: History and Clinical Importance', *Reviews in Urology*, Vol. 6 (Suppl 9), S11–S21.

Marlowe, F., Apicella, C. and Reed, D., 2005, 'Men's Preferences for Women's Profile Waist-Hip-Ratio in Two Societies', *Evolution and Human Behavior*, Vol. 26, pp. 458–468.

Marshall, S., 1947, *Men Against Fire*, New York: William Morrow.

Martínez-Abadías, N., 2011, 'Pervasive Genetic Integration Directs the Evolution of Human Skull Shape', *Evolution*, DOI:10.1111/j.1558-5646.2011.01496.x.

Marwick, M., 1965, *Sorcery in Its Social Setting: A Study of the Northern Rhodesian Cewa*, Manchester: Manchester University Press.

Marx, K. and Engels, F., 2011 [1939], *The German Ideology*, Eastford, CT.:Martino Publishing.
Mason, W., 1976, 'Environmental Models and Mental Modes: Representational Processes in the Great Apes and Man', *American Psychologist*, Vol. 31, pp. 284–294.
Maynard Smith, J., 1971, 'What Use Is Sex?', *Journal of Theoretical Biology*, Vol. 30: 2, pp. 319–335.
Maynard-Smith, J., 1978, 'Evolution and the Theory of Games', in Clutton-Brock and Harvey, 1978.
McCannell, K., 1988, 'Social Networks and the Transition to Motherhood', in Milardo, 1988.
McClelland, J., 1996, *A History of Western Political Thought*, London: Routledge.
McGinnis, P., 1979, 'Sexual Behavior in Free-Living Chimpanzees: Consort Relationships', in Hamburg and McCown, 1979.
McGrew, W., 1979, 'Evolutionary implications of Sex Differences in Chimpanzee Predation and Tool Use', in Hamburg and McCown, 1979.
McGrew, W., 1981, 'The Female Chimpanzee as Human Evolutionary Prototype', in Dahlberg, 1981.
McGrew, W., 1992, *Chimpanzee Material Culture: Implications for Human Evolution*, Cambridge: Cambridge University Press.
McKeganey, N. and Barnard, M., 1996, *Sex Work on the Streets: Prostitutes and their Clients*, Buckingham: Open University Press.
McKinney, M. and McNamara, K., 1990, *Heterochrony*, New York: Plenum Press.
McKnight, J., 1997, *Straight Science: Homosexuality, Evolution and Adaptation*, London: Routledge.
McNamara, K., 1999, 'Embryos and Evolution', *New Scientist*, Inside Science 124 (16 October), pp. 1–4.
Mead, G., 1962, *Mind, Self and Society: From the Standpoint of a Social Behaviorist*, Chicago: University of Chicago Press.
Mead, M., 1942, *Growing Up in New Guinea*, Harmondsworth: Penguin.
Medawar, P., 1969, *The Art of the Soluble*, Harmondsworth: Penguin.
Medvedev, Z. and Medvedev, R., 1971, *A Question of Madness*, New York: Knopf.
Mee, C., 1971, *Gorbachev*, Oxford: Basil Blackwell.
Meikle, J., 2003, 'New Hi-Tech Scans Show Babies Smiling and Crying Before Birth', *The Guardian* (13 September), p. 5.
Mellen, S., 1981, *The Evolution of Love*, San Francisco: W.H. Freeman.
Michels, R., 1962, *Political Parties*, New York: The Free Press.
Midelfort, H., 1972, *Witch Hunting in Southwestern Germany*, Palo Alto: Stanford University Press.
Milardo, R. (ed.), 1988, *Families and Social Networks*, Newbury Park: Sage.
Miller, G., 2000, *The Mating Mind: How Sexual Choice Shaped the Evolution of Human Nature*, London: William Heinemann.
Miller, G., 2010, 'Ulterior Altruists', *New Scientist*, (19/26 December – 2 January), pp. 70–73.
Milton, K., 1988, 'Foraging Behaviour and the Evolution of Primate Intelligence', in Byrne and Whitten, 1988.
Mindlin, B., 2002, *Barbecued Husbands*, London: Verso.
Mitchell, R., 1975, *Depression*, Harmondsworth: Penguin.
Mithen, S. and Maschner, H. (eds.), 1992, *Darwinian Approaches to the Past*, New York: Plenum.
Mithen, S., 2005, *The Singing Neanderthals*, London: Weidenfeld and Nicolson.
Monter, W., 1972, 'The Historiography of European Witchcraft', *Journal of Interdisciplinary History*, Vol. II: 4, pp. 435–451.

Monter, W., 1976, *Witchcraft in France and Switzerland*, London: Cornell University Press.
Moore, J. and Hagedorn, J., 2001, 'Female Gangs: A Focus on Research', Office of Juvenile Justice and Delinquency Prevention, U.S. Department of Justice.
Moore, S., 1994, 'What's Up Doc?', *The Guardian Weekend* (12 February), pp. 16–17.
Moran, L., 2009, 'Richard Dawkins on "Purpose"' (March 10), http://sandwalk.blogspot.co.uk/2009/03
Moran, M., 2006, 'Modern Military Force Structures', Council on Foreign Relations, www.cfr.org/defensehomeland-security
Morgan, E., 1985, *The Descent of Women*, London: Souvenir.
Morgenthau, H., 1946, *Scientific Man versus Power Politics*, Chicago: University of Chicago Press.
Morganthau, H., 1978, *Politics among Nations*, New York: Alfred A. Knopf.
Morris, B., 1991, *Western Conceptions of the Individual*, Oxford: Berg.
Morris, D., 1968, *The Naked Ape*, London: Corgi Books.
Most, W., 2013, 'St. Augustine on Grace and Predestination', www.ewtn.com/library/theology/augustin.htm
MSC, 2014, *Marine Stewardship Council*, www.msc.org/healthy-oceans/the-oceans-today/fish-as-food
Muller, M., *et al.*, 2006, 'Male Chimpanzees Prefer Mating with Old Females', *Current Biology*, Vol. 16, pp. 2234–2238.
Murdock, G., 1957, 'World Ethnographic Sample', *American Anthropologist*, Vol. 59, pp. 664–687.
Murdock, G., 1960, 'The Universality of the Nuclear Family', in Bell and Voget, 1960.
Murdock, G., 1968, 'The Common Denominator of Culture', in Washburn and Jay, 1968.
Murphy, R., 1987, *The Body Silent*, London: Phoenix House.
Myers, P., 2004, 'A Brief Overview of HOX Genes', http://scienceblogs.com/pharyngula
Nachman, M. and Crowell, S., 2000, 'Estimate of the Mutation Rate per Nucleotide in Humans', *Genetics*, Vol. 156: 1, pp. 297–304.
Nadler, R, *et al.*, (eds.), 1995, *The Neglected Ape*, New York: Plenum Press.
Nakamichi, M., Silldorff, A., Bingham, C. and Sexton, P., 2007, 'Baby-transfer and other Interactions between its Mother and Grandmother in a Captive Social Group of Lowland Gorillas', *Primates*, Vol. 45, pp. 73–77.
Nasaw, D., 2007, *Andrew Carnegie*, New York: Penguin Books.
Nekarasov, V., 1975, 'Labour Camps and Asylums: What Hope for My Friends?', *The Observer Supplement* (20 July), p. 33.
Nelson, N., 1987, '"Selling her Kiosk": Kikuyu Notions of Sexuality and Sex for Sale in Mathare Valley, Kenya', in Caplan, 1987.
Nelson, S., 1987, *Incest: Fact and Myth*, Edinburgh: Stramullion.
Neubauer, S., Gunz, P. and Hublin, J., 2009, 'The Pattern of Endocranial Ontogenetic Shape Changes in Humans', *Journal of Anatomy*, Vol. 215: 3, pp. 240–255.
NHS, 2013, 'Symptoms of Autism and Asperger Syndrome', www.nhs.uk/Conditions/Autistic-spectrum-disorder/
Nietzsche, F., 1886, *Human, All Too Human*, in Levy, 1964.
Nietzsche, F., 1961, *Thus Spoke Zarathustra*, Harmondsworth: Penguin.
Nietzsche, F., 1968, *The Will To Power* (trans. W. Kaufmann and R. Hollingdale), New York: Vintage Books.
Nietzsche, F., 1974, *The Gay Science* (trans. Walter Kaufmann), New York: Random House.

Nietzsche, F., 1990, *Beyond Good and Evil*, Harmondsworth: Penguin.
Nietzsche, F., 2010, *Nietzsche, Ultimate Collection*, in Marks, 2010.
Nisbett, R. and Cohen, D., 1996, *The Culture of Honor: The Psychology of Violence*, Boulder CO: Westview.
Nishida, T., 1973, 'The Ant-Gathering Behaviour by the Use of Tools among Wild Chimpanzees of the Mahali Mountains', *Journal of Human Evolution*, Vol. 2, pp. 357–370.
Nishida, T., 1979, 'Social Structure among Wild Chimpanzees of the Mahale Mountains', in Hamburg and McCown, 1979.
Nishida, T., 2003, 'Individuality and Flexibility of Cultural Behavior in Chimpanzees', in de Waal and Tyack, 2003.
Noriuchi, M., Kikuchi, Y. and Senoo, A., 2008, 'The Functional Neuroanatomy of Maternal Love: Mother's Response of Infant's Attachment Behaviors', *Biological Psychiatry*, Vol. 63, pp. 415–423.
Notestein, W., 1955, 'The English Woman 1580–1650', in Plumb, 1955.
Nove, A., 1972, *An Economic History of the U.S.S.R.*, Harmondsworth: Penguin.
Nove, A., 1975, *Stalinism and After*, London: George, Allen and Unwin.
Nyholm, C., 2009, 'Courtship and Dating Valuable Time', Suite, http://suite101.com/a/courtship-and-dating-valuable-time-a92582
Oakley, K., 1958, *Man the Tool Maker*, London: British Museum.
Oakley, K., 1961, 'On Man's Use of Fire with Comments on Tool-Making and Hunting', in Washburn, 1961.
OECD, 2011, 'Health at a Glance 2011; OECD Indicators', OECD iLibrary, www.oecd-ilibrary.org/sites/health_glance-2011
O'Faolain, J. and Martines. L. (eds.), 1973, *Not in God's Image*, London: Virago.
O'Flaherty, W., 1976, *The Origins of Evil in Hindu Mythology*, Berkeley: University of California Press.
O'Grady, S., 2012, *And Man Created God*, London: Atlantic Books.
Okami, P., Olmstead, R. and Abramson, P., 1997, 'Sexual Experiences in Early Childhood: 18-year Data from the UCLA Family Life-Styles Project', *Journal of Sexual Research*, Vol. 34, pp. 339–347.
Olivieri, A., 1985, 'Eroticism and Social Groups in Sixteenth-Century Venice: the Courtesan', in Aries and Bejin, 1985.
O'Neil, D., 2006, Primate Behavior: A Survey of Non-Human Primate Behavior, in 'Social Structure', http://anthro.palomar.edu/behavior/default.htm
Otto, R., 1945, *The Idea of the Holy: An Inquiry into the Non-Rational Factors in the Idea of the Divine and its Relation to the Rational*, reprinted, 1980, New York: Oxford University Press.
Overing, 1985, 'There Is No End of Evil: The Guilty Innocents and Their Fallible God', in Parkin, 1985.
Palsson, G., 1991, 'Hunters and Gatherers of the Sea', in Ingold, Riches and Woodburn, 1991.
Parkin, D. (ed.), 1985, *The Anthropology of Evil*, Oxford: Basil Blackwell.
Parkin, D., 1985a, 'Introduction', in Parkin, 1985.
Parkin, D., 1985b, 'Entitling Evil: Muslims and Non-Muslims in Coastal Kenya', in Parkin, 1985.
Parr, L., 2003, 'Emotional Recognition by Chimpanzees', in de Waal and Tack, 2003.
Parsell, D., 2002, 'Skull Fossil Opens Window into Early Period of Human Origins', *National Geographic News* (11 July).
Parsons, T., 1937, *The Structure of Social Action*, Glencoe: The Free Press.
Parsons, T., 1966, *Societies: Evolutionary and Comparative Perspectives*, Englewood Cliffs, NJ: Prentice-Hall.

Passell, P. and Taylor, J., 1977, 'The Deterrent Effect of Capital Punishment: Another View', *The American Economic Review*, Vol. 67: 3, pp. 445–451.

Patterson, F. and. Linden. E., 1981, *The Education of Koko*, New York: Renchart and Winston.

Pearce, F., 2003, 'Ghosts of the Great eel War', *New Scientist*, (10 March), pp. 58–59.

Penin, X., Berge, C. and Baylac, M., 2002, 'Ontogenetic Study of the Skull in Modern Humans and the Modern Chimpanzees: Neotenic Hypothesis Reconsidered with a Tridimensional Procrustes Analysis', *American Journal of Physical Anthropology*, Vol. 118, pp. 50–62.

Permack, A. and Permack, D., 1972, 'Teaching Language to an Ape', *Scientific American*, Vol. 227: 4, pp. 92–99.

Permack, D. and Permack, A., 1983, *The Mind of an Ape*, New York: Norton.

Perrett, D., *et al*., 1998, 'Effects of Sexual Dimorphism on Facial Attractiveness', *Nature*, Vol. 395, pp. 884–887.

Perry, S., Barrett, H. and Manson, J., 2004, 'White-Faced Capuchin Monkeys Show Triadic Awareness in Their Choice of Allies', *Animal Behaviour*, Vol. 67, pp. 165–170.

Phillips, H., 2003, 'Master Code', *New Scientist* (15 March), pp. 44–47.

Piattelli-Palmarini, M. (ed.), 1980, *Language and Learning: The Debate Between Jean Piaget and Noam Chomsky*, London: Routledge and Kegan Paul.

Pinker, S., 1995, *The Language Instinct*, London: Penguin.

Pinker, S., 1998, *How The Mind Works*, Harmondsworth: Allen Lane.

Pinker, S., 2002, *The Blank Slate*, London: Allen Lane.

Pinker, S., 2007, *The Stuff of Thought: Language as a Window into Human Nature*, London: Allen Lane.

Plant, R., 1991, *Modern Political Thought*, Oxford: Blackwell.

Plato, 1970, *The Laws*, London: Penguin.

Plato, 1974, *The Republic*, Harmondsworth: Penguin.

Plumb, J. (ed.), 1955, *Studies in Social History: A Tribute to G.M. Trevelyan*, London: Books for Libraries Press.

Pomiankowski, A., 2003, 'How Do Mutations Lead to Evolution?', *New Scientist*, Vol. 178: 2399, pp. 34–36.

Power, C., 2005, 'Not the Queen's English', *Newsweek International* (7 March), pp. 62–67.

Power, M., 1988, 'The Collective Foragers: Human and Chimpanzee', in Chance and Omark, 1988.

Price, J., 1988, 'Alternative Channels for Negotiating Asymmetry in Social Relationships', in Chance and Omark, 1988.

Proctor, D., *et al*., 2013, 'Chimpanzees Play the Ultimatum Game', *PNAS*, DOI:10.1073/pnas.1220806110.

Pusey, A., 1979, 'Intercommunity Transfer of Chimpanzees in Gombe National Park', in Hamburg and McCown, 1979.

Pusey, A., 2002, 'Of Genes and Apes: Chimpanzee Social Organization and Reproduction', in de Waal, 2002.

Quiatt, D. and Kelso, J., 1985, 'Household Economics and Hominid Origins', *Current Anthropology*, Vol. 26:2, pp. 207–222.

Rachman, S., 1978, *Fear and Courage*, San Francisco: W.H. Freeman.

Radcliffe-Brown, A. R., 1952, *Structure and Function in Primitive Societies*, London: Cohen and West.

Radcliffe-Brown, A. R. and Forde, D. (eds.), 1950, *African Systems of Kinship and Marriage*, London: Oxford University Press.

Radelet, M. and Lacock, T., 2009, 'Recent Developments: Do Executions Lower Homicide Rates?: The View of Leading Criminologists', *Journal of Criminal Law and Criminology*, Vol. 99: 2, pp. 489–508.

Raffaele, P., 2005, 'Out of Time', *Smithsonian Mag.* (April).

Rancourt, D., 1998, 'Establishment of Spatial Patterns of Gene Expression during Early Vertebrate Development: HOX Genes', http://ucalgary.ca/uofc/eduweb/virtualembryo/hox.hml

Randerson, J., 2003, 'Dawn of Human Race Uncovered', *New Scientist*, (11 June).

Randerson, J., 2003a, 'Iraqis Reclaim Their Ancient Wetlands', *New Scientist*, (4 October), pp. 6–7.

Rands, M., 1988, 'Changes in Social Networks Following Marital Separation and Divorce', in Milardo, 1988.

Rawls, J., 1971, *A Theory of Justice*, Cambridge, MA: Harvard University Press.

Rawls, J., 2001, *Justice as Fairness; A Restatement*, Cambridge, MA: Belknap Press.

Reader, J., 1981, *Missing Links: The Hunt for the Earliest Man*, London: Book Club Associates.

Redfield, B., 1994, *Dictionary of Mythology*, London: Chancellor Press.

Redhead, B., 1984, *Plato to Nato*, London: BBC Books.

Reece, M., et al., 2009, 'Prevalence of Characteristics of Vibrator Use by Men in the United States', *Journal of Sexual Medicine*, Vol. 6, pp. 1867–1874.

Renvoize, J., 1982, *Incest: A Family Pattern*, London: Routledge and Kegan Paul.

Reppetto, T., 2005, *American Mafia: A History of Its Rise to Power*, New York: Owl Books.

Rew, K., 2004, 'Hands off He's Mine', *The Observer Review* (11 July), p. 53.

Richardson, S., Goodman, N., Hastorf, A. and Dornbusch, S., 1961, 'Cultural Uniformity in reaction to Physical Disabilities', *American Sociological Review*, Vol. 26, pp. 241–247.

Ridley, M., 1993, *The Red Queen*, Harmondsworth: Penguin.

Ridley, M., 1999, *Genome*, London: Fourth Estate.

Ritvo, L., 1990, *Darwin's Influence on Freud: A Tale of Two Sciences*, New Haven: Yale University Press.

Roach, J., 2002, 'Controversy Over famed Ancient Skull: Ape Human?', *National Geographic News* (9 October).

Roberts, D., 1980, 'New Perspectives in Human Genetics and Evolution', *Man* (N.S.), Vol. 15: 1, pp. 12–44.

Roberts, D., 1998, *Once they Moved Like the Wind*, London: Pimlico.

Roberts, N., 1992, *Whores in History: Prostitution in Western Societies*, London: HarperCollins.

Roberts, Y., 2005, 'Yes, Yes, Yes ... yesssssssssss: Or Maybe No', *The Observer* (12 June).

Robinson, C., Lockard, J. and Adams, R., 1979, 'Who Looks at a Baby in Public?', *Ethology and Sociobiology*, Vol. 1, pp. 87–91.

Rodger, N., 2004, *The Command of the Ocean: A Naval History of Britain Vol. II, 1649–1815*, London: Allen Lane.

Rogers, A., et al., 2004, 'Genetic Variation at the MC1R Locus and the Time since Loss of Human Body Hair', *Current Anthropology*, Vol. 45, pp. 105–108.

Rogers, S. and Turner, C., 1991, 'Male-Male Sexual Contact in the U.S.A.: Findings from Five Sample Surveys, 1970–1990', *Journal of Sex Research*, Vol. 28, pp. 491–519.

Rosenberg, K. and Trevathan. W., 1996, 'Bipedalism and Human Birth: The Obstetrical Dilemma revisited', *Evolutionary Anthropology*, Vol. 4, pp. 161–168.

Rosenberg, K. and Trevathan, W., 2001, 'The Evolution of Human Birth', *Scientific American*, Vol. 285: 5, pp. 72–77.

Ross, C. and Henneberg, M., 1995, 'Basicranial Flexion, Relative Brain Size, and Facial Kyphosis in Homo Sapiens and Some Fossil Hominids', *American Journal of Physical Anthropology*, Vol. 98, pp. 575–593.

Ross, D., 1995, *Aristotle*, London: Routledge.

Rossiaud, J., 1985, 'Prostitution, Sex and Society in French Towns in the Fifteenth Century', in Aries and Bejin, 1985.

Roth, G. and Dicke, U., 2005, 'Evolution of the Brain and Intelligence', *Trends in Cognitive Science*, Vol. 9:5 (May), pp. 250–257.

Rothblatt, B. (ed.), 1968, *Changing Perspectives on Man*, Chicago: University of Chicago Press.

Rowell, T., 1972, *The Social Behaviour of Monkeys*, Baltimore: Penguin Books.

Rowell, T., 1988, 'What Do Male Monkeys Do Besides Competing?', in Greenberg and Tobach, 1988.

Rull, K., Nagirnaja, L. and Maris, L., 2012, 'Genetics of Recurrent Miscarriage: Challenges, Current Knowledge, Future Directions', *Frontiers in Genetics*, Vol. 3, 10.3389/fgene.2012.00034 online.

Rumbaugh, D., et al., 1975, 'The Language Skills of a Young Chimpanzee in a Computer-Controlled Training Situation', in Tuttle, 1975.

Rumbaugh, D., et al., 1973, 'Computer-Controlled Language Training System for Investigating the Language Skills of Young Apes', *Behavior Research Methods and Instrumentation*, Vol. 5, pp. 385–392.

Runciman, W. G., Maynard Smith, J. and Dunbar, R. (eds.), 1996, *Evolution of Social Behaviour: Patterns in Primates and Man*, Oxford: Oxford University Press (for the British Academy).

Ruse, M., 1988, *Homosexuality*, Oxford: Basil Blackwell.

Ruse, M., 2012, *The Philosophy of Human Evolution*, Cambridge: Cambridge University Press.

Russell, J., 1988, *The Prince of Darkness: Radical Evil and the power of good in History*, Ithaca: Cornell University Press.

Russo, K., 2009, 'Double Trouble? India's Mysterious Twin Village', ABC World News (20 May).

Rutherford, A., 2013, *Creation: The Origin of Life. The Future of Life*, London: Penguin.

Ryan, A. (ed.), 1993, *Justice*, Oxford: Oxford University Press.

Saayman, G., 1975, 'The Influence of Hormonal and Ecological Factors Upon Sexual Behavior and Social Organization in Old World Primates', in Tuttle, 1975.

Safilos-Rothschild, C., 1970, *The Sociology and Social Psychology of Deformity and Rehabilitation*, New York: Random House.

Sagan, C., 1977, *The Dragons of Eden*, New York: Hodder and Stoughton.

Saint-Simon, H., 1964 [1925], *Social Organization, The Science of Man, and other Writings* (F. Markham ed.), New York: Harper Torchbooks.

Sandars, N., 1964a, 'Introduction', in Sandars, 1964.

Sandars, N. (ed.), 1964, *The Epic of Gilgamesh*, Harmondsworth: Penguin.

Sanford, N. and Comstock, C. (eds.), 1971, *Sanctions For Evil*, San Francisco: Jossey-Bass.

Sapolsky, R., 1994, *Why Zebras Don't Get Ulcers*, New York: Freeman and Co.

Savage-Rumbaugh, S. and McDonald, K., 1988, 'Deception and Social Manipulation in Symbol-using Apes', in Byrne and Whiten, 1988.

Scarf, M., 1981, *Unfinished Business*, Glasgow: Fontana.

Schaik, C. and Deaner, R., 2003, 'Life History and Cognitive Development in Primates', in De Waal and Tyack, 2003.

Schell, C., et al., 2007, 'Socioeconomic Determinants of Infant Mortality: A Worldwide Study of 152 Low-, Middle-, and High-Income Countries', *Scandinavian Journal of Public Health*, Vol. 35: 3, pp. 288–297.

Schillaci, M., 2006, 'Sexual Selection and the Evolution of Brain Size in Primates', DOI:10.1371/journal.pone.0000062.
Schmidt, K., 2002, 'Sugar Rush', *New Scientist* (26 October), pp. 34–38.
Schmitt, D. and Buss, D., 2001, 'Human Mate Poaching: Tactics and Temptations for Infiltrating Existing Mateships', *Journal of Personality and Social Psychology*, Vol. 80, pp. 894–917.
Schultz, A., 1949, 'Sex differences in the Pelvis of Primates', *American Journal of Physical Anthropology*, Vol. 7, pp. 887–964.
Schultz, A., 1963, 'Age Changes, Sex Differences, and Variability Factors in the Classification of Primates', in Washburn, 1963.
Schultz, A., 1968, 'The Recent Hominoid Primates', in Washburn and Jay, 1968.
Schultz, A., 1978, 'Illustrations of the Relations between Primate Ontogeny and Phylogeny', in Washburn and McCown, 1978.
Science, 2009, *Science*, Vol. 326: 5949 (October 2), pp. 36–106.
Scott, G., 1994, *The Power of Fantasy: Illusion and Eroticism in Everyday Life*, New York: Carol Publishing Group.
Seafarer, 2013, Anglo-Saxons.net, 2013, www.anglo-saxons.net/hwaet/? do=get&type=text&id=Sfr
Sedgwick, E., 1985, *English Literature and Homosexual Desire*, New York: Columbia University Press.
Seligman, M., 1975, *Helplessness: On Depression, Development, and Death*, San Francisco: Freeman and Co.
Sellin, T., 1959, 'The Death Penalty', in Sellin, 1967.
Sellin, T. (ed.), 1967, *Capital Punishment*, New York: Harper and Row.
Seminole, 2013, Seminole Tribe of Florida, 'History', www.semtribe.com/History/Bibliography.aspx
Senut, B., *et al.*, 2001, 'First Hominid from the Miocene (Lukeino Formation, Kenya)', *Comptes Rendus de l'Académie des Sciences, Série Iia*, Vol. 332, pp. 137–144.
Service, E., 1963, *Profiles in Ethnology*, New York: Harper and Row.
Seyfarth, R. and Cheney, D., 2003, 'The Structure of Social Knowledge in Monkeys', in De Waal and Tyack, 2003.
Shakespeare, W., 1977, *The Complete Works of William Shakespeare*, London: Abbey Library.
Sheehan, J. and Sosna, M., 1991, *The Boundaries of Humanity*, Berkeley: University of California Press.
Shepher, J., 1971, 'Mate Selection among Second Generation Kibbutz Adolescents and Adults: Incest Avoidance and Negative Imprinting', *Archives of Sexual Behaviour*, Vol. 1, pp. 293–307.
Shepher, J., 1978, 'Reflections on the Origin of the Human Pair Bond', *Journal of Sociological and Biological Structure*, Vol. 11, pp. 253–264.
Shepher, J., 1979, *Incest: The Biosocial View*, Cambridge: Cambridge University Press.
Silk, J., Alberts, B. and Altmann, J., 2003, 'Social Bonds of Female Baboons Enhance Infant Survival', *Science*, Vol. 302, pp. 1231–1234.
Silver, L., 2007, 'The Year of Miracles', *Newsweek International* (15 October), pp. 38–43.
Singer, B., Kancebaum, L. and Thomas, R., 2004, 'Men and Women in Therapy 2004: Inside the Gender Gap', *PacifiCare Behavioral*, www.napabipolardepression.org/images/men_women_therapy.pdf
Singer, E., 2003, 'Brain Chemistry Could Be the Key to Autism Severity', *New Scientist* (6 September), p. 1.

Singh, D., 1993, 'Adaptive Significance of Female Physical Attractiveness: Role of Waist-to-Hip Ratio', *Journal of Personality and Social Psychology*, Vol. 65: 2, pp. 293–307.

Singh, D. and Luis, S., 1995, 'Ethnic and Gender Consensus for the Effect of Waist-to-Hip Ratio on Judgment of Women's Attractiveness', *Human Nature*, Vol. 6:1, pp. 51–65.

Singh, D. and Young, R., 2001, 'Body Weight, Waist-to-Hip Ratio, Breasts, and Hips: Role in Judgments of Female Attractiveness and Desirability for Relationships', *Ethology and Sociobiology*, Vol. 16: 6, pp. 483–507.

Skinner, B. F., 1953, *Science and Human Behaviour*, New York: Macmillan/Free Press.

Skinner, B. F., 1971, 'A Behavioral Analysis of Value Judgments', in Tobach, *et al.*, 1971.

Skinner, B. F., 1988, 'Genes and Behavior', in Greenberg and Tobach, 1988.

Sloan, A., 1972, *My Years with General Motors*, New York: Anchor Books.

Sloane, S., Baillargeon, R. and Premack, D., 2012, 'Do Infants Have a Sense of Fairness?', *Psychological Science*, Vol. 23:3, pp. 196–204.

Small, M., 2002, 'The Happy Fat', *New Scientist* (24 August), pp. 34–37.

Smillie., S., 2002, 'Killer Cats Hunted Human Ancestors', *National Geographic News* (20 May).

Smith, G. E., 1927, *The Evolution of Man*, Oxford: Oxford University Press.

Smith, H., 1958, *The Religions of Man*, New York: Mentor Books.

Smithsonian, 2013, 'What Does It Mean to Be Human?', National Museum of Natural History, http://humanorigins.si.edu/

Smithsonian, 2013a, 'Climate Effects on Human Evolution', National Museum of Natural History, http://humanorigins. si.edu/research/climate-research/effects

Smuts, B. and Gubernick, D., 1992, 'Male-Infant Relationships in Nonhuman Primates: Parental Investment or Mating Effort?', in Hewlett, 1992.

Snowdon, C., 2002, 'From Primate Communication to Human Language', in de Waal, 2002.

Soble, A., 2002, *Philosophy of Sex: Contemporary Readings*, Boulder, CO: Rowan and Littlefield Publishers.

Soble, A., 2002a, 'Masturbation: Conceptual and Ethical Matters', in Soble, 2002.

Solomon, E., 1984, in *Psychology Today*, (March).

Sommerstein, A. H., 2008, *Aeschylus*, Cambridge, MA: Loeb Classical Library (Harvard University Press).

Spellman, W., 1997, *John Locke*, Basingstoke: Macmillan.

Spencer, H., 1967, *The Evolution of Society: Selections from Herbert Spencer's Principles of Sociology*, Chicago: University of Chicago Press.

Spoor, F., *et al.*, 2007, 'Implications of New Early Homo Fossils from Ileret, East of Lake Turkana, Kenya', *Nature*, Vol. 448, pp. 688–691.

Srinivas, M., 1952, *Religion and Society among the Coorgs of South India*, Oxford: Clarendon Press.

Sroufe, L., 1977, 'Wariness of Strangers and the Study of Infant Development', *Child Development*, Vol. 48, pp. 1184–1199.

Stanford, C., 2002, 'The Ape's Gift: Meat Eating, Meat Sharing and Human Evolution', in De Waal, 2002.

Steiner, F., 1967, *Taboo*, Harmondsworth: Penguin Books.

Stepan, A., 1971, *The Military in Politics: Changing Patterns in Brazil*, Princeton: Princeton University Press.

Sternberg, R. and Kaufman J. (eds.), 2002, *The Evolution of Intelligence*, London: Lawrence Erlbaum.

Steward, J., 1968, 'Causal Factors and Processes in the Evolution of Pre-Farming Societies', in Lee and DeVore, 1968.

Stone, L., 1977, *The Family, Sex and Marriage in England, 1500–1800*, London: Weidenfeld and Nicolson.
Stouffer, S., *et al.*, 1965, *The American Soldier Vol. II.*, New York: John Wiley.
Strauch, B., 2003, *Why Are They So Weird? What's Really Going on in a Teenager's Brain?*, London: Bloomsbury.
Strier, K., 2002, 'Beyond the Apes: Reasons to Consider the Entire Primate Order', in De Waal, 2002.
Summit, R., 1982, 'Beyond Belief', in Kirkpatrick, 1982.
Suzuki, A., 1975, 'The Origin of Hominid Hunting: A Primatological Perspective', in Tuttle, 1975.
Symonds, J., 1893, *A Study of Walt Whitman*, London: John C. Nimmo.
Symonds, J., 1907, *Essays, Speculative and Suggestive*, London: Smith Elder.
Symons, D., 1979, *The Evolution of Human Sexuality*, London: Oxford University Press.
Symons, D., 1992, 'On the Use and Misuse of Darwinism in the Study of Behavior', in Barkow, Cosmides and Tooby, 1992.
Szalavitz, M., 2002, 'Love Is the Drug', *New Scientist* (23 November), pp. 38–40.
Szasz, T., 1973, *The Manufacture of Madness*, St Albans: Paladin.
Tanner, N., 1981, *On Becoming Human*, Cambridge: Cambridge University Press.
Tattersall, I., Delson, E. and Van Couvering, J., 1988, *Encyclopedia of Human Evolution and Prehistory*, London: St. James Press.
Taylor, D., 1985, 'Theological Thoughts about Evil', in Parkin, 1985.
Taylor, S., 2002, *The Tending Instinct: How Nurturing Is Essential to Who We Are and How We Live*, New York: Holt.
The Telegraph, 2009, 'Indian village with 250 sets of twins', *The Telegraph* (11 May).
Thomas, K., 1971, *Religion and the Decline of Magic*, London: Weidenfeld and Nicolson.
Thor, 2011, 'German Rifle Squad (Schützengruppe)', History Wars Weapons, http://historywars weapons.com/german-rifle-squad-schutzengruppe/
Thornhill, R. and Gangestad, S., 1993, 'Human Facial Beauty: Averages, Symmetry and Parasite Resistance', *Human Nature*, Vol. 4, pp. 237–269.
Thorpe, S., 2006, 'Walking the Walk: The Evolution of Human Bipedalism', Darwin Day, School of Medicine, University of Birmingham, UK (10 February).
Tiger, L., 1969, *Men in Groups*, New York: Random House.
Tilly, L. and Tilly, C., 1981, *Class Conflict and Collective Action*, Los Angeles: Sage.
Titaley. C., *et al.*, 2008, 'Determinants of Neonatal Mortality in Indonesia', *BMC Public Health*, Vol. 8, p. 232.
Tobach, E., Aronson, L. and Shaw, E. (eds.), 1971, *The Biopsychology of Development*, New York: Academic Press.
Tooby, J. and Cosmides, L., 1992, 'The Psychological Foundations of Culture', in Barkow, Cosmides and Tooby, 1992.
Tooby, J. and Cosmides, L., 1995, 'Forward', in Baron-Cohen, 1995.
Toups, M., *et al.*, 2010, 'Origin of Clothing Lice Indicates Early Clothing Use by Anatomically Modern Humans in Africa', *Molecular Biology and Evolution*, http://mbe.oxfordjournals.org content/28/1/29.full
Trinkaus, E. (ed.), 1989, *The Emergence of Modern Humans: Biocultural Adaptations in the Late Pleistocene*, Cambridge: Cambridge University Press.
Trinkaus, E. and Howells, W., 1979, 'The Neanderthals', *Scientific American*, Vol. 241, pp. 118–133.
Tripp, C., 1977, *The Homosexual Matrix*, London: Quartet Books.

Trivers, R., 1972, 'Parental Investment and Sexual Selection', in Campbell, B., 1972.
Trivers, R., 1985, *Social Evolution*, Menlo Park, CA: Benjamin/Cummings Pub. Co.
Turke, P. and Betzig, L., 1985, 'Those Who Can Do: Wealth, Status, and Reproductive Success on Ifaluk', *Ethology and Sociobiology*, Vol. 6:2, pp. 79–87.
Turnbull, C., 1965, *Wayward Servants; The Two Worlds of the African Pygmies*, Garden City, NY: Natural History Press.
Turnbull, C., 1976, *The Forest People*, London: Pan Books.
Turner, L., et al., 1999, 'Preliminary Research on Plasma Oxytocin in Normal Cycling Women: Investigating Emotion and Interpersonal Distress', *Psychiatry*, Vol. 62, pp. 97–113.
Turner, R. and McGuinness, T., et al., 1999, 'Hormone Involved in Reproduction May Have Role in the Maintenance of Relationships', http://www.ucsf.edu/news/1999/07/5079/
Tutin, C., 1975, 'Exceptions to Promiscuity in a Feral Chimpanzee Community', in Kondo, Kawai and Ehara, 1975.
Tutin, C., 1979, 'Mating Patterns and Reproductive Strategies in a Community of Wild Chimpanzees', *Behavioral Ecology and Sociobiology*, *Vol.* 6, pp. 29–38.
Tuttle, R. (ed.), 1975, *Socioecology and Psychology of Primates*, The Hague: Mouton.
Twitchell, J. 1987, *Forbidden Partners: The Incest Taboo in Modern Culture*, New York: Columbia University Press.
Tyack, P., 2003, 'Dolphin Communication about Individual-Specific Relationships', in de Waal and Tyack, 2003.
Uglow, J., 2002, *The Lunar Men*, London: Faber and Faber.
Ulick, J., 2005, 'What to Watch for', *Newsweek International* (28 February), pp. 46–47.
UN, 2014, Food and Agricultural Organization, www.fao.org/FOCUS/E/fisheries/intro.htm
Unger, R., 1979, *Female and Male: Psychological Perspectives*. New York: Harper and Row.
UNICEF, 2013, 'Childinfo: Monitoring the Situation of Women and Children', www.child.org
Urban Dictionary, 2013, 'Diesel Dyke', www.urbandictionary.com/define.php?term=Diesel
USGS, 1998, *Status and Trends of the Nation's Biological Resources*, Vol. 1–2, National Wetlands Research Centre, USGS, www.nwrc.usgs.gov/sandt/
Utley R., 1998, *A Life Wild and Perilous: Mountain Men and the Paths to the Pacific*, New York: Henry Holt.
Van Couvering and Van Couvering, 1976, 'Early Miocene Mammal Fossils from East Africa: Faunistics and Paleoecology', in Isaac and McCown, 1976.
Van Hooff, J. and Preuschoft, S., 2003, 'Laughter and Smiling: The Interaction of Nature and Culture', in de Waal and Tyack, 2003.
Van Sommers, P., 1988, *Jealousy: What Is It and Who Feels It?*, Harmondsworth: Penguin.
Van Wyk P. and Geist, C., 1995, 'Biology of Bisexuality: Critique and Observations', *Journal of Homosexuality*, Vol. 28: 3–4, pp. 357–373.
Vannelli, R., 1977, *The Politics of Danger*, PhD Thesis, University of Reading, UK.
Vannelli, R., 2001, *Evolutionary Theory and Human Nature*, Boston: Kluwer Academic Press.
Vellacott, P. (ed.), 1956, *Aeschylus: The Oresteian Trilogy*, Harmondsworth: Penguin.
Veyne, P., 1985, 'Homosexuality in Ancient Rome', in Aries and Bejin, 1985.
Vignaud, P. et al., 2002, 'Geology and Paleontology of the Upper Miocene Toros-Menalla Hominid Locality, Chad', *Nature*, Vol. 418, pp. 152–155.
Voltaire, 1759, *Candide*, in Aldington, 1928.
Voltaire, 1972, *Philosophical Dictionary*, Harmondsworth: Penguin.
Waber, D., et al., 1981, 'Behavioral Correlates of Physical and Neuromotor Maturity in Adolescents from Different Environments', *Developmental Psychobiology*, Vol. 14: 6, pp. 513–522.

Waber, D., 1976, 'Sex Differences in Cognition: A Function of Maturation Rate?', *Science*, Vol. 192, pp. 572–573.
Wallman, J., 1992, *Aping Language.*, Cambridge: Cambridge University Press.
Walter, E., 1969, *Terror and Resistance: A Study in Political Violence*, Oxford: Oxford University Press.
Wang, et al., 2003, 'Conception, Early Pregnancy Loss, and Time to Clinical Pregnancy: A Population-Based Prospective Study', *Fertility and Sterility*, Vol. 79: 3 pp. 577–584.
Ward, L., 1897, *Dynamic Sociology*, New York: Appleton.
Ward, L., 1906, *Applied Sociology*, Boston: Ginn.
Washburn, S., 1960, 'Tools and Human Evolution', *Scientific American* (September).
Washburn, S. (ed.), 1961, *Social Life of Early Man*, Chicago: Aldine Pub. Co.
Washburn, S. (ed.), 1963, *Classification and Human Evolution*, Chicago: Aldine.
Washburn, S., 1963a, 'Behavior and Human Evolution', in Washburn, 1963.
Washburn, S., 1968, 'Speculations of the Problem of Man's Coming to the Ground', in Rothblatt, 1968.
Washburn, S., 1968a, 'Discussion', pp. 245–247, in Lee and DeVore, 1968.
Washburn, S., 1973, 'Human Evolution: Science or Game?', *Yearbook of Physical Anthropology*, Vol. 17, pp. 67–70.
Washburn, S., 1978, 'The Evolution of Man', *Scientific American*, Vol. 239; 3, pp. 194–208.
Washburn, S. and DeVore, I., 1961, 'The Social Life of Baboons', *Scientific American*, Vol. 204, pp. 62–71.
Washburn, S. and Jay, P. (eds.), 1968, *Perspectives on Human Evolution*, New York: Holt, Rinehart and Winston.
Washburn, S. and Lancaster, C., 1968, 'The Evolution of Hunting', in Lee and DeVore, 1968.
Washburn, S. and McCown, E. (eds.), 1978, *Human Evolution: Biosocial Perspectives*, Menlo Park, CA: Benjamin/Cummings.
Watson, J., 1913, 'Psychology as the Behaviorist Views It', *Psychological Review*, Vol. 20, pp. 158–177.
Watts, A., 1962, *The Way of Zen*, Harmondsworth: Penguin.
Weatherburn, P., et al., 1990, 'Heterosexual Behaviour in a Large Cohort of Homosexually Active Men in England and Wales', *AIDS Care*, Vol. 2: 4, pp. 319–324.
Weber, M., 1958, 'Politics as Vocation', in Gerth and Mills, 1958.
Weber, M., 1978, *Economy and Society*, Vol. I. Berkeley: University of California Press.
Weber, M., 1981, *General Economic History*, New Brunswick: Transaction Books.
Weinberg, M., Williams, C. and Pryor, D., 1994, *Dual Attraction: Understanding Bisexuality*, New York: Oxford University Press.
Welch, J. (with Stekler, P.), 1994, *Killing Custer*, New York: Norton.
Wellings, K., Field, J., Johnson, A. and Wadsworth, J., 1994, *Sexual Behaviour in Britain*, London: Penguin.
Wheeler, P., 1984, 'The evolution of bipedality and loss of functional body hair in hominids', *Journal of Human Evolution*, Vol. 13: 1, pp. 91–98.
White, E. (ed.), 1981, *Sociobiology and Human Politics*, New York: Lexington Books, D.C. Heath.
White, M., 2005, 'Source List and Detailed Death Tolls for the Twentieth Century Hemoclysm', http://users.erols.com/mwhite28/warstat1.htm
White, P. and Speiser, P., 2000, 'Congenital Adrenal Hyperplasia due to 21-Hydroxylase Deficiency', *Endocrine Reviews*, Vol. 21:3, pp. 245–291.

White, T., et al., 2009, *Ardipithecus Ramidus* and the Paleobiology of Early Hominids, *Science*, Vol. 326: 5949, pp. 64, 75–86.
White, T., Suwa, G. and Asfaw, B., 1994, '*Australopithecus ramidus*, a New Species of early Hominid from Aramis, Ethiopia', *Nature*, Vol. 371, pp. 306–312.
Whiten, A., 1997, 'The Machiavellian Mindreader', in Whiten and Byrne, 1997.
Whiten, A. and Byrne, W., 1997, *Machiavellian Intelligence II: Extensions and Evaluations*, Cambridge: University of Cambridge Press.
Whybrow, P., 2004, 'Miocene Vertebrate Fossils of Arabia', The Natural History Museum London: www.nhm.ac.uk/science
Wikipedia, 2013, 'Squad; Military Organization'; Inca army, http://en.wikipedia.org/wiki/
Wikipedia, 2014, 'Military organization', http://en.wikipedia.org/wiki/Military_organizat
Wilcox, A., Baird D. and Weinberg, C., 1999, 'Time of Implantation of the Conceptus and Loss of Pregnancy', *New England Journal of Medicine*, Vol. 340: 23, pp.1796–1799, doi:10.1056/NEJM199906103402304. PMID 10362823.
Wilkinson, M., 2009, 'Raskolnikova: Rodion Romanovich's Struggle with the Woman Wit', *Gen* Issue 50.
Wille, J., 2012, 'Occurrence of Fibonacci Numbers in Development and Structure of Animal Forms: Phylogenetic Observations and Epigenetic Significance', *Natural Science*, Vol. 4, pp. 216–232.
Williams, G., 1966, *Adaptation and Natural Selection: A Critique of Some Current Evolutionary Thought*, Princeton, NJ: Princeton University Press.
Williams, R., 1968, *Textbook of Endocrinology*, London: Saunders.
Wills, C., 1993, *The Runaway Brain*, London: HarperCollins.
Wilson, M. and Daly, M., 1992, 'The Man Who Mistook His Wife for a Chattel', in Barkow, Cosmides and Tooby, 1992.
Wilson, E. O., 1975, *Sociobiology: The New Synthesis*, Cambridge, MA: The Belknap Press.
Wilson, E. O., 1978, *On Human Nature*, Cambridge, MA: Harvard University Press.
Wingert, P. and Brant, M., 2005, 'Reading Your Baby's Mind', *Newsweek International*, (15 August), pp. 48–55.
Winkvist, A., Mogren, I. and Hogberg, U., 1998, 'Familial patterns in birth characteristics: impact on individual and population risks', *International Journal of Epidemiology*, Vol. 27: 2, pp. 248–254.
Winn, D., 1974, *Prostitutes*, London: Hutchinson.
Winnicott, D., 1965, *The Family and Individual Development*, London: Tavistock.
Winter, H., et al., 2001, 'Human type I Hair Keratin Pseudogene PhihHaA Has Functional Orthologs in the Chimpanzee and Gorilla', *Human genetics*, Vol. 108: 1, pp. 37–42.
Wiseman, R., 2007, 'The Truth about Lying and Laughing', *The Guardian Weekend* (21 April), pp. 42–53.
Wispé, L., 1991, *The Psychology of Sympathy*, New York: Plenum Pub. Corp.
WoldeGabriel, G., et al., 1994, 'Ecological and Temporal Placement of Early Pliocene Hominids at Aramis, Ethiopia', *Nature*, Vol. 371, pp. 330–333.
Wolfe, T., 1980, *The Right Stuff*, New York: Bantam Books.
Wood, B., 1992, 'The Origin and Evolution of the Genus Homo', *Nature*, Vol. 355, pp. 783–790.
Wood, B., 1994, 'The Oldest Hominid yet', *Nature*, Vol. 371, pp. 280–281.
Wood, B, 2002, 'Paleoanthropology: Hominid Revelations from Chad', *Nature*, Vol. 418, pp. 133–135.
Wood, B., 2002a, 'Who Are We?', *New Scientist* (26 October), p. 44.
Woodburn, J., 1982, 'Egalitarian Societies', *Man* (NS), Vol. 17, pp. 431–451.

World Science, 2005, 'Wired for war? Killer Chimps Fuel Debate on How War Began', www.world-science.net/exclusives/050209_warfrm.htm

Wrangham, R., *et al.*, 1994, *Chimpanzee Cultures*, Cambridge, MA: Harvard University Press.

Wrangham, R., 1999, 'Evolution of Coalitionary Killing', *American Journal of Physical Anthropology*, Vol. 110: 29, pp. 1–30.

Wrangham, R., 2002, 'Out of the Pan, into the Fire: How Our Ancestors Evolved Depended on What They Ate', in de Waal, 2002.

Wrangham, R., 2009, *Catching Fire: How Cooking Made Us Humans*, London: Profile Books.

Wray, A. (ed.), 2002, *The Transition to Language*, Oxford: Oxford University Press.

Wright, *et al.*, 2012, 'Human Responses to Unfairness with Primary Rewards and Their Biological Limits', *Scientific Reports*, 2 DOI:10.1038/srep00593.

Wynn, T., 1979, 'The Intelligence of Later Acheulean Hominids', *Man*, (NS) Vol. 14: 3, pp. 171–191.

Wynn, T., 1988, 'Tools and the Evolution of Human Intelligence', in Byrne and Whitten, 1988.

Zaehner, R. (ed.), 1966, *Hindu Scriptures*, New York: Dutton.

Zahavi, A., 1975, 'Mate Selection – A Selection for a Handicap', *Journal of Theoretical Biology*, Vol. 53, pp. 205–214.

Zeh, J. and Zeh, D., 1996, 'The Evolution of Polyandry I: Intergenetic Conflict and Genetic Incompatibility', *Proceedings of the Royal Society, London, Series B*, 263, 1711–1717.

Zeh, J. and Zeh, D., 1997, 'The Evolution of Polyandry II: Post-Copulatory Defenses Against Genetic Incompatibility', *Proceedings of the Royal Society, London, Series B*, 264: 69–75.

Zelzer, E., 2009, 'Tendons Shape Bones during Embryonic Development', in *ScienceDaily* (14 December).

Zihlman, A., 1981, 'Women as Shapers of Human Adaptation', in Dahlberg, 1981.

Index

21-hydroxylase (congenital adrenal hyperplasia), 77
5-aR (5 alpha-reductase deficiency – type 2), 40–41

abnormality, 23, 40, 225–226, 228–229, 231, 246, 250–251
abortion, 37, 40, 93, 95, 117, 232, 252, 265–269
abstinence, 74, 93, 95, 101, 237, 267
accusations, 125, 130–132, 178, 218, 230, 232, 235–236, 239
Ache, 185
adaptation, 16–17, 24, 32, 48, 63
addicted, 95, 169–170, 175
addiction, 74, 95
administration, 1, 204, 258, 265
adolescence, 14–15, 29, 53, 55, 59, 62, 65, 67, 74, 76–77, 79, 82–83, 86–88, 90, 95–96, 98, 104, 130–131, 152, 158–159, 168, 176, 209, 218, 220, 249, 251
adultery, 30, 52, 91, 152, 154, 221, 231, 238–242, 249, 251
Adventures of Huckleberry Finn, 173
Aeneid, 172
Africa, 25–27, 33–34, 56, 118, 244
agonistic behaviour, 58, 113, 155–157, 159, 178, 200, 202–203
agriculture, 33, 135, 143, 147, 180, 185, 188
AIDs, 34, 210
Alice (in Wonderland), 227
alleles, 39, 40, 85
alliances
 alliance behaviour, 61, 97, 164, 167, 175, 178, 185, 199, 250
 alliance groups, 46, 57–58, 128, 198
 alliance mates, 46, 123, 133, 141, 144, 154, 165, 172, 199–200, 203, 213, 220–221, 249
 alliance politics, 58, 60, 178, 183, 185
 alliance systems, 180
alms, 200, 232, 236
altruism, 31, 151, 156, 239

ambiguity, 17, 78, 86, 88–89, 97, 103, 107, 121, 125, 128, 131, 149, 160, 181, 210, 222, 238, 249–250, 269
ambition, 75, 106–107, 132, 190, 202, 223, 247
ambivalence, 15, 21–22, 88, 96, 107, 121, 213–214, 226, 238, 260
analytical intelligence, 123
androgens, 77
anthropology, 2, 4, 66, 135, 142, 153, 199, 230, 245, 267
Antigone, 5
anxiety, 8–9, 15, 22–23, 29, 83, 88, 93, 95, 99, 106–107, 125, 134, 164, 177, 191, 193, 206, 213–218, 223–224, 238, 243, 252
aquatic environments, 33, 112
aquatic flora/fauna, 27, 35, 117
Aquinas, St Thomas, 4, 173, 240,
Ardipithecus kadabba, 25
Ardipithecus ramidus, 26, 28, 112
aristocracy, 159–160, 211
Aristotle, 3–4, 172–173, 242,
arrogance, 121, 185, 232, 234–236, 238, 243, 247, 255
arts, the, 253–254
Asperger's, 121
attachment, 93, 102, 115, 134, 140, 215, 241
Augustine, St, 3–4, 173, 223
aunt behaviour, 60, 81, 196
Austen, Jane, 215
Australopithecus, 25, 27, 35, 61, 64
autism, 125
Azande, 185
Aztecs, 185

baboons, 10, 49–50, 52, 54, 56, 144, 157, 194
baby
 baby talk, 94, 104
 baby-face, 90
 baby-like, 84, 90, 195
 baby-ness, 83, 94–95
balance of power, 58, 178, 180, 191, 271
balanced sex ratio, 152

bands, 4, 46, 50, 65, 133–134, 138, 141–142, 145–147, 151–153, 155, 161, 163, 191, 221, 270
  band societies, 141–142, 145, 150, 153
Barrie, J.M., 224, 227
Baum, L., 41, 227
Beach, F., 168
beauty, 9, 67, 69, 246
behaviourism, 12, 16
Bhagavad Gita, 241
Bible, 5, 223, 241
bioelectric, 2, 10, 20, 107, 140, 155, 209
biological determinism, 16
bipedal, 24–30, 35–36, 38, 56, 59, 61, 63, 64, 70, 72, 77, 85, 91, 100, 111–116, 124, 144
Birdsell, J., 147, 151–152
birth
  birth canal, 100, 112, 115, 117–118
  birth control, 95, 197, 252, 267–268
  birth cycle, 72, 103, 116
  birth intervals, 29–30, 62
  birth rate, 76–78, 93
birth control, 93
bi-sexuality, 195
Bishop Jewel, 253
Bodin, J., 222
bonding, 44–45, 54, 94, 113, 123, 141, 148–149, 151, 153, 156, 164, 166, 177, 204, 225, 258
  female bonding, 194, 198
  male bonding, 166, 170–171, 174, 175, 199
  pair bonding, 54, 73, 89, 120, 127, 146, 166, 250
Book of the Dead, 239
Boulton, M., 174
boundaries, 96, 120, 140, 177–178, 203–204, 206, 224, 227–228, 237, 239, 243, 248, 255–256, 258
boyfriend, 46, 193, 202, 218
brachiation, 56
Brahmins, 188
brain
  brain bioelectrics, 2
  brain case, 38, 116–117, 124
  brain size, 34, 91, 111, 114–115, 122, 124, 143, 145
  brain/body ratio, 145
  computer-like (mind/brain), 12, 29
  limbic-hypothalamic brain, 117
  neocortex, 28, 55, 78, 117, 119, 143–146
  pre-knowledge, 15, 107, 140, 207
  visual cortex, 115
breasts, 12–13, 73–74, 77, 79, 85, 113, 131, 167, 228, 246
  external breasts, 62, 66, 84–86
brother/sister love, 81
brotherhood, 133, 176, 193, 204, 221, 254, 258
Buddhism, 5, 241
buddyships, 164, 167

buddy stories, 172
buddy system, 147
buttocks, 66, 84–85, 246

camaraderie, 106, 151, 166, 172, 188
Candide, 6
Canterbury Tales, 223
capital punishment, 237, 263, 265
Carnegie, Andrew, 201
Carroll, J., 215–216, 218
Carroll, Lewis, 227
caste, 137, 176, 186–187
castration, 55, 78, 209, 215, 222, 231, 246
Catherine the Great, 228
cause and effect, 120, 260
celebrations, 153, 155, 160
ceremonies, 140, 153, 155
chance, 1, 5, 7, 17, 32, 42, 44, 48, 67, 119, 130, 157, 179, 184, 186, 190, 194, 213, 218, 236, 255, 259, 270
Chang and Eng, 228
charisma, 6, 156, 161, 184, 216, 227, 234
cheap sperm, 42, 44, 47, 103, 220
child
  child care, 95, 114, 117, 124, 156, 199, 208, 268
  child rearing, 14–15, 60, 63, 65, 81, 87, 93, 100, 120, 151–152, 164, 194, 199–200, 258
  child sexuality, 13, 30, 79
  childhood, 13, 38, 61, 72, 86, 95, 125–126, 145, 151, 227
  child-like, 30, 66, 68, 82, 94, 102, 105, 109, 112, 166–167, 169, 184, 195, 224, 246, 249–250
  child-ness, 13, 94–95
  child-oriented, 86–87, 103, 196
China, 18, 27, 118, 137, 201, 261
chivalry, 186
Christian, 212, 240–241
chromosomes, 39, 65
Church, 240–241
Church, The, 143, 187, 235, 240–241
churches, 186–187, 201, 241
Churchill, Winston, 132
Cicero, 242
circumcision, 209–210
cities, 1, 4, 22, 135, 144, 146, 150, 155, 171, 173, 183, 188, 221, 235
citizen, 111, 150, 179, 188, 189, 202, 240, 264
civil society, 153, 188, 270
civilization, 15–16, 17, 21, 30–31, 74–75, 107, 140, 175–177, 183, 197, 216–218, 229, 231–232, 234, 239, 244, 256, 258
clans, 50, 143, 153, 156, 175, 201
classification, 20–21, 119, 121, 138, 146, 247, 255
clothing, 20, 64, 66, 73–74, 84, 89, 91, 96, 119, 134, 165, 189, 196, 209–210, 248
coalitions, 49, 58, 63
Code of Hammurabi, 239

codes of honour, 75, 106, 124, 186, 245
cognitive psychology, 16
communication, 13, 34, 49–50, 60, 63, 67–69, 75, 108, 114, 125–127, 131, 133, 135–140, 143, 145, 152, 165, 180
companionship, 170–171, 223, 249–250
compromise, 48, 109, 200, 203–204, 253, 255, 258, 260, 269–271
comradeship, 147, 149, 153, 164, 166, 171–174, 177–178, 187, 199, 240
Comte, A., 3
conception, 40, 48, 228, 265
confession, 168, 213, 230, 232, 237, 239, 241, 264
Confucius, 5, 242, 263
conscious, 2, 9, 12, 17, 21, 49, 76, 87, 99, 103, 109, 111, 139, 146, 165, 177, 206, 208, 213
consciousness, 6–9, 11–12, 16, 42, 49, 83, 92, 103, 109, 167, 208–209, 215, 240, 265
consortship, 52–53, 59, 62, 87, 90, 109, 156
conspiracy, 121, 131, 146, 148, 157, 178, 187–188, 204, 231, 234, 238, 258
  conspiracy theorists, 121
constitutions, 20, 133, 137, 139, 150, 161
contracts, 133–135, 137, 141, 161, 173, 203, 239–240, 243, 247–248, 269
conversation, 69, 128, 130–132
cooking, 24, 119
Coolidge effect, 104
cooperation, 29, 31, 68, 91, 97, 106, 114, 120, 123, 130, 134, 164, 166, 181, 199, 253–254, 260
copulation, 41, 46, 51–55, 59–62, 70, 72–73, 83, 88, 90, 116, 126–127, 156, 167, 169, 192, 227, 246
  copulatory courtship, 51
  ventral/dorsal, 52, 59
  ventral/ventral, 52, 62, 70
corruption, 191, 271
cost-benefit, 22, 29, 31, 87, 103, 107, 148, 166
counting coups, 65, 177
court life, 75, 137, 173, 181, 186–188, 203, 219
courts (judicial), 180, 182, 233
courtship, 14, 41, 68–69, 75, 88, 90, 102, 104, 106, 120, 127–128, 131, 137, 145, 198, 217–218, 224, 251
cousins, 79, 81, 88, 101, 152, 187, 196
  cousin marriage, 40, 80
coyness, 13, 70, 87, 94–95, 105, 125, 127
crime, 160, 176, 181–182, 221, 264
Crime and Punishment, 223
cuckold, 69, 102, 105, 165, 179, 199, 221, 246, 249
cults, 139, 173, 176, 188, 201, 253
culture, 21, 25, 73, 75, 85–86, 124, 139, 147, 173, 188, 200, 211, 221, 224–225, 250, 269
  cultural determinism, 16
  multi cultural, 226
customs, 142, 189, 229, 270
cybersex, 170

Dante, 240
Darwin, Charles, 1, 4, 7–12, 17, 19, 23–24, 30–31, 42, 71, 91, 258, 259
Darwin, Erasmus, 174
daughters, 5, 31, 48, 74, 81, 100–101, 136, 152, 158, 184–185, 196, 200, 214
Dawkins, Richard, 17
daydreams, 20, 22, 70–72, 74, 78, 83, 98, 108, 141, 183, 224, 254
de Waal, F, 104
deals, 111, 134, 136, 139–140, 149, 153, 157, 161, 173, 203–204, 258, 270
death, 29, 80, 161, 165, 172, 176, 178, 190, 191, 194, 210, 214, 221, 225–226, 232–233, 237, 240, 254, 260, 265, 267
deception, 20–21, 23, 29, 44, 54, 58, 63, 87, 89, 116, 129, 136–137, 141, 145, 157, 231, 238, 240
deference, 99, 159–162, 186–187
deferred gratification, 13, 72, 96, 148, 164, 192, 235, 238, 248, 255
deformity, 67, 80, 82, 226, 237–238, 246
delusion, 5, 204, 258, 271
democracy, 15, 189–191, 253
demonology, 231, 233
dependency, 106, 215, 224, 246, 254–255
  creative dependency, 141, 146
  creative subordination, 51, 75, 99, 106, 109, 141, 150, 158
  dependency signals, 99
  dependency trap, 105, 123
  emotional dependency, 59, 106, 108–109, 128, 164, 175, 193
  infant dependency, 59, 68, 72, 76, 87, 92, 94, 96, 103, 113, 116, 192, 194, 249
depression, 20, 22–23, 29, 31, 91, 93, 103, 120, 140, 177, 206, 213–217, 220–221, 223, 230, 244–245, 248, 266–267
deprivation, 214
Descartes, Rene, 3
deserts, 26, 33, 118, 135, 145
design, 1, 2, 7, 9, 16–17, 21, 32, 37, 258
  Grand Design, 7
despair, 22, 90, 144, 148, 206, 243, 245
developmental processes, 37, 42, 48, 77–78
Devil, 222, 231–234, 237, 260
dialect, 130, 138–139, 142, 161, 225
  dialectical groups, 138–139
  dialectical tribes, 142, 152–154, 157, 160–161, 175
Diamond, Jared, 35
differentiation, 24–25, 28, 30, 32–33, 48, 63, 77, 90, 92, 112, 116–117, 126, 163, 167, 245
dilemmas, 173, 189, 213, 222–224, 254
  the human dilemma, 215–216, 223, 254
  the master ontological conflict for humans, 108

disability, 225–226, 248, 267
  disabilities, 67, 81, 210
  disabled, 225–226
disease, 34, 121, 207, 210–212, 224, 231, 233, 236–238
disgust, 22, 90, 93, 161, 206–207, 209, 223–224, 229, 245, 249
dishonour, 249
display, 13, 29, 45, 50–52, 61, 73, 75, 98, 105, 108, 113, 115–116, 123, 132–133, 140, 157, 160, 163–164, 175, 186, 188, 190, 192, 197, 246–247, 252
division of labour, 36, 106, 129, 137, 146
divorce, 20, 30, 182, 229, 266
DNA, 36, 39, 50, 55, 78, 112, 220, 264
Dollimore, Jonathan, 20
dominance, 43–45, 50–52, 58, 61–62, 107–108, 111, 116, 125, 127, 157–159, 163, 184
domination, 31, 44, 58, 61, 102, 150, 160, 178, 194–195, 223
dopamine, 15, 78, 166
Dostoevsky, F., 5, 223
Dragon Bone Hill (China), 27
dreams, 5, 11–12, 20, 22, 29, 79, 89, 141, 207, 251, 254
drives, 2, 12, 17, 21, 25, 42, 48, 66, 76, 92–93, 97, 107–109, 111, 122–123, 146, 155, 157, 213, 244–246, 256, 258
Dunbar, R., 142–143, 147–148
Durkheim, E., 242
duty, 96, 186–187, 198, 199, 201, 238–239, 241, 249
dwarfs, 227–229

Ecclesiastes, 4–5
economics, 6
Eemian interglacial, 134
Einstein, Albert, 3
ejaculation, 43, 52–53, 62, 70, 88, 209, 267
elders, 31, 62, 93, 142, 153, 159, 176, 179–180, 182–183, 189, 201, 217, 232
Eliot, George, 5
embarrassment, 9, 14, 20, 73, 83, 87–88, 96–97, 99, 101, 103, 120, 134, 140, 159, 164, 177, 209, 247–250
embryo, 39, 42, 86, 117, 267–268
embryology, 30, 48, 78, 80
emotional-cognitive motivators, 30, 83, 92, 93
emotions
  emotional expressions, 8, 63, 67–68
  emotional power, 193
  emotional separation, 134
  emotionality, 15–16, 20, 25, 63, 122, 146, 191, 242
empathy, 20, 22, 49, 63, 115, 121–123, 146, 165, 193, 211, 252, 262, 271
endogamy, 187
enforcement, 178, 182, 189, 270

engulfment, 29, 101, 213, 215, 222–223, 248–249, 254
Enlightenment, 1, 12, 19, 25, 190, 229, 258, 261
entertainment, 85, 111, 131
environments, 37, 42, 44, 48, 60, 80, 92, 118–120, 124, 133, 138, 142, 216, 248, 250, *See* aquatic environments
  biochemical environments, 37
  social environments, 31, 153, 255
envy, 19–20, 22, 29, 31, 88, 91, 97, 99, 115, 120, 123, 140–141, 148, 154, 165, 189–190, 206, 213, 229, 231–232, 238, 240–242, 244–245, 247, 252, 255
Epic of Gilgamesh, 172
equality, 107, 125, 127, 156, 189–190, 255
erection, 56, 59, 62, 73, 79, 84, 114, 163, 165, 239
erogenous zones, 13, 83
erotic, 197
estrus, 44, 47, 51–52, 54, 59, 62–63, 66, 72, 87, 101, 113, 196
ethnicity, 137, 139, 162, 179–180, 188, 260
  ethnic cleansing, 1, 139, 146
Europe, 152, 173, 184–185, 187, 222, 230–233, 235–237, 253
evaluate, 2, 20–22, 42–43, 51, 67, 71, 102, 108, 115, 117, 127–128, 131–132, 141, 146, 158, 161, 196, 246, 255
even-up behaviour, 181
evolutionary psychology, 12, 16, 30, 31
Evolutionary Theory and Human Nature, 2
exaggerate, 21, 69, 128, 133, 137, 146, 186, 217, 225
excitement, 20, 56, 73, 83–84, 119, 155, 169, 209, 219, 238, 244
exclusion, 132, 216, 221, 229
execution, 19, 263–265
exogamy, 153, 156
expectations, 96, 106, 141, 175, 203, 205, 213, 220, 235–236, 254–255
extended families, 139, 145, 150, 187
eyes, 8, 39, 63, 66, 90, 109, 113, 125–126, 130–131, 135, 197, 215, 229

faces
  face power, 109
  facial expressions, 8–9, 49, 54, 67–68, 73, 90, 96–97, 108, 115, 125, 127, 134
  human faces, 9, 193, 246
  infant faces, 67–68
failure, 22–23, 97, 99, 149, 160, 178, 190, 201, 205, 208–209, 213, 216–218, 220, 244, 247, 254, 260
fairness/unfairness, 3, 23, 181–182, 238, 259, 261–266, 268–271
false consciousness, 190, 243
false witness, 240

fangs, 38, 115–116
fantasy, 5–6, 11, 20–21, 31, 69, 70–72, 74, 78, 83, 89, 96, 98, 107–108, 122–123, 131, 141, 161, 170, 171, 183, 217, 220–221, 223–224, 227–228, 250–252, 254, 271
farming, 98, 135, 250
Fascist, 173
fatty acids, 117
FDR (Franklin Delano Roosevelt), 227, 244
fear of strangers, 134
federations, 134, 142, 153–154
femininity, 41, 68, 85, 167, 169, 195, 196, 229, 246, 250
femmes fatales, 31, 197, 232, 249–250
fertility, 41, 44, 51, 55, 62, 67, 71, 76, 79, 82, 85, 87, 90, 96, 158, 169, 196
feudalism, 186–187, 235
Feynman, Richard, 19
Fiedler, L, 227,
film noir, 222
Final Cause, 3, 7, 32
fish, 27, 28, 33–37, 64, 80, 117
  fishing, 27, 33, 36, 81, 113–114, 119, 123, 130, 135, 184–185, 187
fission-fusion, 46, 50, 58, 61, 133, 138, 142
  fission, 65, 80, 133, 142, 151, 156
  fusion, 65, 142
flight, 20, 22, 104, 135, 155, 157–158, 206, 245
foetus, 39, 77, 265, 268
foraging, 24, 27, 29, 50, 81, 142–143, 145, 155, 156
forests, 24, 26–28, 33–35, 46, 49, 56, 60, 63, 112, 116, 135, 145, 154
forgiveness, 181, 254, 263
Fortune, 4, 6
  fate, 4–5, 144, 165, 229, 234, 260
  luck, 1, 6, 253, 272
fossils, 24–28, 33, 35, 40, 77, 90–92, 117, 119
FOX P2 gene, 127
fraternity, 189–190
freaks, 40, 48, 79–80, 119, 206, 226–229, 237, 247
free will, 4, 16
freedom, 189–190, 215, 224, 248, 252, 254, 257, 260, 262, 268–269
Freud, S., 4, 11–17, 30, 74, 83, 97, 104, 208
friendship groups, 65, 100, 155, 173, 196
fundamentalists, 178, 271

Galileo, 3, 19
games theorists, 181, 242
gangs, 97, 139–140, 148, 163–164, 166–167, 170–171, 173–174, 176–180, 187–188, 190, 202, 227
gaze avoidance, 99
genes
  genetic capture, 44

genetic drift, 82, 86
genetic processes, 2
genome, 37
HOX genes, 37–38, 112
selfish gene, 11, 15
genitals, 13–15, 51, 58, 69–70, 73, 79, 84, 86, 89, 131, 134, 165, 227
genital play, 13, 50–51, 54, 62, 79
genocide, 80, 183, 238, 259
Germany, 18, 173
giants, 227–228
gibbons, 46, 52, 143
gift-giving, 44–45, 105, 108, 258
Gilgamesh, 172, 185
glory, 134, 222, 231, 254, 256
God/gods, 1–4, 7, 19, 29, 93, 101, 122, 139, 149, 160, 173, 175, 185–186, 189–190, 205, 211, 219, 221, 225, 228–232, 238–241, 253–254, 257
golden rule, the, 261
Goodall, J, 112
gorillas, 27, 51–54, 92, 143
gossip, 65, 69, 72, 74, 89, 111, 125, 131–133, 139–140, 153, 171, 175, 182, 199–200, 204, 213, 230–232, 236–237, 248, 258, 260, 271
government, 34, 150, 154, 170, 175, 188, 189, 191, 201–202, 243
grandchildren, 79, 101, 152, 228, 266
grandparents, 152–153
greetings, 50, 58, 62, 139, 155
grooming, 13, 44, 50–51, 54, 58, 60, 79, 84–85, 88, 147
group size, 50, 58, 143–145, 151, 154
groupthink, 147–149, 153
guilt, 5, 14–15, 20, 22, 29, 73, 83, 86, 88, 97–99, 101–103, 136, 140–141, 148, 159–160, 164, 166, 177, 179, 190, 193, 199, 201, 206, 209–210, 213–216, 218, 223, 229, 236, 238, 243, 250–251, 254, 263–264, 266–267
  depressive guilt, 216
  dispositional guilt, 215

Hadar (Ethiopia, Hetro people), 27
hair, 52, 77, 84–86, 88–90, 167, 225, 228, 246, 248
Hamilton, W.D., 37
Hamlet, 214, 216, 221–222
handicap, 225
happiness, 8, 21, 29, 90, 99, 172, 175, 256–257
Haraway, D., 25
hate, 19–20, 22, 29, 66, 88, 93, 99, 115, 120, 140–141, 144, 151, 206, 214, 216, 229, 241, 263–264
hedonic, 155–157, 160, 199, 203–204, 257
  hedonic motivators, 156–157
  hedonic politics, 200
Hegel, G., 3, 99, 242
heightened sensitivity, 20, 93

Heisenberg, W., 7
helplessness, 98–99, 109
herd mammals, 53, 144, 256
heretics, 21, 192, 260
hermaphrodites, 227–228
Herod Effect, 164
Herodotus, 242
heroes, 69, 89, 97, 106, 123–124, 131, 151–152, 155, 159, 166–167, 170, 172, 174–175, 183, 186, 190–191, 203, 221, 223, 225, 239, 246–247, 249, 251, 253–254
    hero worship, 20
heroines, 69, 97, 131, 247
heterochrony, 38–39, 78
heterosexual, 168–170, 196, 198–199
heterozygous, 34, 39, 47, 169–170, 174, 177, 195, 197
HGH (human growth hormone), 77, 112
hierarchy, 18, 44, 58, 141, 147, 155, 157, 159, 167, 186–188, 212, 224, 236, 255–256, 259, 270
higher calling, 73, 98, 248
Hindu epics, 241
Hinduism, 5, 241
Hobbes, T., 4, 6–7
Homo erectus, 27–28, 34, 77, 118–119, 124, 145, 154
Homo ergaster, 119
Homo floresiensis, 34, 77, 124
Homo georgicus, 25
Homo habilis, 25, 27, 119
Homo heidelbergensis, 27, 119
Homo neanderthalensis, 118
Homo sapiens, 11, 27, 78, 118, 135, 143, 145, 154, 205
homosexuality, 167–171, 173–174, 197–198, 246, 251
    camp homosexuals, 171
homozygous, 39, 47, 169, 174, 196
honour, 99, 160–161, 165, 171, 173–174, 176, 178–179, 182, 193, 199, 221–222, 224, 238, 247, 249
hormones, 15, 37, 41, 55, 77–78, 85, 89, 166
households, 24, 64–65, 147–148, 152, 172, 184, 200, 221
human condition, the, 108, 244
humanists, 12, 16, 173
Hume, David, 4, 6, 7, 21
humiliation, 23, 70, 123, 159, 179, 182, 234, 248, 254
humour, 125, 127
    sense of humour, 67, 69, 127
hunting and gathering, 4, 21, 55, 65, 93, 135, 138, 141, 143, 145, 147–148, 153, 156, 180, 185, 218
    hunting, 27–29, 36, 91, 98, 113, 119, 123, 132, 145, 148, 151, 164, 184, 247, 250
husbands, 46, 52, 71, 93, 101, 107, 132, 151–152, 170, 193–195, 198, 199, 202, 216, 218–219, 222–223, 236, 242, 246–247, 251–252, 258

Huxley, T.H., 24
hyper sexuality, 197
hypocrisy, 149, 243, 252, 270

ice ages, 118–119
idealism, 17–18, 121, 148, 191–192
identity, 50, 88, 108, 133, 137, 189, 205, 224, 226
ideology, 189, 239, 269
Iliad, 172
immature births, 76–77, 82, 112
immaturity, 15, 127
impotence, 222–223, 231
impregnate, 42, 79, 103, 199, 233–234, 266
impurity, 3, 161, 192, 209, 214, 237, 239, 257
Inca, 80, 185
incest, 5, 14–15, 30, 47–48, 80–82, 86, 89, 101, 187, 213, 231, 237–238, 246
    incest avoidance, 48, 55, 82, 88, 127, 163
    incest taboo, 14–15
inclusion, 129, 132, 159, 215–216
inclusive fitness (Darwinian fitness), 30
independence, 31, 65, 108–109, 213, 215, 219, 224, 235–236, 249–251
India, 76, 187
individual
    individual rights, 191
    individualism, 20, 155–156, 189, 235
    individuality, 63, 108, 227, 229
industrial societies, 19, 55, 147, 189
infant mortality, 60, 92, 100, 267
infanticide, 43, 45, 93, 95, 144, 196, 231, 234, 237, 252, 267–268
infertility, 82, 246, 260
infidelity, 22, 129, 148
information processing, 16, 143
innocence, 13, 74, 86, 90, 94–95, 98, 138, 176, 213, 229, 259–260, 264–265
instincts, 2, 3, 12, 17, 21, 25, 42, 48, 56, 92, 97, 101–102, 144, 155, 157, 185, 245, 256, 258
intelligence, 21, 25, 29, 36, 60, 63, 67, 69, 108, 111, 115, 118–124, 143, 145–146, 177
interest groups, 162
intolerance, 190, 268
Islam, 241
isolation, 34, 36, 40, 124, 232
    genetic, 30
    geographical, 33–34, 40, 80
    sexual, 30, 34, 40, 51, 80
    social, 34, 80, 154

James, William, 10–11
Jane Eyre, 218
Janis, Irving, 148
Java, 118
jaw
    ape jaw, 38, 115
    human jaw, 30, 38, 66, 117, 119–120

jealousy, 19–20, 22, 29, 31, 49, 66, 72, 88, 91, 96–97, 99, 102–103, 105–107, 115, 120, 128, 140–141, 165, 189–190, 206, 213, 215, 232, 238, 240–241, 245, 252, 255
Jencks, C., 6
Jesus, 173
Jocasta, 5
jokes, 11, 69, 75, 89, 127, 136, 139, 213, 252
 joking, 19, 69–70, 74, 96, 127–128, 131–132, 136, 209, 247, 250
Jones, Ernest, 12
joy, 5, 8, 30, 90, 93, 103, 120, 125, 155, 233, 244
judges, 153, 176, 230, 232
junk DNA, 37, 40, 47
justice, 3–4, 23, 149–150, 178, 180–181, 183, 217, 235, 244, 248–249, 258–265, 268–271
juveniles, 38, 50, 53, 55, 62, 65, 77, 79–80, 82–83, 96–97, 114, 145, 158, 164, 166, 202

k selection, 15, 45–47, 103
Kanapoi and Allia Bay (Kenya), 27
Keith, Sir Arthur, 24
Kennedy, J. F., 201
kindness, 66, 69, 90, 248
King Arthur, 174, 221
kings, 160, 180, 184–187, 201–202
Kinsey, A., 51, 74, 168–169, 197
kinship, 15, 30–31, 55, 58, 61, 65, 73, 75, 150, 154, 156, 174, 188, 199, 204, 222, 230, 232, 235, 241, 258
kissing, 13, 62, 70, 84, 125
knights, 172, 174, 176, 186–187, 249
knowledge, 5, 63, 125, 130–131, 133, 191, 198, 226, 241, 253, 256, 268, 269
Kwakiutl, 35

labour, 5, 19, 30, 36, 66, 99, 152, 157, 171, 184, 187, 190, 204, 211, 256, 261
Lake Chad, 26
lakes, 26–28, 32–33, 63, 116, 118, 135, 154
Lamarckian, 11
language
 non-emotional language, 128, 134, 136–137, 139
laughter, 8, 68, 70, 94–95, 98, 104, 125–127, 134
law and order, 120, 166, 180
laws, 3–4, 6–7, 14, 23, 32, 64, 106, 133–134, 137, 139, 150, 153, 161, 175, 178, 180, 183, 186, 189–191, 203, 213, 230, 239, 241–242, 250, 257–259, 270
leaders, 18, 106, 123–124, 129, 138, 159–160, 175–176, 180, 186, 189–191, 200–202, 234, 241–242
Lear, 5
LeDoux, J., 16
legends, 121, 124, 221
lesbians, 202

lice, 34, 84–85
lies, 129–133, 140, 144, 148, 179, 192, 204, 235, 239, 243, 252, 257
 white lies, 129, 132
life history, 145
Lincoln, Abraham, 124, 271
live-and-let-live, 160–161
Locke, John, 4, 242
lord of the manor, 184
loss, 22–23, 97, 100, 103, 108, 134, 149, 159, 160, 163, 164, 174–175, 178, 205–206, 209–211, 214–218, 220, 222, 224–225, 227, 239, 244, 245–248, 250, 254–255, 262–263, 266
love making, 127
loyalty, 15–16, 73, 102, 105, 107, 122, 129, 137, 157, 164–165, 172–175, 187–188, 198, 199, 203, 221, 224, 238, 241, 247, 249
Lucy, 27
Lunar Men, 174
lust, 15, 19, 20, 22, 29, 31, 46, 47, 71, 75, 82, 89, 91, 96, 98, 106, 108, 110, 115, 120, 140–141, 153, 164, 165, 166, 177, 185, 199, 218, 219, 221–223, 238–239, 246–247

Machiavelli, N., 4, 6–7
 Machiavellian, 21–22
Macmillan, Harold, 6
Mafia, 173, 184
magic, 227, 253
magistrates, 176, 180, 230, 232, 234, 236
maleness, 196
mammals, 24, 27–28, 35–36, 42, 45, 50, 55, 65, 70, 77, 93, 118, 133, 164, 194, 207
Mao Zedong, 18
marriage, 64–65, 75, 124, 133, 135, 137, 142, 149, 151–153, 156, 174, 187, 193–194, 196, 214, 218, 224, 235, 249, 266
Marx, K., 3–4, 242
Marxist, 18
masculinity, 41, 78, 85, 171, 174, 195, 202, 246, 249
mass media, 191
master behaviours, 24, 28, 255
masturbation, 29, 50, 60, 62, 69, 71, 74, 79, 131, 169–170, 174, 209–210, 246, 250–251
mate competition, 29, 44–46, 178
mate guarding, 44, 105, 116, 132
Mead, M., 81
meaning, 2, 5, 7–9, 17, 21, 32, 125–126, 133, 137, 240, 253, 257
meat eating, 24, 119, 207
memory, 22, 37, 95, 118–121, 126, 132, 145–146, 154, 177, 254, 267
menstruation, 55, 211
Middle Ages, 34, 184–185
midwives, 100, 232

migrate out (out migration), 50, 58, 61, 65, 82, 157, 163
military, 34, 139, 143, 146–147, 162, 171, 198, 204, 258
   military squads, 149, 174, 198
Miller, G., 52, 130
miniaturization, 77
moderation, 238–239, 242
modesty, 12, 14, 73, 134, 164, 217–218, 229, 242, 246
molly houses, 173
momism, 215
monasteries, 189, 198, 201
monkeys, 46, 49–50, 52–57, 60, 77, 93, 114, 116, 143, 145, 156, 159, 177
   New World monkeys, 54, 70, 143
monogamy, 52–55, 64–66, 68, 70, 72, 88, 106, 140, 145, 151–152, 156, 163, 165, 187
   serial monogamy, 46, 65
mood states, 20, 97
morality, 1, 2, 19, 121, 182, 197, 221, 224, 231, 240, 249, 261–262, 264–266, 271
   moral aggression, 242
   moral claims, 268, 271
   moral codes, 18, 74, 260, 269
   moral fundamentalists, 271
   moral guilt, 215
   moral obligations, 73, 201, 236
morphology, 28, 30, 37, 77, 91, 110, 112
mother
   mother love, 13, 15, 101, 109, 214, 224
   motherhood, 101, 106
   mothering, 46, 81, 87, 89, 94, 98, 101–106, 128, 137, 159, 194, 198, 218, 223–224, 249–252
mother/son bond, 61, 81, 94, 100, 104
motivation, 7, 11–12, 16, 21, 31, 36, 41, 102, 122, 133, 138, 146, 157, 165, 170, 181, 203, 210, 223, 245, 250, 252
motivational, 10, 21, 89, 92, 120, 122–123, 171, 257, 258
multi-mating, 46, 52–54, 65, 145
murder, 29–30, 237–240, 242, 263–265, 268
mutations, 30, 32, 34, 37–40, 47–48, 78–80, 82, 85, 117, 119
mythology, 1, 4, 11, 21, 69, 74, 124, 127, 155, 174, 182, 186, 198–199, 204, 215, 218, 221, 223, 227, 234, 238–239, 253, 258

nakedness, 12, 72–73, 84–85, 94, 250, 254
naming, 137–138
narcissism, 15, 95–97,
national interests, 189
natural human group/grouping, 142, 144, 146, 151, 153

natural selection, 11, 17, 22, 34
Nazi, 18, 261
needs, 150, 157, 218, 255, 260
neighbours, 132, 153, 175, 183, 187, 200, 221, 230–232, 240, 250, 253, 255
   neighbourhood, 131, 162, 189, 199, 202, 235, 258, 270
Neolithic revolution, 135
neoteny, 38, 68, 77–78, 84, 110, 112, 114, 116–117, 145, 195
nepotism, 150
nests, 14, 36, 56, 63–64, 72, 75, 81, 83, 94, 98, 104, 108–109, 113–114, 116, 207–208, 247, 250, 258, 271
neurons, 14, 61, 95
   neural networks, 8
   neurotransmitters, 8, 78, 85
Newton, Isaac, 3, 224
N-glycolylineuraninic, 117, 119
Nietzsche, F., 4, 6, 11, 183
non-verbal communication, 98, 108–109, 114, 126–127, 131, 134–136, 139
noradrenaline, 78
novels, 5, 127, 170, 182, 215, 222–223, 250
Nuer, 185

oaths, 134, 137, 199, 237
obedience, 109, 183, 186–187, 201, 216
obligations, 75, 98, 124, 135, 137, 160–161, 191, 201, 203, 223, 235, 238, 240, 259, 263
obsession, 20, 29, 74, 166, 169, 223, 243, 271
oedipal, 17, 150, 176, 216, 217
Oedipus, 5, 206, 216
Okefenokee, 34, 36
oral sex, 13, 62, 79, 250
orangutans, 35, 51, 52–54, 143,
order-in-process, 2, 7, 20–21, 23
Oresteia, 214, 261
Orestes, 214, 216
organizations, 25, 139, 142, 144, 153, 176, 188, 191, 200, 204, 257–258
orgasm, 51, 59, 62, 70–71, 74, 104, 168, 196–197, 210, 224
original sin, 241
Orrorin tugenensis, 26
orthograde feeding, 35, 112, 154
ovulation, 46, 51, 54, 59, 62, 72, 87, 113, 134
oxytocin, 89, 166

paedomorphism, 38, 78, 112, 116, 145
paedophiles, 169
pain, 99–100, 172, 210, 226
pair-bonding, 46, 54, 63, 73, 134, 138, 152, 245
palaces, 140, 186, 188
palaeontology, 24, 27, 29–30, 91–92
panic, 20, 22, 206–207, 210, 226
parasites, 34, 48, 82

parents
  parent/offspring conflicts, 120
  parental investment, 45–46, 48, 76, 103
  parenthood, 98
  parenting, 47, 97–98, 101–103, 106, 109–110, 120, 123, 132, 153, 181, 202, 213, 249, 257
partnerships, 94, 128–129, 163, 167, 171–172, 175, 178, 181, 193–194, 199, 225
passions, 6, 21, 182, 211, 218, 223, 229, 235, 238, 241–244, 247
patriarch, 66, 251
patronage, 50, 75, 99, 139, 141–142, 149–150, 158–159, 175, 180–181, 183–189, 191–193, 199–204, 247, 249, 258
  patron/client, 106, 167, 184, 189
  patron/client politics, 99
patterns of behaviour, 20, 92, 181
Paul, St., 223
Peking Man, 27
pelvis, 26, 100
penis, 12, 40, 41, 43, 51, 53, 55, 59, 60, 62, 70–71, 77–78, 84, 120, 131, 228–229, 246
  penile display, 50–51, 163
  penis fencing, 50, 62
peramorphosis, 38
perfection, 2–4, 16, 18–19, 23, 31, 107, 124, 173, 225, 249, 253, 256, 259, 270
  perfectibility, 2, 19
  perfecting, 3, 29, 74
persecutions, 230, 259, 261
Peter Pan, 227
phenotype, 30, 34, 36–40, 42, 72, 78, 90, 225
philandering, 46–47, 104–105, 132, 200, 246
philanthropy, 200
philosophy, 1, 4, 6, 17–18, 121, 124, 149, 172, 190–191, 201, 238–239, 242, 244, 253–254, 256
phylogenetic, 8–10, 12, 30, 47, 49, 56, 68, 71, 78, 207
physics, 3, 4, 7, 120, 258
Plato, 3, 242
play
  playfulness, 88, 102, 164, 166, 168
  play-like, 82, 90, 126
  playmates, 61, 88, 93, 95, 97, 101–102, 126, 140, 163–164, 185, 196, 248
Pleistocene, 35, 118, 121
Pliocene, 26–27, 35
Plio-pleistocene, 26
policing, 18, 108, 131, 176–178, 189, 199, 261
politics
  alliance politics, 58, 60, 178, 180, 183, 185, 188, 202
  political factions, 178, 188, 203
  political networks, 167
  political parties, 162, 188–189, 191

political society, 4, 99
political states, 111, 139, 146, 180, 183, 189, 191
politicians, 121–122, 132, 160, 176, 191–192, 201–203, 252
politicking, 106, 122, 125, 131, 177–178, 191, 255
primate politics, 49, 183
polluted, 209–212, 236–238, 240, 243, 247
polyandry, 52, 65, 87
polygenetic, 39, 47, 123–124, 166, 168
polygyny, 44, 52–53, 55, 65–66, 70–71
polymorphic, 39, 121, 167
polymorphous perversion, 12–13, 88, 166, 213
populists, 191
pornography, 30, 69, 74, 86, 169–170, 250–251
Portnoy, 215
post-natal, 45–46, 55, 77, 82, 103
  post-natal depression, 93
  post-natal development, 38
prayer, 12, 161, 190, 212, 230, 241, 248, 253–254
predation, 24, 28, 45–46, 144, 154, 156, 207
pregnancy, 40, 42–43, 46, 53–54, 59, 85, 87, 99, 101, 103–104, 128, 169, 193, 196, 198–199, 212, 266–268
prejudices, 197, 254, 261
premature births, 231
premature deaths, 231
prestige, 99, 108, 150, 247
pride, 14, 22, 45, 96, 99, 120, 136, 148, 212, 230
prisons, 169–171, 176, 182, 189, 264, 269
professions, 100, 153, 162, 189, 236
progenesis, 38
Progress, 2–3, 16–18, 74–76, 140, 150, 205, 244, 256, 259, 261
promiscuity, 15, 43–46, 51–53, 59, 65, 70, 73–74, 101–105, 115–116, 134, 145, 169, 197–198, 213, 217–218, 246, 251–252, 265, 269
prosimians, 52, 143
prostitution, 70, 91, 149, 152, 157, 169–171, 219, 223–224, 234, 236, 251
pseudo sex, 166, 168
psychiatry, 18, 190, 216
psychology, 10–11, 15–16, 18, 22, 30, 92, 101, 199, 213, 220, 245, 261, 264
psychosexual, 13, 29, 61, 71–72
puberty, 40–41, 78, 86, 96–97, 127, 152, 158, 164, 166–168, 196, 227, 266, 268
  post-puberty, 84, 158, 163, 166–167, 169, 196
pubic hair, 84–85
purity, 74, 122, 190, 209, 211–213, 229, 237, 239, 241
purpose, 1–4, 7, 16–17, 19, 21, 32, 76, 121–122, 129, 136, 140, 190, 201, 231, 241, 247–248, 250, 253, 257–258, 266
pygmies, 34,

quantum electrodynamics (QED), 7
Qur'an, 5

r selection, 15, 45, 103
race, 18, 46–47, 173, 179, 243, 261
randomness, 1–3, 5, 7, 20, 23, 29, 31, 39, 48, 121–122, 258
rape, 30, 217, 246
rational
    rational animal, the, 30, 245
    rational man, 24–25, 90
    rationalistic, 125, 143, 146, 148–149, 181, 192, 245, 260, 262, 264, 271
    rationalists, 12, 16, 191, 271
    rationality, 1–3, 6–8, 11–12, 16–18, 21, 23, 25, 28–29, 32, 76, 92, 111, 121, 190–191, 204, 229, 242–243, 253–254, 259–262
rationalize, 87, 98, 107, 120, 142, 156, 182, 254
Rawls, J., 242
rebellion, 34, 75, 96, 187, 213
redeeming, 264
reflection, 243, 256
rejection, 14, 39, 52, 80, 95, 97, 103, 123, 128, 181, 205, 208, 214, 216, 218, 223, 230, 239, 241, 246–247, 254–255, 258
religions, 1, 124, 139, 173, 179, 187, 189, 200, 203, 225, 230, 241, 243, 253–254, 256, 262
REM sleep, 78–79
repression, 14–15, 124, 141, 146, 214, 243, 245, 252
    repressed aggression, 141
    repressed sexuality, 15, 214, 217–218, 250
reproductive propensities, 166, 197
repulsion, 22, 138, 206–208, 219, 224, 226, 229, 243, 245, 250
responsibility, 15, 178, 186, 199, 200–201, 239, 251, 254, 260
restorative justice, 181
reverted escape, 104, 155, 158
revolutions, 18, 173, 175–177, 188, 201, 261
righteous indignation, 242
rights, 106, 160–161, 164–165, 178–180, 182, 187, 189, 191, 200–204, 252, 255, 258, 265, 268–269
rivers, 26–27, 32–33, 56, 60, 63, 81, 116, 118–119, 135, 145, 154, 173
Roman, 147
romance, 69, 72, 75, 89, 96, 111, 123, 127, 131, 198, 213, 218, 220, 224, 247, 249, 257
    romantic love, 14, 146, 220
Rosemary's Baby, 233
Rousseau, J-J., 242
runaway
    runawayevolution, 116
    runaway sexual selection, 115

sacrifice, 148, 223, 239, 247, 254, 267
Sahelanthropus tchadensis, 26

savannahs, 24–29, 33, 35, 63, 76, 144–145
scapegoats, 149, 236
scavenging, 24, 207, 236
Schopenhauer, A., 11, 20
science
    scientific, 2, 7, 12, 16–19, 34, 71, 76, 92, 133, 201, 211, 213, 237, 256, 259, 261–262, 265, 266, 268–269
    scientists, 7, 12
Scientific Socialism, 18
Second World War, 202
secondary sexual characteristics, 41, 55, 73, 77–78, 82, 86–87, 89, 101, 168, 246
seduction, 69, 73–75, 106, 125, 128, 132–133, 135–136, 140, 164–165, 169, 176, 185, 204, 232, 234, 246, 257
seed eating, 24, 204, 256
self
    presentations of self, 22, 89, 148, 171, 195, 206, 211–212, 227, 248–249,
    self-awareness, 10, 13, 29, 49, 60, 63, 83, 101, 108, 117, 133, 141, 143, 146, 178, 189, 211, 214–215, 239, 243, 254, 262
    self-blame, 206, 220
    self-centred, 95, 146, 238
    self-control, 123, 177, 224, 235, 243, 248, 254
    self-deception, 20, 75, 125, 129, 136–137
    self-definition, 208
    self-delusion, 108
    self-discovery, 208
    self-esteem, 140, 216, 224, 248
    self-harm, 206, 269
    self-indulgence, 212
    self-serving, 123
    self-value, 108, 165, 229, 255
Seminole, 34
sensual sucking, 13, 15
separation anxiety, 218, 223
separations of powers, 191
serfs, 180, 187, 200
serotonin, 78
servants, 64, 95, 157, 160, 184, 186, 199, 239, 267
sex
    sex play, 54, 70, 74, 79
    sex talk, 174
sexual
    consummatory (pleasure), 41, 51, 167, 174, 195–196,
    proceptivity, 41, 51, 70, 78, 114, 167–169, 174, 195–197, 218
    receptivity, 13, 29, 41–43, 42, 46, 51, 59, 73, 85, 90, 114, 134, 167–169, 171, 174, 192, 195–197
    sexual attractivity, 41, 43, 87, 102, 167, 196
    sexual dimorphism, 70, 166–167

sexual (cont.)
  sexual drives, 21, 46, 51, 73, 107, 115, 198, 212, 220–221, 250
  sexual identities, 88, 108
  sexual inactivity, 82, 127
  sexual politics, 29, 31, 60, 202
  sexual secrecy, 14, 69–70, 72–74, 83, 87–89, 126, 250
Shakespeare, W., 5, 222
shame, 5, 9, 14, 20, 73, 83, 87–89, 97, 101, 103, 120, 134, 140–141, 159, 164, 177, 199, 201, 206, 209, 215, 218, 243, 250
shyness, 9, 14, 83, 88, 101, 103, 134, 164
Siamese twins, 228–229
siblings, 15, 31, 49, 75, 79, 82, 88, 94–97, 100, 106, 113, 132, 140–141, 151, 158, 184, 192, 199, 248, 252
  sibling rivalry, 15, 29, 83, 88, 91, 96, 212
sign language, 60, 126
sin, 160, 190, 211, 222, 240–241, 253
Sioux, 157
sisterhoods, 198
Skinner, B.F., 16
skull, 25, 30, 38–39, 77, 100, 114–117, 119–120
Sloan, A.P., 147–148
Small, William, 174
smiling, 8, 11–12, 68, 75, 87, 90, 94, 104, 109, 113, 125–126, 193, 252
Smith Woodward, Sir Arthur, 24
SNPs (single nucleotide polymorphisms), 37, 77, 112
soap operas, 1, 128, 182, 214, 253
social
  social constructionists, 12, 16–17, 243
  social display, 75, 108
  social mobility, 157
  socialization, 150
  socialness, 133
social sciences, 3, 6, 12, 16, 19, 23, 149, 243
socialism, 18, 190, 201, 260–261
sociobiology, 30–31, 242
sociology, 4, 6
sociosexual activities, 50, 55, 62
Socrates, 242
Solomon, Ezra, 6
Solomon, King, 138
Sophists, 4
Sophocles, 5
soul, 172, 212, 241, 248
Soviet Union, 18, 261
species-typical
  species-typical desires and fears, 2, 3, 20–21, 23, 32, 162, 204, 245, 255, 258–259, 263, 269, 270–271
  species-typical emotions, 2, 7, 245–246
sperm competition, 43–45, 47, 53, 59, 70, 105, 116, 145

spirits, 1, 211, 230, 237, 239, 260
spirituality, 29, 172, 241, 253
spiritualize, 3, 101, 140, 153
sports teams, 147–148, 164, 174, 188, 198, 203
Stalin, J., 18
stalkers, 169
stereotyping, 20–21, 121, 136–137, 140–141, 146, 149, 153–155, 157, 159–161, 191, 193, 216, 247, 254–255, 271
sterility,
  infant/child sterility, 127, 196
  juvenile/adolescent sterility, 55, 82, 164, 166
stigmatizing, 168, 225, 250
stories
  story tellers, 122
  storytelling, 71, 131, 247, 251
strange, 80, 82, 84, 128, 161, 194, 227–228, 238
strangers, 21, 50, 55, 64, 66, 69, 73, 87–88, 94, 96, 98, 101, 122, 133–134, 139–140, 146, 148, 155, 157, 194, 208, 226, 237, 247, 251–252, 259, 263
stress, 23, 56, 95, 164, 194, 220
structures, 6, 16, 38, 112, 121, 150, 176
sublimation, 14–15, 21, 106, 140, 170, 175, 183, 199, 204, 215–216, 245, 251–252, 256–258
submission, 1, 8, 44, 50, 58, 62, 68, 125, 129, 135, 140–141, 158–160, 178, 201, 247, 254
subordination, 31, 51, 54, 56, 58, 79, 104, 106, 157–159, 160, 183
suffering, 2, 206, 211, 259, 262
Swetnam, J., 222
Swift, Jonathan, 4, 6
symbols, 69, 126, 140, 153, 155, 160–161, 215
symmetry, 37, 67, 246, 249
sympathy networks, 148–149

Taboo nests, 63
taboos, 14, 23, 69, 73, 75, 83, 87–89, 106, 120, 140, 142, 153, 165, 177–178, 186, 189, 203–204, 206, 210, 212–213, 224, 229, 235, 237, 239, 248, 255–259
Taung-hwa, S., 201
team mates, 129, 140, 171, 183
team spirit, 174
teasing, 9, 69, 70, 74, 83, 89, 104, 127, 132, 136, 174, 209, 250, 252
technologies, 18–19, 33, 118, 138–139, 175, 188, 203, 261
teenagers, 74, 91, 98–99, 213, 267
teleology, 2–4, 11, 15–17, 19, 21, 25, 31–32, 63, 92, 150, 254, 257, 259, 271
Tennyson, A., 172
terminal branch feeding, 35
testicles, 43, 51, 53, 59–60, 62, 70
testosterone, 40, 55, 77–78, 96, 175–176, 188, 196, 221
The Brothers Karamazov, 215

The Expression of the Emotions in Man and Animals, 8
The Origin of Species, 7
the rule of law, 183, 189, 191
The Selfish Gene, 17
The Ugly Duckling, 216
theory of mind, 143
Thorpe, S., 35
threat, 8, 68, 90, 96, 104, 157–158, 180, 182–183, 186, 193, 210, 221, 233, 244
toddlers, 81–83, 94–95
toilet awareness, 208
tolerance, 269, 271
tool making, 24, 113, 119
tool use, 8, 25, 28–29, 63, 91, 111, 113, 142–143
torture, 1, 230, 232, 255, 259
toys, 60, 75, 95, 151, 189, 200, 256
trade unions, 139, 149, 153, 162, 188
tranquillity, 29, 93, 255
tribes, 34, 130, 139, 142, 147, 157, 161, 194
Troilus and Cressida, 222
tropics, 26–27, 33, 35–36, 49, 56, 63, 112, 118, 154
trust, 16, 31, 81, 90, 105–106, 108–109, 120, 141, 146, 175, 187–188, 193, 197, 199, 213–214, 220, 224, 234, 237–240, 242–243, 249, 255, 259, 270
    trustworthiness, 9, 67, 69, 87, 90, 109, 164, 173, 177, 204, 224, 247, 257
Turkana basin (Kenya), 27
twins, 76, 225, 228–229
tyranny of the majority, 191

unconscious, 5, 9–12, 14, 17, 21–22, 71, 76, 87, 97, 99, 103, 109, 111, 206, 213, 215–216, 223, 226, 245, 250
unforeseen consequences, 1, 203, 270
us-groups, 141
utopias, 2, 16, 18–19, 23, 217, 261, 270

vagina, 12, 43, 53, 71
vasopressin, 89, 166
vegetarians, 49, 57
vengeance, 58, 130, 139–140, 148, 152–153, 163, 172, 176–177, 179, 180, 181, 182, 183, 186, 188–189, 203, 221, 234, 263
    seeking vengeance, 106, 114, 140, 148, 179, 181, 192, 199, 200, 202, 245
    vengeance feuds, 180, 188
    vengeance groups, 129–130, 151, 162–163, 176
verbal courtship, 128

victim, 86, 148, 183, 210, 219, 263–264
vigilance, 10, 102, 106, 240, 243, 251
violence, 29, 46, 73, 95, 131, 160, 169, 179, 183, 234, 239, 248, 268
Virgil, 172
virtues, 74, 120, 165, 173–174, 176, 183, 187, 238, 242, 247, 250, 254
Voltaire, 4–5, 7, 260
vulva, 61, 84

waist to hip ratio, 66, 85
war, 6–7, 19, 22, 28–29, 91, 131–132, 153, 174, 176–177, 179–180, 183, 186, 202–203, 222, 247–248, 253–254, 256–257, 261
    warriors, 101, 109, 114, 128, 130, 132, 174, 176, 180, 185–186, 195, 221, 223, 253
Washburn, S.L., 91,
Weber, Max, 6, 7, 165
welfare, 103, 200–202, 204, 258, 260
Whitman, Walt, 172
whores, 170, 197, 219, 224, 249, 251
widowers, 64, 152
widows, 48, 64, 152, 194, 236
will to power, 6, 29, 86, 91, 99, 108–109, 141, 162, 176, 215, 262
Winnicott, D., 223
witchcraft
    common tradition, 230–233
    learned tradition, 230–235, 237
Wolfe, Tom, 130
woodland, 27, 33–35
worship, 1, 99, 140, 151, 155, 159, 166–167, 175, 183, 187, 191, 229, 242, 247, 254–255, 257
writing, 3, 130, 133, 136–137, 139, 161, 180, 189
Wylie, Philip, 215
Wyoming, 33

Xun Qing, 5

Yanomamo, 80, 180
youth
    youth force, 176–177, 180, 183, 193, 200, 203
    youth gangs, 166, 176, 178–180, 187
    youth-apes, 79–82, 87–90, 114, 118, 158, 163–164, 166, 168

Zulu, 185
zygote, 34, 37, 40, 42, 68

Printed in the United States
By Bookmasters